Introduction to the Physics of Landslides

Fabio Vittorio De Blasio

Introduction to the Physics of Landslides

Lecture Notes on the Dynamics
of Mass Wasting

Fabio Vittorio De Blasio
NHAZCA s.r.l., spin-off "Sapienza" University
Rome

ISBN 978-94-007-1121-1   e-ISBN 978-94-007-1122-8
DOI 10.1007/978-94-007-1122-8
Springer Dordrecht Heidelberg London New York

Library of Congress Control Number: 2011926632

© Springer Science+Business Media B.V. 2011
No part of this work may be reproduced, stored in a retrieval system, or transmitted in any form or by any means, electronic, mechanical, photocopying, microfilming, recording or otherwise, without written permission from the Publisher, with the exception of any material supplied specifically for the purpose of being entered and executed on a computer system, for exclusive use by the purchaser of the work.

Printed on acid-free paper

Springer is part of Springer Science+Business Media (www.springer.com)

Preface

A book like this should certainly start emphasizing the death toll and destruction potential of landslides and the consequent need for a better understanding of these geohazards. Nevertheless, the interesting and inspiring aspect of natural catastrophes should also be stressed. Much physics, mathematics, engineering, technology, and even chemistry and biology form the foundation of the study of natural hazards. It is a subject that will become more physically oriented in the near future, and from which the motivated researcher will enjoy material for study, research, and discovery.

There are numerous books, reports, and Internet documents about landslides, and no scientist knows even a small fraction of them. This book is not an attempt to review the field of landslides, as this would be an impossible task for one person, an effort which besides would soon go outdated. During my teaching at the University of Oslo I have realized the need for better understanding of the basic physics necessary for understanding landslides, and the geological phenomena in general. Sometimes students, investigators, and practitioners who use conceptual and numerical models in the study of landslides pay little attention to the basic physical laws, and ignore the mathematics necessary to describe it. The understanding of physical processes may become fuzzy; formulas and computer programs are arcane and their range of validity is not tested. One of the negative consequences of the rapid progress in science is the extreme specialization, which entails a deep knowledge of a narrow subject, limited understanding of a close topic, and complete ignorance of distant disciplines. Although pigeonholing scientific knowledge is useful for learning, it must be reminded that cultural compartments are fictitious. Strengthening the basis of a certain subject appears as an appropriate treatment to reduce this problem.

Quantitative methods in the study of landslides are not novel. However, geotechnics and geological engineering mostly deal with problems of slope stability. Apart from Erismann and Abele's book only dealing with rock avalanches, I am not aware of any book on this subject at the introductory level, although it is possible that some books do exist in language other than English like my own short introduction in Italian. Although this is also a topic on which many contributions continuously appear on specialized journals, the field of landslide dynamics is still far from a state of maturity. Perhaps this is reflected by the lack of a name defining the subject of landslide dynamics like seismology is for the science of earthquakes (probably a name like "ruinology" could be appropriate, from ruina=landslide in Latin).

This book will especially deal with the physics and dynamics of landslides. In short, the book aims at: (1) Informing about the physical basis of the mass wasting phenomena. (2) Stand as a reference of some basic physics needed for working with landslide modeling. (3) Help to work out physical models of landslides. (4) Influence the students to become more curious than erudite.

It is easier to list what the book is not about. It is not intended as an updated report packed with references. Nor does it treat the subject in a systematic and thorough way. The emphasis is instead on simple models rather than up-to-date complex analytical and numerical techniques. The book does not try to be comprehensive and updated; it rather aims at promoting physical reasoning. Important subjects have been left out for different reasons: because too complicated, too novel, very far from the author's expertise (a good number of books represent a more personal view than the author would admit), or simply for lack of space. To cite only one of the several missing topic: a systematic treatment of numerical simulations of landslides. I anticipate that many scientists who contributed a great deal to the field will be disappointed to see their contribution underrepresented.

A list of the subjects treated is the following.

Chapter 1 is an introduction to the subject of landslides. The reader will not find much novel information but rather an ordering of the subject useful for later use.

Chapter 2 opens with the laws of friction and cohesion. It then introduces very shortly some problems of slope stability. The chapter is perhaps the least novel in the book, as the subject is treated in a more thorough way in several other textbooks, but an introduction to slope stability appeared to be necessary.

Chapter 3 is an introduction to fluid mechanics useful for landslide studies. Some basic concepts of fluids are initially brought in heuristically. In a second step, the laws of fluid mechanics are introduced in the proper conceptual framework.

Chapter 4 initially introduces the subject of non-Newtonian fluids. In the second part, the chapter deals with rheological flows such as mudflows, lahars, and debris flows.

Chapter 5 is a general introduction to the physics of granular media, both dry and wet. It forms the bases for Chap. 6 and for part of Chap. 9.

Chapter 6 is the longest of the book. It deals with granular flows and rock avalanches. It is partly descriptive, with a tendency to become more quantitative in some particular topics. Special subjects such as rocks generated by frictional heat or vapor lubrication are also discussed.

Chapter 7 considers landslides in peculiar environments: water reservoirs, glaciers, and mass wasting on the surface of other planets.

Chapter 8 is a plain introduction to rock falls, which are single falling boulders. Even if much smaller in volume than rock avalanches, they are a very common threat in mountain environment.

Chapter 9 is a primer to the physics of submarine landslides. The contact forces with water become important, and thus the subject strongly builds on concepts developed in Chap. 3.

Chapter 10 considers a number of gravity mass flows, not necessarily landslides, and draws a comparison to other landslides examined in the book. The reader will also be able to appreciate how previous knowledge can be applied to other forms of hazardous phenomena.

To help (or even to encourage) the reader jumping from different parts of the book, the reference to other parts of the book are signaled with arrows. A right pointing arrow ($\rightarrow$Sect. 3.2.1) indicates that particular concept will be developed in a later part of the book, in this case in section 3.2.1. A left-pointing arrow ($\leftarrow$Sect. 1.2.3) signals that the subject has been treated earlier.

Boxes are scattered throughout the book. Some are introductive of concepts that not all readers may have (they are termed "one step back"). The more advanced ones or the ones not central to the discussion but interesting or even enjoyable have been labeled as "one step forward." Both can be skipped in a first reading, for opposite reasons. Other boxes are very short introductions to some famous landslides and gather basic information. They are called "brief case study." Another kind of box is "external link," dealing with disciplines other than ruinology but in which landslides play an interesting role. A few boxes of the series "simple views" illustrate easy theoretical schemes or uncomplicated experiments to make a particular subject clearer. Finally, other boxes are not categorized.

Three appendices complete the book: a mathematical appendix, also called *MathApp* throughout the book, a Geological appendix called *GeoApp*, and a physical appendix *PhysApp*. These can be useful as quick look-up reference for some data or equations. Most of the photographs are original and unpublished.

For providing comments, diagrams, articles, or figures, the following persons deserve my gratitude: Stein Bondevik, Hedda Breien, Giovanni Battista Crosta, Graziella Devoli, Anders Elverhøi, Birgitte Freiesleben De Blasio, Ulrik Domaas, Peter Gauer, Lisetta Giacomelli, Carl Harbitz, Kaare Høeg, Dieter Issler, Ole Mejlhede Jensen, Paolo Mazzanti, Marco Pilotti, Roberto Scandone, May-Britt Sæter, Maarten Venneste, Hassan Shaharivar, Alessandro Simoni, Roger Urgeles, and the numerous students who during these years have been the test subjects of the lecture notes this book grew from. Needles to say, the author is the sole responsible for any misconception or inaccuracy that plagues every book, and this probably more than others.

Milano and Oslo Fabio Vittorio De Blasio
December 2010

Contents

1 Introduction and Problems 1
   1.1 Landslides: An Overview 2
      1.1.1 What Is a Landslide? 2
      1.1.2 Landslides as a Geological Hazard 3
      1.1.3 Landslides as a Geomorphic Driving Force 5
      1.1.4 Physical Aspects of Landslides 9
   1.2 Types of Landslides 11
      1.2.1 Geometrical Characteristics of a Landslide 11
      1.2.2 Description of the Seven Types of Movements 12
   1.3 A Physical Classification of Gravity Mass Flows 15
      1.3.1 Rheological Flows (Cohesive) (→Chap. 4) 15
      1.3.2 Granular Flows (Frictional) (→Chap. 6) 16
      1.3.3 Rock Falls and Topples 17
      1.3.4 Slow Landslides and Creep 18
      1.3.5 Other Gravity Mass Flows Similar to Landslides 19

2 Friction, Cohesion, and Slope Stability 23
   2.1 Friction and Cohesion 25
      2.1.1 Normal and Shear Stresses 25
      2.1.2 Friction 25
      2.1.3 Cohesion 30
   2.2 Slope Stability 31
      2.2.1 A Few Words on Slope Stability 31
      2.2.2 An Example: Layered Slope 33
      2.2.3 A Few Basics Concepts of Soil Mechanics
            and an Application to Slumps 40
      2.2.4 Other Factors Contributing to Instability 49

3 Introduction to Fluid Mechanics ... 53
- 3.1 Introduction ... 55
  - 3.1.1 What Is a Fluid? ... 55
- 3.2 Fluid Static ... 56
  - 3.2.1 Isotropy of Pressure ... 56
  - 3.2.2 Pressure Increase with Depth ... 57
  - 3.2.3 Pressure Exerted by a Suspension: Total Pressure and Water Pressure ... 58
  - 3.2.4 Force Exerted on a Dam by Water in a Reservoir ... 60
- 3.3 Simple Treatment of Some Topics in Fluid Dynamics ... 60
  - 3.3.1 Fluid Flow (Key Concept: Velocity Field, Streamlines, Streamtubes) ... 60
  - 3.3.2 Fluid Flow in a Pipe with a Constriction (Key Concepts: Continuity, Incompressibility) ... 61
  - 3.3.3 Lift Force on a Half-Cylinder (Key Concept: Energy Conservation and the Bernoulli Equation) ... 62
  - 3.3.4 Flow of a Plate on a Viscous Fluid (Key Concepts: No-Slip Condition, Viscosity, Newtonian Fluids) ... 65
  - 3.3.5 Fluid Pattern Around a Cylinder (Key Concepts: Reynolds Number, Turbulence) ... 67
- 3.4 Microscopic Model of a Fluid and Mass Conservation ... 69
  - 3.4.1 The Pressure in a Gas Is Due to the Impact of Molecules ... 69
  - 3.4.2 Viscosity ... 71
- 3.5 Conservation of Mass: The Continuity Equation ... 73
  - 3.5.1 Flux ... 73
  - 3.5.2 Continuity Equation in Cartesian Coordinates ... 74
- 3.6 A More Rigorous Approach to Fluid Mechanics: Momentum and Navier–Stokes Equation ... 76
  - 3.6.1 Lagrangian and Eulerian Viewpoints ... 76
  - 3.6.2 Momentum Equation ... 78
  - 3.6.3 Analysis of the Forces: The Momentum Equation ... 79
  - 3.6.4 Adding up the Rheological Properties: The Navier–Stokes Equation ... 82
- 3.7 Some Applications ... 84
  - 3.7.1 Dimensionless Numbers in Fluid Dynamics ... 84
  - 3.7.2 Application to Open Flow of Infinite Width Channel ... 85

4 Non-Newtonian Fluids, Mudflows, and Debris Flows: A Rheological Approach ... 89
- 4.1 Momentum Equations, Rheology, and Fluid Flow ... 91
- 4.2 Dirty Water: The Rheology of Dilute Suspensions ... 91

4.3	Very Dirty Water: Rheology of Clay Slurries and Muds		93
	4.3.1	Clay Mixtures	93
	4.3.2	Interaction Between Clay Particles	94
	4.3.3	Rheology of Clay Mixtures and Other Fluids	94
	4.3.4	Bingham and Herschel-Bulkley	96
	4.3.5	Shear Strength as a Function of the Solid Concentration	97
	4.3.6	Relationship Between Soil Properties and Fluid Dynamics Properties	99
4.4	Behavior of a Mudflow Described by Bingham Rheology: One-Dimensional System		100
4.5	Flow of a Bingham Fluid in a Channel		103
	4.5.1	Calculation for a Cylindrical Channel	103
	4.5.2	Triangular Channel	104
4.6	Rheological Flows: General Properties		105
	4.6.1	Introduction	105
	4.6.2	Geological Materials of Rheological Flows	107
	4.6.3	Structure of a Debris Flow Chute and Deposit	108
	4.6.4	Examples of Rheological Flows	108
4.7	Debris Flows: Dynamics		117
	4.7.1	Velocity	117
	4.7.2	Dynamical Description of a Debris Flow	121
	4.7.3	Impact Force of a Debris Flow Against a Barrier	125
	4.7.4	Quasi-Periodicity	126
	4.7.5	Theoretical and Semiempirical Formulas to Predict the Velocity	127

5 A Short Introduction to the Physics of Granular Media 131

5.1	Introduction to Granular Materials		133
	5.1.1	Solid Mechanics: Hooke's Law, Poisson Coefficients, Elasticity	133
	5.1.2	Angle of Repose	134
	5.1.3	Force Between Grains	135
5.2	Static of Granular Materials		136
	5.2.1	Pressures Inside a Container Filled with Granular Material	136
	5.2.2	Force Chains	140
5.3	Grain Collisions		142
	5.3.1	Grain-Wall Collisions	142
	5.3.2	Grain–Grain Collisions	144
5.4	Dynamics of Granular Materials: Avalanching		148
	5.4.1	General	148
	5.4.2	Dynamics of Granular Materials at High Shear Rate: Granular Gases and Granular Temperature	149

		5.4.3	Haff's Equation	150
		5.4.4	Fluid Dynamical Model of a Granular Flow	151
	5.5	Dispersive Stresses and the Brazil Nuts Effect		153
		5.5.1	Dispersive Pressure	153
		5.5.2	Brazil Nuts and Inverse Grading	156
6	**Granular Flows and Rock Avalanches**			159
	6.1	Rock Avalanches: An Introduction		161
		6.1.1	Historical Note	161
		6.1.2	Examples of Rock Avalanches: A Quick Glance	163
		6.1.3	The Volumes of Rock Avalanches	164
	6.2	Rock Avalanche Scars and Deposits		169
		6.2.1	Rock Avalanche Deposits: Large-Scale Features	170
		6.2.2	Rock Avalanche Deposits: Intermediate-Scale Features	170
		6.2.3	Rock Avalanche Deposits: Some Small-Scale Features	177
	6.3	Dynamical Properties of Rock Avalanches and Stages of Their Development		178
		6.3.1	Velocity of a Rock Avalanche	178
		6.3.2	Stages in the Development of a Rock Avalanche	180
	6.4	Simple Lumped Mass and Slab Models for Rock Avalanches		183
		6.4.1	A Simple Model of Landslide Movement	184
		6.4.2	Use of Energy Conservation (1): Runout of a Coulomb Frictional Sliding Body	187
		6.4.3	Use of Energy Conservation (2): Calculation of the Velocity with Arbitrary Slope Path	188
		6.4.4	A Slab Model	191
	6.5	Application of the Models to Real Case Studies		193
		6.5.1	Elm	193
		6.5.2	The Landslides of Novaya Zemlia Test Site	194
	6.6	The Fahrböschung of a Rock Avalanche		196
		6.6.1	The Importance of the Centre of Mass of the Landslide Distribution	196
		6.6.2	Fahrböschung of a Rock Avalanche	197
	6.7	How Does a Rock Avalanche Travel?		201
		6.7.1	Shear Layer as an Ensemble of High-Speed Particles	202
	6.8	The Problem of the Anomalous Mobility of Large Rock Avalanches		208
		6.8.1	Statement of the Problem	208
		6.8.2	A List of Possible Explanations	209
		6.8.3	Explanations That Do Not Require Liquid or Gaseous Phases	210

		6.8.4	Explanation of the Anomalous Mobility of Rock Avalanches Invoking Exotic Mechanisms and New Phases	212
	6.9	\multicolumn{2}{l	}{Frictionites, Frictional Gouge, Thermal Effects, and Behavior of Rocks at High Shear Rates}	215
		6.9.1	Frictionite, Melt Lubrication, and the Kofels Landslide	215
		6.9.2	Vapor or Gas at High Pressure	221
7	\multicolumn{3}{l	}{**Landslides in Peculiar Environments**}	223
	7.1	\multicolumn{2}{l	}{Landslides Falling into Water Reservoirs}	224
		7.1.1	General Classification	224
		7.1.2	Limit $C \ll 1$ (Mass of the Landslide Much Greater Than the Water Mass)	225
		7.1.3	Mass of the Landslide Comparable to the Water Mass, $C \approx 1$	232
	7.2	\multicolumn{2}{l	}{Coastal Landslides and Landslides Falling onto Large Water Basins, $C \gg 1$}	233
		7.2.1	General Considerations	233
		7.2.2	Lituya Bay	235
		7.2.3	Landslides Propagating Retrogressively from the Sea to Land	238
		7.2.4	Landslides Falling on a Tidal Flat	239
		7.2.5	Generation and Propagation of the Tsunami in Lakes and Fjords	243
	7.3	\multicolumn{2}{l	}{Landslides Traveling on Glaciers}	245
		7.3.1	General Considerations	245
		7.3.2	Dynamics of Landslides Traveling on Glaciers	247
	7.4	\multicolumn{2}{l	}{Landslides in the Solar System}	253
		7.4.1	Landslides on Planets and Satellites, Except Mars	253
		7.4.2	Landslides on Mars	255
8	\multicolumn{3}{l	}{**Rockfalls, Talus Formation, and Hillslope Evolution**}	263
	8.1	\multicolumn{2}{l	}{Introduction to the Problems and Examples}	265
		8.1.1	General	265
		8.1.2	Physical Processes During a Rock Fall	267
	8.2	\multicolumn{2}{l	}{Simple Models of a Simple Object Falling Down a Slope}	268
		8.2.1	Simple Models of Rolling, Bouncing, Gliding, and Falling	268
	8.3	\multicolumn{2}{l	}{Simple Rockfall Models}	273
		8.3.1	A Simple Lumped Mass Model	273
		8.3.2	The CRSP Model	275
		8.3.3	Three-Dimensional Programs	277

		8.4	The Impact with the Terrain	277
		8.4.1	The Physical Process of Impact Against Hard and Soft Ground	277
		8.4.2	Coefficients of Restitution and Friction	278
		8.4.3	Block Disintegration and Extremely Energetic Rockfalls	281
	8.5	Talus Formation and Evolution		282
		8.5.1	Kinds of Talus and Their Structure	283
		8.5.2	Physical Processes on Top of Taluses	283
	8.6	Topple		293
9	**Subaqueous Landslides**			295
	9.1	Introduction and Examples		297
		9.1.1	Some Examples in Brief	297
	9.2	Peculiarities of Subaqueous Landslides		301
		9.2.1	Types of Subaqueous Landslides	301
		9.2.2	Differences Between Subaerial and Subaqueous Landslides	301
		9.2.3	The H/R-Volume Diagram for Submarine Landslides	303
	9.3	Triggering of Subaqueous Landslides (Especially Submarine)		304
	9.4	Forces on a Body Moving in a Fluid		309
		9.4.1	General Considerations	309
		9.4.2	Drag Force	310
		9.4.3	Skin Friction	313
		9.4.4	Added Mass Coefficient	314
	9.5	Block Model for Subaqueous Landslides		319
		9.5.1	Equation of Motion for the Block	319
		9.5.2	Application of the Block Model to Ideal Cases	321
	9.6	Tsunamis		324
		9.6.1	Introduction	324
		9.6.2	Propagation of Tsunami Waves in the Ocean	331
		9.6.3	Tsunamis Generated by Submarine Landslides	334
	9.7	More Dynamical Problems		343
		9.7.1	Outrunner Blocks	343
		9.7.2	Debris Flows	347
		9.7.3	Theories for the Mobility of Submarine Landslides	349
10	**Other Forms of Gravity Mass Flows with Potentially Hazardous Effects**			353
	10.1	Lava Streams		355
	10.2	Ice Avalanches		358
		10.2.1	Fall of an Ice Avalanche	358
		10.2.2	Stability Condition of an Ice Avalanche	360

10.3	Catastrophic Flood Waves		361
10.4	Snow Avalanches		364
10.5	Slow Landslides and Soil Creep		366
	10.5.1	Sackungs and Lateral Spreads	366
	10.5.2	Soil Creep and Other Superficial Mass Movements	368
10.6	Suspension Flows: Turbidites and Turbidity Currents, and Relationship with Submarine Landslides		368
	10.6.1	Turbiditic Basins	370
	10.6.2	Ancient Turbidites	370
	10.6.3	Flow of a Turbidity Current	373

Appendix GeoApp (Geological and Geotechnical) 377

Appendix PhysApp (Physical) 383

Appendix MathApp (Mathematical) 385

References 395

Index 405

Chapter 1
Introduction and Problems

On May 31, 1970, a strong quake shakes the Peruvian Andes, causing the collapse of 50 million cubic meters of rock and ice from the mountain called the Nevados Huascaran. After 500 m of free fall, the material collapses against the glacier 511; disintegrating at once, it generates a shock wave. An ominous black cloud composed of pulverized material develops in the area of the impact, obscuring part of the Nevados. Traveling rapidly across the glacier, the material reaches astounding velocities, perhaps greater than 90 m/s. Huge boulders are cast in the air. Bombarding locations some kilometers away, boulders kill people and cattle and devastate a large area. The intake of water and ice rapidly transforms the solid material in a debris flow, a lethal and rapid river of dense fluid capable of carrying huge boulders with colossal devastating power. The towns of Yungay and Rahnrahirca are shattered and 8,000 of the inhabitants killed. The area is completely covered with a heavy muddy deposit. Reaching the river Rio Santa, the debris flow increases its death toll in the town of Matacoto.

The disaster of the Nevados Huascaran represents one of the worst landslide disasters in historic times. It is emblematic at the start of this book in conveying several physical-dynamical phenomena related to landslides: the huge energy acquired during the initial fall, the travel on the glacier, the transformation into a debris flow, and its destructive potential. It is disturbing that the catastrophe was not unforeseeable. Similar events had occurred previously, the latest in 1962.

Landslides are not only a significant geohazard; they also contribute to the geomorphic reshaping of the landscape. Rivers and glaciers are among the most effective geomorphic agents sculpting the mountain areas. However, they erode slowly and selectively the bedrock and soils that happen to lie along their path. So doing, they make slopes steeper and steeper. Suddenly, the gravity takes over. In the time span of a mere minute, a landslide transforms the local topography, redistributing the material to long distances from the source. Landslides should be envisaged as a natural and common phenomenon in the geological history, but they are rarely perceived as such by the community. Only because of our biased perception of the geological times and conditions, do we watch landslide catastrophes as exceptional and aberrant.

Landslides are very variable in terms of material involved, size, velocity, and destructive potential. They may involve rock, soil, mud, water, and ice. The physical behavior of landslides is much variable depending on the material involved.

The figure below shows the terminal branch of the debris flow lobe formed by the Nevados Huascaran debris avalanche of May 31, 1970. Yungay is visible in the lower part of the picture, partly covered by the secondary lobe of the debris flow. Ranrahirca is on the opposite side. On the right side, the torrent Rio Santa meanders on the debris; at the extreme right the town of Matacoto. Photograph taken in June 1970. Image pla00012 (USGS) of public domain. Author: Plafker.

1.1 Landslides: An Overview

1.1.1 What Is a Landslide?

We can define a landslide as the movement of rock, detritus, or soils caused by the action of gravity. To distinguish landslides from other forms of gravity mass flows, we require in the definition that the bulk of the moving material should have density at least 10% greater that the density of water. Most landslides are very small: Every year mountain roads need removal of blocks fallen from the flanks. Larger collapses may affect local watercourses and influence the activity of local communities; greater slides may provoke disasters and change the geomorphologic setting of several square kilometers of land. Some landslides evolve very slowly, and special instruments may be necessary to become aware that they are in fact moving. Others may travel faster than 100 km/h. And several ones start with a creeping, imperceptible movement, to suddenly accelerate and degenerate in a catastrophic debris avalanche. Some landslides travel similar to a fluid, resembling the flow of water. Others are akin to granular flows. Many landslides come to a halt without affecting vast areas beyond the immediate surroundings; others plunge into the sea and cause damage hundreds of kilometers away.

In this book, the denomination of gravity mass flow is also adopted. It stands for a somehow broader class comprising any catastrophic movement due to the action of gravity, irrespective of the material involved and density. Thus, snow avalanches, catastrophic flood water waves, hyperpycnal flows (caused by variation in salinity or temperature that affect the density of water), as well as suspension flows (due to suspension of solid material in air or water, such as pyroclastic flows and turbidity currents) are gravity mass flows but not landslides. The division seems at first artificial, and indeed it is partly, but it is justified by the different physics, means of investigation, and scientists involved in the analysis.

1.1.2 Landslides as a Geological Hazard

The basic motivation behind landslide studies is the prevention and mitigation of disasters and reduction of risk (Fig. 1.1). Much has been published on this theme, so very little will be added in the present book. Most landslides are small and their killing potential is limited. However, large landslides may be very catastrophic. Table 1.1 reports some of the deadliest landslides in the twentieth century.

Delayed consequences to landslides may also tragically contribute to the death toll. In 1786, a strong earthquake in the province of Sichuan in China released a landslide that dammed a local river for 10 days. As many as 100,000 people were drowned when the dam failed inundating an area 1,400 km downstream. The Vaiont tragedy (Northern Italy, 1963) was the result of water spilling over a dam, thrust by the body of a large landslide that invaded the artificial reservoir. Submarine landslides may generate devastating tsunamis. Accounting for both the small but frequent and for the large and rarer catastrophic landslides, it has been calculated that during one average year, landslides kill about 5–7 people in Norway, 18 in Italy, 25–50 in USA, 186 in Nepal, 170 in Japan, and 140–150 in China (Sidle and Ochiai 2006).

Fortunately, this is not an extremely heavy death toll compared to other natural disasters or car accidents. However, the negative consequences of landslides are not limited to loss of life, but include the destruction of houses and infrastructures (Fig. 1.1), loss of productivity in the area affected, unpredictable changes in the local watercourse, and reduction of arable or habitable land. Estimated costs of landslide damage (including both the direct costs caused by destruction, and

Fig. 1.1 *Top*: In addition to the death toll, landslides often interrupt railways and roads and devastate infrastructures. (**a**) A boulder has interrupted the railway in Maccagno (Varese, Italy). It is part of a larger landslide occurred in 2004 in Varenna (Northern Italy). (Photograph courtesy of G. B. Crosta.) (**b**) The village of La Conchita in California is frequently affected by killing landslides. (Photograph USGS of public domain.) *Bottom* (**c**) A small landslide in Cortenova (northern Italy). (Photograph courtesy of G. B. Crosta.)

1.1 Landslides: An Overview

Fig. 1.1 (continued) (**d**) Higher devastation is carried by landslides of high volume and mobility. The picture shows the pathway of the lahar from the Nevado del Ruiz volcano that in November 1985 wiped out the town of Armero (Colombia). Armero was located in the center of the picture (Photograph USGS of public domain)

indirect costs due to long-time effects in the local economy) are about 4 billion USD in Japan, 70 million USD in Canada, 2.6–5 million USD in Italy (Sidle and Ochiai 2006).

1.1.3 Landslides as a Geomorphic Driving Force

Landslides contribute significantly to the geomorphic evolution of the natural environment. As soon as a slope is steepened by tectonic uplift or by river and glacial erosion, gravity will tend to redistribute the rock or soil and smooth out the terrain. The instability that frequently ends with the observable, catastrophic event (a rock fall, a rock slide, topple, or soil movement) is thus a normal result of natural phenomena. For example, a glacial valley appears as "U" shaped also as a consequence of rock falls and slope adjustments that after glacier retreat have smoothed the steep valley walls. It is difficult to assess the amount of landslide material and the total volume of rock and soil derived from landslides and its distribution, because ancient landslide deposits may have been covered by more recent sediments and go unnoticed or misinterpreted, for example, as moraines. Landslides may affect the geomorphology of vast areas, creating new local topography. Landslide bodies often interrupt the river

Table 1.1 Some of the deadliest landslides of the twentieth century. A few of them are described in more details in the book (From different sources)

Location	Country	Date	Killed	Landslide type and notes
Gansu province	China	December 16, 1920	180,000	675 loess flows affected an area of 50,000 km^2
Armero (Nevado del Ruiz)	Colombia	November 13, 1985	25,000	Lahars formed by eruptions of ice-capped Nevado del Ruiz volcano
Yungay (Nevados Huascaran)	Peru	May 31, 1970	18,000	Rock avalanche mixed with ice and water, caused by earthquake along the path of a glacier
Khait	Tajikistan	1949	12,000	Rock avalanche detached by a strong quake
Huaraz	Peru	December 1941	4,000–6,000	Debris flow caused by a morain dam break impounding a glacial lake (GLOF); more events in 1962 and 1970
Kelud volcano	Indonesia	1919	5,160	Drainage of the Crater lake
Yungay (Nevados Huascaran)	Peru	January 10, 1962	4,000–5,000	Rock avalanche mixed with ice and water, caused by earthquake
Colima	Honduras	September 20, 1973	2,800	
Ranrahirca (Nevados Huascaran)	Peru	January 10, 1962	4,000	Rock avalanche mixed with ice and water, and debris flow
Longarone (Belluno)	Italy	October 9, 1963	2,000	Rock avalanche onto artificial water reservoir; killing due to water splash
Bihar, Bengal	India	October 1, 1968	1,000	The region is prone to large quakes
Villa Tina (Medellin)	Colombia	September 27, 1987	217	Small and shallow slide in lateritic residual soils
La Conchita	California (USA)	January 10, 2005	10	Debris flow mobilized previous landslide deposits due to intense rainfall

networks and give origin to new lakes, like the Fernpass Lake in the Austrian Alps or the Molveno Lake in the Italian Alps. Normally, lakes dammed by rock avalanches may last for thousands of years, while those created by debris flows are ephemeral. By reducing ablation, landslides falling on glaciers act as insulating layers changing positively the mass balance of the glacier, as documented for example for the Sherman landslide in Alaska ($\rightarrow$Box 7.3).

> **Box 1.1** External Link: Biological and Paleobiological Consequences of Landslides
>
> Landslides are capable of changing at once the local geomorphology of a vast area. One example of ancient lake created by landslide deposits is the Pianico Lake of Riss-Mindel interglacial age in Northern Italy. The lake has hosted a community of animals and plants thriving for thousands of years. The lake varves contain a spectrum of macrofloral tree genera like *Acer*, *Pinus*, or *Prunus*, now extinct for that area. In addition to the thin layers deposited in tranquil lake environment, the Pianico deposits also revealed tephra and turbidite deposits. Documentation of this kind allows a reconstruction of paleoclimatology and paleobiology of the area.
>
> Another example of biological aspect related to landslides is the colonization of landslide deposits by different organisms. The landslide may change locally the kind of soil, promoting some plant species and stunting the growth of others, thus changing also the local ecosystem of animal species. In the case of the Sherman landslide ($\rightarrow$Box 7.3) the deposit is surrounded by ice. Thus, the landslide body represents a small isolated environment that allows testing the validity of the Island Biogeography theory, according to which the local species equilibrium derives from continuous colonization by new species and possible extinction of other species inhabiting the deposit.
>
> Sometimes landslides may uncover interesting sediments of geological and paleontological interest. A remarkable discovery was made in the area of the Lavini di Marco landslide, in Northern Italy. The analysis of landslide deposits reveals a sequential deposition from different events, dating from ancient to historical. In May 1988 an amateur naturalist, L. Chemini, was on a mountain trip on top of the detachment niche of the landslide. He noticed a series of crater-like holes on the upper layers of the lower Jurassic limestone, each some decimeter across. These were the first dinosaur footprints ever found in Italy. Before that discovery, most specialists rejected the idea that dinosaurs could inhabit Italy during the Mesozoic, as most of the peninsula was submerged. However, the richness of these footprints clearly defied that simplistic concept demonstrating varied dinosaur communities. Since that finding, several other dinosaur remains and traces have been found in Italy. Figure 1.2 shows some footprints from the landslide area.
>
> (continued)

Box 1.1 (continued)

Fig. 1.2 Dinosaur footprints on the scar of the Lavini di Marco landslide (Trento, Italy). *Upper photograph*: footprint of a biped dinosaur. *Lower*: sauropod footprint track with view on the Adige valley. It is interesting that the footprints have not been erased by the rock avalanche that slid on them

1.1.4 Physical Aspects of Landslides

Many researchers have treated the dynamics of landslides using the language and methods of mechanics and physics. In principle everything from the Universe to the human psychology embodies a physical system. However, it is difficult to apply the reductionism character of physics when systems are too complex. Although landslides are in fact a complex phenomenon, understanding their behavior requires in principle a few basic laws of mechanics. Their apparently complex and unpredictable behavior is due more to lack of knowledge on the physical conditions and materials characteristics, rather than to a lack of knowledge for the basic laws. The motion of landslides can consequently be described in terms of well-known laws of friction, cohesion, and gravity (Fig. 1.3a). There are several reasons why a physical description of landslides is desirable.

1. The need to quantify the observable facts typical of exact sciences urges the geologist to go beyond the mere description of the phenomenon. It compels to think in terms of physical laws, breading the perspective of the geological phenomenon. One could also cite the science fiction writer Robert Heinlein: "If it can't be expressed in figures, it is not science; it is opinion." If on the one hand this is not always possible with geological problems, on the other hand the process of thinking quantitatively is often rewarding in terms of gaining a much deeper insight into observable facts.
2. Secondly, as people continue to move to hazardous zones, the practical need to predict future run-out, damage, energy, and path of landslides or other gravity mass flows will increase. Good predictive models may result in saving human life and properties. The writing of the equations of motion of a landslide is a prerequisite to any computer modeling.
3. Part of the physics and mathematics needed to describe landslides belong to the standard curriculum of a physicist and engineer: fluid mechanics, elasticity theory. Others, like the mechanics of granular media and rheology, are often

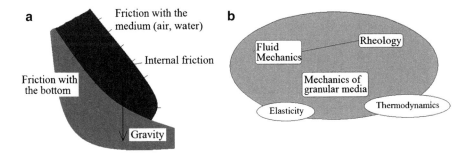

Fig. 1.3 (**a**) Very simplified scheme of the forces acting on a moving landslide. (**b**) Physics and mechanics needed for a physical approach to landslides

skipped in regular courses (Fig. 1.3b). In any case, the knowledge acquired by studying the physical aspects of gravity mass flows can be applied to other areas of physics, applied mathematics, and earth sciences. The study of landslides thus gives the possibility to tackle interesting problems in other disciplines.

Box 1.2 One Step Back: The Gravity Pull

There would not be landslides in the absence of the gravity force. In contrast to other forces acting on a landslide that take on different forms depending on the material involved and on the geometry, the gravity force is a body force and has a very simple form. Suppose dropping two coins of very different weight from the same height. The coins will reach the ground at the same time (with perhaps a slight difference if the two coins have different shape and size). Deceived by common sense and by the authority of Aristotle, scholars throughout the centuries were persuaded that the heavier body would reach the ground first. We know from Newton's law of gravitation that the force of gravity between two bodies has the form

$$P = G\frac{Mm}{R^2} \tag{1.1}$$

where in the present example M is the Earth's mass, m is the mass of one coin, R is the Earth radius and G is Cavendish gravitational constant. However, Newton's second law of dynamics states that to each force there corresponds an acceleration

$$g = \frac{P}{m} \tag{1.2}$$

from which substituting Eq. 1.1, one obtains that the acceleration is independent of the mass of the coin

$$g = G\frac{M}{R^2} \tag{1.3}$$

The value of gravity acceleration along the equator at sea level is

$$g = 9.81 \text{ ms}^{-2}. \tag{1.4}$$

> **Box 1.3** One Step Back: Velocity and Displacement of a Body Subject to Constant Acceleration
>
> It is useful to get acquainted with the kinematical laws of a freely falling body. A complete calculation is presented in the PhysApp; here we pick up some essential and very basic results.
>
> If the body is dropped at zero velocity and the air resistance is neglected, the velocity acquired is proportional to the time t from start
>
> $$v = gt \tag{1.5}$$
>
> and the space is proportional to the time squared
>
> $$z = \frac{1}{2}gt^2. \tag{1.6}$$
>
> The velocity as a function of the position and vice versa are so given as
>
> $$v = \sqrt{2gz}; \quad z = \frac{v^2}{2g}. \tag{1.7}$$
>
> As an example, Eq. 1.7 gives a velocity of 44.29 m/s after 100 m of free fall in the Earth gravity field.
>
> The force is a vector, i.e., it can be decomposed into three components acting along three arbitrary Cartesian directions of space x, y, z. In the study of gravity mass flows, the gravity acceleration is often projected along one vector perpendicular to the local slope $g_\perp = g\cos\beta$ and one vector parallel to slope, $g_= = g\sin\beta$ where β is the slope angle. The equation of motion of a body sliding along a surface with constant inclination β and without the effect of friction can be treated as the free-falling body with g substituted with $g_= = g\sin\beta$. In this case the velocity represents the component parallel to the slope.

1.2 Types of Landslides

1.2.1 Geometrical Characteristics of a Landslide

Figure 1.4 reproduces an idealized block diagram of a landslide. The failed material starts from a zone of depletion and deposits in the accumulation zone. The crown of the landslide identifies the region adjacent to the highest parts of the failed mass. The scarp is the steep rupture surface between the failed body and the terrain. Several minor scarps due to internal shearing may also punctuate

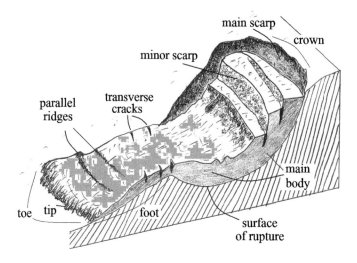

Fig. 1.4 Main geometrical elements of a landslide (From Varnes 1978, redrawn and simplified)

the main landslide body. The surface of rupture identifies the interface at the base of the landslide where the material has slid. In this example the surface appears curved while in other cases it may be planar of complex. The foot is the material deposited in the accumulation zone, beyond the surface of rupture. The landslide deposit ends with a toe, which is the line (usually bent) between the accumulated material and the untouched terrain. The tip is the point of the landslide deposit farthest from the crown. Some landslides exhibit transverse tension cracks in the region of the foot closest to the scarp. Closer to the toe, transverse ridges are sometimes formed. The right and left flanks of the landslide are identified by standing with shoulders to the crown.

1.2.2 Description of the Seven Types of Movements

Various systems of landslide classification have been proposed, such as the system by Varnes (1978), Hungr et al. (2001), and Hutchinson (1988). Various authors have contributed to the EPOCH classification system, which stems from the Hutchinson system (1998). The EPOCH system recognized seven classes and three material types, for a total of 21 possibilities. These are listed in Table 1.2.

1.2.2.1 Fall

A fall is the movement of material from a stiff headwall or cliff. It generally involves limited volumes of material, most usually rock. The material falls en

1.2 Types of Landslides

Table 1.2 Classification of landslides adopted from the EPOCH system (Dikau et al. 1996)

	Material		
Movement	Rock	Debris	Soil or earth
Fall	**Rockfall**	Debris fall	Soil fall
Topple	**Rock topple**	Debris topple	Soil topple
Translational slide	**Rock slide**	Debris slide	Soil slide
Rotational slide or "slump"	Rock slump	**Debris slump**	**Soil slump**
Lateral spreading	**Lateral rock spreading**	Lateral debris spreading	**Lateral soil spreading**
Flow	**Rock flow or *Sackung***	**Debris flow**	**Soil flow**
Complex			

In boldface the most common types

masse, moving freely in the gravity field. The contact with the terrain occurs especially in the last part of the trajectory, where the material becomes frequently shattered (Fig. 1.5a).

1.2.2.2 Topple

A topple is the rotation of a vertical slab about a pivoting point located at the base (Fig. 1.5b). Topple is typical of compact vertical slabs (usually but not exclusively rock) lying on soft, unconsolidated terrain. The movement may be extremely slow for long periods, culminating with a catastrophic fall of the slab.

1.2.2.3 Translational Slide

A slide is defined as the movement of material along a shear surface. For a translational slide, this surface is planar (Fig. 1.5c). The identity of the shear surface is somehow preserved and distinguishes a slide from a flow.

1.2.2.4 Rotational Slide

In a rotational slide, the detachment surface is roughly circular, spoon-like. In contrast with translational slides, where the planar surface often originates from a weakness zone, the circular shape of a rotational slide is created by the failure itself and derives from the geometrical distribution of the shear stress (Fig. 1.5d).

1.2.2.5 Flow

According to Dikau et al. (1996, p. 149), a flow is "a landslide in which the individual particles travel separately within a moving mass. They involve whatever

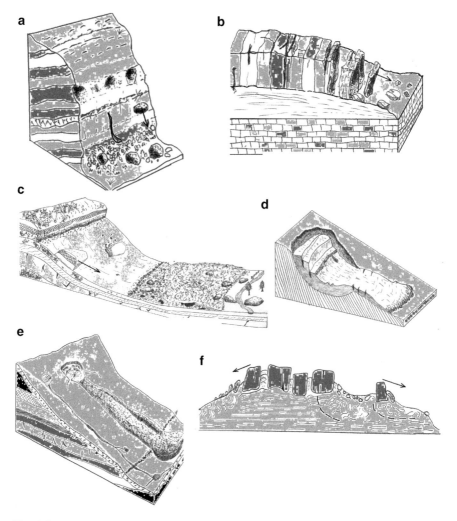

Fig. 1.5 Six landslide types: (**a**) fall, (**b**) topple, (**c**) translational slide, (**d**) rotational slide, (**e**) flow, and (**f**) lateral spreading (Some drawings inspired from Dikau et al. 1996)

material is available to them and may therefore be highly fractured rock, clastic debris in a fine matrix or a simple, usually fine, grain size. Flow in its physical sense is defined as the continuous, irreversible deformation of a material that occurs in response to applied stress." A flow is thus characterized by a fluid-like movement, in which the information on the detachment surface has been lost. A slide may evolve into a flow if the energy and/or the run-out are sufficient to rework completely the material (Fig. 1.5e). In this work, the word "rock avalanche" will be preferred to "rock flow" to denote a catastrophic landslide mostly composed of rock, usually very fast and mobile. Rock avalanches, however, fall into the type "Complex" in the EPOCH classification scheme.

1.2.2.6 Lateral Spreading

It consists of a lateral movement of rock or soil, often of large extension (Fig. 1.5f). In the case of rock spreading, the rate is often slow (from one tenth of mm to 10 cm per year) and is generally caused by deep-seated viscoplastic material underneath the rocky slabs. Soil spreads, like the ones involving quick clays, can move extremely fast, with speed in the range of several meters per second.

1.2.2.7 Complex

It is a generic name used when a landslide changes behavior during the movement. In addition to the rock avalanche (rock slab turned into a granular flow) the EPOCH system considers the flow slide as a member of this class (Dikau et al. 1996). A flow slide consists of a portion of soil loosing cohesion during the flow, to the point of becoming a completely fluidized mass.

1.3 A Physical Classification of Gravity Mass Flows

It seems convenient for the purpose of this book to introduce a classification of gravity mass flows based on the kind of predominant physical behavior and physical description that appears more appropriate. The classes are listed as follows.

1.3.1 Rheological Flows (Cohesive) (→Chap. 4)

A mudflow is a kind of landslide where large amount of clay-rich soil mix with water and runs downslope, often canalized along river beds. Such kind of flow represents the epitome of a rheological flow. A rheological flow is ideally portrayed as a fluid with complex properties (Figs. 1.1d and 1.5e). Its physical description will be based on fluid mechanics, with rheological properties specified in terms of a non-Newtonian fluid. The basic building blocks of a rheological flow are the fine particles (especially clay or silt) that when mixed with water confer a cohesive behavior to the flow. Rheological flows in nature will normally include sand, gravel, and boulders, which can possibly give a frictional component to the rheological behavior. Depending on composition, rheological flows may be called

Fig. 1.6 An example of the product of a rheological flow: lahar deposits of the Cotopaxi volcano, Ecuador (Photograph courtesy of Roberto Scandone and Lisetta Giacomelli)

debris flows (if the composition has certain variability in clast size), mudflows (if mostly composed by fines as in the previous example), or hyperconcentrated flows (if the water is present in higher degree, resulting in a Newtonian fluid behavior like in a flooding river).

Rheological flows affect particular sedimentary environment such as moraines, loess, pyroclastic deposits, river banks, or subaqueous environments dominated by clastic deposition. Water greatly affects the mechanical behavior of debris flows, altering the rheology and diminishing the effective shear strength and viscosity (Fig. 1.6).

1.3.2 Granular Flows (Frictional) (→Chap. 6)

Granular flows are rapid movement of granular material where friction, described by the classical laws of Amontons-Coulomb, plays an important role in the dynamics (Fig. 1.7). A small-scale analog of a granular flow is the superficial movement of sand inside a jar gently tilted around the axis. Sand avalanches on a large scale, however, are rare. In nature, large-scale granular flows normally develop from the collapse of an initially unbroken rock slab. High-energy impact, internal deformation and crushing determine a rapid disintegration of the rocky mass.

1.3 A Physical Classification of Gravity Mass Flows

a

Fig. 1.7 (a) The hummocky deposit of a small rock avalanche in the Elbortz mountains, Iran

1.3.3 Rock Falls and Topples

A rockfall is characterized by the fall of blocks in isolation in the gravity field, in which the phase of free fall represents a significant part of the path (Fig. 1.8). The rock fall is essentially described by a rigid body motion, but the impact with the terrain may be a much complicated problem. The block may bounce against hard rock, and then collide and roll on soft ground. The soil deformation involves complex interactions between a cohesive and the frictional component of the soil. Usually the process is described in terms of empirical coefficients of restitution and rolling friction, but this level of description neglects the complexity of block-terrain interaction. However, the rigid body dynamics coupled with coefficients of restitution to describe the impact is in a first approximation considered satisfactory to describe the process and to provide valuable information for mitigation measures in rockfall-prone areas.

In this class we include topple failures, where sub-vertical slabs revolve at high speed around a pivoting point at the base, much like a book resting on an edge and pushed at the top. Upon impact with the ground, toppling slabs may partly disintegrate, without much horizontal movement.

Fig. 1.7 (continued) (**b**) A small rock avalanche in Valdossola (Alps) (Original photograph courtesy of Giovanni B. Crosta)

1.3.4 Slow Landslides and Creep

Certain slides move very slowly and often their presence is revealed only by clefts and fallen blocks (Fig. 1.9). The behavior of slow landslides is often poorly known. Slow slides may respond to the presence of clay layers

1.3 A Physical Classification of Gravity Mass Flows

Fig. 1.8 An example of rock fall. The large bluish boulder in the torrent is a porphyrite block fallen from the side of a steep gorge. Val di Scalve (Italy)

underneath the rock formation, while others may be associated to fractures along a weak plane. They are often affected by karst phenomena or freeze-thaw activity.

1.3.5 Other Gravity Mass Flows Similar to Landslides

Suspension flows, in which solid grains travel upheld by air or water, are not properly landslides. The flow along slope is promoted by the presence of a solid fraction in the medium, which locally increases the density. The speed acquired by the mass flow in turn increases the velocity and turbulence in the medium, thus aiding the suspension. The process in therefore self-maintaining, waning in proximity of a slope break. Examples of subaerial suspensions are the pyroclastic flows, where the upward flow of turbulent air is assisted by hot gases. Also pulverized rock from rock avalanches may form suspensions. Turbidity currents are well-known suspension flows in the subaqueous environment. They will be considered in some more detail, as they are close relatives of submarine landslides. The physics of suspension flows is the result of a quite complex interplay between the solid suspension and the turbulent fluid, and requires sophisticated understanding of fluid mechanics.

Fig. 1.9 The La Verna monastery in Italy is built on cliffs standing on a clayey basement. The slow creep causes the opening of clefts

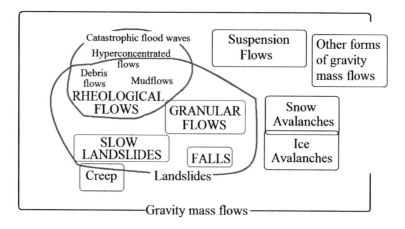

Fig. 1.10 Landslides as members of the broader class of gravity mass flows. For brevity, not all kinds of gravity mass flows are cited

1.3 A Physical Classification of Gravity Mass Flows

Ice and snow avalanches are not landslides either, but have many characteristics in common and may be partly associated to landslides, like for example in rock-ice mixed avalanches. Figure 1.10 summarizes some gravity mass flows and their relationship with landslides.

It is apparent that the present classification is somehow artificial, and has limits. Firstly, there are often cases in which one type of gravity flow transforms into another. In the example of the Nevados Huascaran, a granular flow changed into a rheological flow. Secondly, different typologies can coexist during the same event. For example, a rock avalanche may fall on soft soil with fluid properties, and slide on it. A submarine rock avalanche may transform in a debris flow and at the same time generate a turbidity current. Thirdly, the physical classification may be in some cases problematic. For example, there may be uncertainty as to whether soil flows with low amount of fines is rheological or frictional.

General references Chapter 1: Sidle and Ochiai (2006); Dikau et al. (1996); Turner and Schuster (1996).

Chapter 2
Friction, Cohesion, and Slope Stability

Every solid or liquid mass on Earth is influenced by gravity. A mass of soil or rock remains stable if the gravity force is counterbalanced by the reaction forces exerted by the adjacent bodies and the terrain. Rock masses and soils on the surface of the Earth appear steady at first sight. However, this impression is often deceiving, as the masses may slowly creep, terminating with a sudden collapse. Natural buttressing of a potential landslide may be removed of weakened, causing portions of the mass to fall. Change in stability conditions may be consequent to a variety of causes such as river undercutting or ice melting. Earthquakes can instantly change the local force equilibrium, anticipating the fall. The process of mountain building continuously overloads rock masses with renewed stress throughout time scales of several million years. Newly produced deposits may also become unstable. For example, volcanic eruptions deposit enormous amounts of pyroclastic materials, which may subsequently be mobilized by rain.

A landslide starts as consequence of terrain instability, and for this reason it is important in geotechnical practice to ascertain the stability conditions of soils or rocks. Owing to the significance in the prevention of disasters, slope stability has been the subject of much effort. There exist numerous numerical models, textbooks, and computer programs for assessing the stability on different kinds of terrain. Here the problems of instability and the initiation phase of a landslide are very briefly considered, limiting ourselves to only a few basic concepts.

The chapter starts with the basic laws of friction and cohesion, of fundamental importance not only for the problems of slope stability, but also for the dynamics of landslides.

The figure shows parallel tension cracks in soil, indicative of instability. The whole area is subject to creep, which may culminate in a catastrophic landslide. The barren surface visible in the background is part of the detachment niche of a landslide that on June 20, 1990, cost the life of at least 170 people. Most of the bodies were never found. Fatalak (Iran), April 2003.

2.1 Friction and Cohesion

2.1.1 Normal and Shear Stresses

Consider an object like a book resting on a plane inclined with angle β. Because we assume the book to be static, according to the laws of dynamics the gravity force must be counterbalanced by the reaction force exerted by the table. The gravity force can be decomposed into the components normal $F_\perp$ and parallel $F_=$ to the plane: $F_\perp = Mg\cos\beta$ and $F_= = Mg\sin\beta$. The component of gravity $F_=$ is equal and opposite to the reaction force, so that there is no net force perpendicular to the plane (Fig. 2.1). The force balance parallel to the plane is more complex and requires introducing the friction force ($\rightarrow$Sect. 2.2).

We first define normal stress σ and shear stress τ the force, divided by the area S of the surface in contact, respectively, normal and parallel to the plane. Expressing the magnitude of the weight force as $\rho g D'S$, we can write

$$\sigma = \frac{F_\perp}{S} = \frac{\rho g D' S \cos\beta}{S} = \rho g D' \cos\beta = \rho g D \cos^2\beta$$
$$\tau = \frac{F_=}{S} = \frac{\rho g D' S \sin\beta}{S} = \rho g D' \sin\beta = \rho g D \sin\beta \cos\beta. \quad (2.1)$$

where $D' = D\cos\beta$ is the thickness of the object and D is the vertical projection of the thickness (Fig. 2.1). The stress is measured in pascals (Pa).

2.1.2 Friction

2.1.2.1 Basic Laws of Friction

The friction force is universally present in everyday life. The very actions of walking, running, or driving a car are possible because friction acts between the surfaces of solid bodies. A pond of oil on the road reduces the friction between the car tires and the road, causing a dangerous loss of grip. Friction is perhaps better known for its negative implications: parasitic resistance between different parts

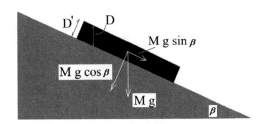

Fig. 2.1 Elementary geometrical elements for the identification of normal and shear stress

of engines, overheating of joints and machines, wear, and loss of efficiency in industrial processes.

The physics of friction was explored for the first time by Leonardo da Vinci (1452–1519). Leonardo established that the friction force is proportional to the total weight but is independent of the mass distribution. However, like many other discoveries of the renaissance genius, these studies on friction were lost for centuries. At the end of the seventeenth century, Guillaume Amontons (French inventor and physicist, 1663–1705) rediscovered the same principles, and further noticed that, at least for the materials he experimented, the friction force was about one third of the load. Nearly one century later another Frenchman, Charles-Augustine de Coulomb (1736–1806), famous for his studies on electricity, attempted a physical explanation of the laws of friction described empirically by Amontons.

To study the friction forces, the experimenter applies a force parallel to a horizontal plane as shown in Fig. 2.2. The force F_A in correspondence of which the body begins to move is the friction force.

In modern terminology, the properties of the friction force as found by Leonardo da Vinci, Amontons, and Coulomb, can be stated as follows:

1. The friction force is independent of the contact area between the two surfaces. For example, if the body is shaped as a parallelepiped with different faces, the friction force is independent on which face it rests.
2. The friction force F_A is found to be proportional to the body weight $P = Mg$, where M is the body mass. The ratio $\mu = F_A/P$ between the two forces is thus independent of the mass and of the gravity field; it is called the static friction coefficient. In shorthand,

$$F_A = \mu P. \tag{2.2}$$

3. Typically, the magnitude of the friction force is comparable for materials of similar properties. For metals it is about one third of the weight, and for rocks it is about one half of the weight (which means friction coefficients of 1/3 and 1/2, respectively).
4. As stated earlier, a body remains static if the applied tangential force is smaller than the friction force. Let a horizontal force F'_A slightly exceeding the friction force be applied to the center of mass of the body. The body starts moving along

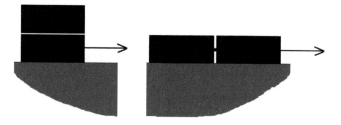

Fig. 2.2 Whether the weights are on top of each other or besides, the friction force is the same

the direction of the applied force, and with constant acceleration given by Newton's law of dynamics $a = (F'_A - F_A)/M$. Hence, when the body is moving, the friction force is collinear to the tangential force but has opposite direction.
5. At a closer inspection, point 4 results to be an oversimplification. This is because the friction force necessary to commence sliding is greater than that measured during sliding. Distinguishing a static from a dynamic friction force $F_{A,DYN}$, one should properly write $a = (F'_A - F_{A,DYN})/M$ where $F_{A,DYN} < F_A$ is the dynamic friction force. Experiments, however, show that the static and dynamic frictions do not differ much, and for practical purposes this difference may be neglected, or $F_{A,DYN} \approx F_A$. Dealing with landslides, many effects like the presence of pore water or the variability of rock behavior will influence the friction coefficient in a more substantial manner. The description in terms of two different coefficients is an oversimplification, anyhow. A full analysis would require considering the whole process from the static condition to full sliding. In fact, the friction coefficient decreases continuously as a function of the time of contact between the two surfaces, reaching a constant value after fractions of a second.
6. The friction force is only weakly dependent on the velocity. For many practical purposes it can be considered as independent of it.

2.1.2.2 Inclined Plane

It has been shown (←Sect. 2.1) that the gravity force parallel and perpendicular to an inclined plane are, respectively, $Mg \sin \beta$ and $Mg \cos \beta$. From point (2) it follows that in the presence of friction, the force necessary to set a body in motion is $Mg\mu \cos \beta$. The condition of instability becomes $Mg \sin \beta > Mg\mu \cos \beta$ from which it follows $\tan \beta > \mu$. Thus, increasing the inclination angle the body starts to glide once a threshold angle ϕ is reached. This angle is called the friction angle. Equating the friction coefficient to the ratio between friction and normal force, it is obtained that

$$\mu = \frac{Mg \sin \phi}{Mg \cos \phi} = \tan \phi \tag{2.3}$$

showing that the friction coefficient is the tangent of the friction angle.

2.1.2.3 Microscopic Interpretation of Friction

In the following discussion we consider in particular rock for our analysis, though many of the concepts apply to other kinds of materials as well, such as metals. Even if the surface of polished rock appears smooth, at the microscopic level it reveals an irregular outline. Thus, the regions of contact between the surfaces of two bodies are irregularly distributed; as a consequence the effective surface of contact is much smaller than the geometric contact area of the two bodies.

According to the adhesion theory of friction, two solid bodies in contact yield off in correspondence of the areas where asperities come in contact, a little like loaded spring. Partial welding takes place around these areas, forming so-called *junctions*. Adhesion is explained as the resistance due to microscopic welding around asperities. We can anticipate a proportionality relationship between the effective area of contact A_r and the loading pressure P in the following way

$$\frac{P}{p} = A_r \tag{2.4}$$

where a material property p called the *penetration hardness* accounts for the strength of the material. It represents the efficiency of indentation between the two solid surfaces: a high value indicates a small deformation at the junctions. The penetration hardness is linked to the yield stress of the material. For example, for metals the hardness is about three times the yield stress (Rabinowicz 1995).

So far, we have examined the behavior of the two bodies during compression. To test the adhesion theory in predicting the properties of friction force, we need to consider the role of shear force applied between the two surfaces. For slippage to occur, the shear strength of the rocks must be overcome in correspondence of the junctions. Calling s the shear strength of the material, the friction force F is predicted by the theory to be

$$F = sA_r. \tag{2.5}$$

From (2.4) and (2.5) it follows that ratio F/P between the tangential and the normal force is constant and dependent on the properties of the bodies. We can identify this ratio with the friction coefficient, and so

$$\mu = \frac{s}{p}. \tag{2.6}$$

The relation (2.6) explains the independence of the friction force on the load and on the total area of contact. The fact that μ does not depend on the velocity is a consequence of the velocity independence of the bulk properties.

Among the possible critiques of the adhesion theory, one is particularly relevant. For elastic materials (like most hard rocks) the deformation at the asperities should be elastic. However, in the elastic limit, Hertz's theory predicts a nonlinear relationship between the area of contact and the load

$$A_r = kP^{2/3} \tag{2.7}$$

from which it follows that the friction coefficient should decrease with the load; moreover, the relationship between friction and load is nonlinear as well

$$\mu = skP^{-1/3}. \tag{2.8}$$

2.1 Friction and Cohesion

Whereas diamond does obey a relationship like (2.8), most rocks follow the linear behavior predicted by adhesion theory. The reason why an incorrect microscopic model of elastic compression predicts the correct macroscopic behavior is bewildering; it has probably to do with the geometrical arrangement of indenters. It has been shown that a model where many indenters of different sizes are distributed in hierarchies (a large indenter supports more smaller indenters of the same shape, each of which in turn supports the same number of smaller indenters, and so on) reproduces a linear friction in the limit of the number of hierarchies tending to infinite, although each indenter satisfies Hertz's law (Archard 1957; Scholz 2002).

2.1.2.4 Friction Coefficients for Rocks

Remarkably, friction coefficients for rocks turn out to depend little on the lithology. Data collected by Byerlee (1978) show that at overburden pressures lower than 200 MPa the average friction coefficient is

$$\mu \approx 0.85 \quad (\sigma < 200 \text{ MPa}). \tag{2.9}$$

This value should be considered as indicative. For example, some granites have friction coefficient between 0.6 and about twice as much, while some kinds of limestone have $0.70 < \mu < 0.75$. Other values of the friction coefficient are reported in the GeoApp. It is interesting to note that for higher pressures, $\sigma > 200$ MPa

$$\mu \approx \frac{50}{\sigma} + 0.6 \quad (\sigma > 200 \text{ MPa}) \tag{2.10}$$

showing that the friction coefficient slightly decreases with pressure, even though these values are beyond the pressure range of interest for landslides (200 MPa correspond to some 5–10 km of overburden rock).

Data reported in the tables and the fitting relations Eqs. 2.9 and 2.10 refer to polite surfaces. For asperities lengths less than some *mm*, the friction coefficient is independent of the roughness. However, interlocking between asperities greatly improves with surface roughness, with the effect of increasing the friction coefficient. In this case the effective friction coefficient is given by the Barton empirical formula (Barton 1973)

$$\mu_{\text{EFF}} = \tan\left[\tan^{-1}(\mu) + \text{JRC} \, \log_{10}\left(\frac{\sigma_j}{\sigma}\right)\right] \tag{2.11}$$

where σ_j is the compressive strength of rock ($\rightarrow$GeoApp) and JRC is called the roughness coefficient. Typical values for JRC range between 0 and 20 from very smooth to rough surfaces. In the field, the roughness can be measured by comparing the surface profile of rock with standard profiles. Roughness may be very important for assessing rock stability in the presence of joints, but is of limited interest in dynamical studies of landslides, where rock is fragmented.

2.1.2.5 Final Form of the Friction Force to be Used in the Calculations

To summarize, the total horizontal force (gravity plus friction) acting on a body at rest on a plane inclined with angle β can be written as

$$F_= = 0$$
(if $\tan \beta < \tan \phi$ and $U = 0$)
$$F_= = Mg(\sin \beta - \cos \beta \tan \phi) \quad (2.12)$$
(if $U \neq 0$ or $U = 0$ and $\tan \beta \geq \tan \phi$)

2.1.2.6 Work Performed by Friction Forces

Let us consider again a block resting on a horizontal table. A force of magnitude greater than the static friction force, $Mg\mu = Mg \tan \phi$, is now applied to the body. Thrust by the external force, the block moves from an initial point A to a final position P during a certain time interval. The work performed by the friction force between A and P is

$$L(A \to P) = Mg \tan \phi \overline{AP} \quad (2.13)$$

where $\overline{AP}$ is the curvilinear distance measured along the table (i.e., the trajectory length). For simplicity, we consider rectilinear trajectories. If the table is inclined with an angle β, a factor $\cos \beta$ has to be accounted for in the friction force, and the work becomes

$$L(A \to P) = Mg \tan \phi \cos \beta \overline{AP} = Mg \tan \phi R \quad (2.14)$$

where $R = \cos \beta \overline{AP}$ is the horizontal displacement. The last equation only follows if the displacement occurs along the slope direction.

If the inclination or the friction coefficient changes with the position, it is necessary to perform an integration

$$L(A \to P) = Mg \int_A^P \tan \phi \cos \beta \, dl \quad (2.15)$$

where dl is the line element along the trajectory. The integral returns again the total horizontal length, Eq. 2.14.

2.1.3 Cohesion

If a shear force is applied to a cube of muddy soil or rock at zero normal pressure, the resulting shear deformation is accompanied by a measurable resistance.

The resistance force per unit area is termed cohesion, and is measured in pascals (Pa). In natural soils, cohesion results from electrostatic bonds between clay and silt particles ($\rightarrow$Chap. 4). Thus, soils devoid of clay or silt are not cohesive except for capillary forces arising when little water forms bridges between sand grains, resulting in negative pore pressure (or "suction"). Values of soil cohesion typically are of the order of some kPa. In contrast, rocks normally exhibit much greater cohesion, thousands of times larger than soils.

At finite normal stresses, soils and rocks normally display both cohesive and frictional behavior. The shear strength of a soil is thus the sum of the cohesive and frictional contributions. Let us consider a slab of cohesive-frictional soil with constant thickness resting on a plane inclined with angle β. The resistive force is given by the combined effect of friction and cohesion in the following way

$$F_{\text{res}} = Mg \cos \beta \tan \phi + CwL \tag{2.16}$$

where w is the width of the slab and L is its length. If this combined force is lower than the gravity component along slope, the slab will not move. Because the mass is

$$M = \rho D w L \cos \beta \tag{2.17}$$

it is found that $F_{\text{res}} = \rho g D w L \cos^2 \beta \tan \phi + CwL$, or also

$$F_{\text{res}} = \sigma w L \tan \phi + CwL \tag{2.18}$$

where $\sigma = \rho g D \cos^2 \beta$ is the normal stress.

Finally, note that cohesion is also responsible for the finite value of tensile strength in both soils and rocks. The tensile strength, i.e., the tensile force per unit area, is normally a fraction of the cohesion.

An introduction to cohesion in soils is given by Selby (1993).

2.2 Slope Stability

2.2.1 A Few Words on Slope Stability

Gravity would tend to flatten out slopes, if it was not for the cohesion and friction forces of rocks and soils. However, the stability conditions may change due to temporary adjustments of equilibrium or because of external perturbations. In this case, a landslide may be triggered. There are numerous books and articles on slope stability. Here only a few basic examples are discussed to illustrate stability problems without any pretence of completeness.

The stability of a slope depends on several factors:

1. The kind of material involved. For example, recent volcaniclastic material may become very unstable and collapse into debris flows and lahars following

Fig. 2.3 A rock overhang unsupported at the base is an unstable condition that may lead to the detachment of portions of rock. Southern Norway

intense precipitation. In contrast, a hard and compact rock like intact gneiss is normally very stable.
2. The geometry of the material. Layers of rocks dipping toward slope are particularly unstable (Fig. 2.3). The slope angle is another important variable. The Frank landslide in Canada was probably due to instability along a bedding plane (Cruden and Krahn 1973).
3. The distribution of weight along slope. Loading the top of a slope may have great influence on stability. Likewise, cutting the slope at its base diminishes the buttressing of the lower layers underneath and promotes sliding conditions. This was particularly evident with the Betze-Post mine, where a mass of 3–10Mm^3 of unconsolidated deposits showed a slow creep of some cm/day. Transferring some of the material from the top of the heap to the foot proved of immediate effect in diminishing the creep rate (Rose and Hungr 2007).
4. Water is one of the most important instability factors. It decreases cohesion in soils and increases weight and pore water pressure in granular media. The rate at which water seeps into to the slope may also be critical. Some slopes may become unstable if even small amounts of water penetrate fast; others are more sensitive to the amount of water fallen in a long time span. The earthflow near Honolulu, Oahu, Hawaii, is a shallow (7–10 m) landslide that is periodically reactivated but only after massive precipitation. Recorded displacements do

2.2 Slope Stability

not exceed some centimeters, however. More dramatic are the rapid flows that take place in many areas of the world where rock is blanketed by a thick layer of soil. Following intense rain, several landslides may be created at once, forming a characteristic barren landscape, like in the San Francisco area in 1982, or in the Sarno region in southern Italy in 1997 and 1998.

5. External impulsive forces such as earthquakes, waves, and volcanic eruptions. In July 1888, a swarm of strong earthquakes shook Mount Bandai, in Japan. A series of volcanic explosions, partly phreatic, destabilized a large portion of the summit, which collapsed in a debris avalanche covering an area of 3.5 km^2. Better known is the eruption of the St. Helens of March 1980. A flank of the volcanic edifice slowly bulged during the 1980 activity following more than a century of dormancy. The progressive deformation finally resulted in a giant collapse and a debris avalanche with approximate run-out of 30 km. Following the landslide, the pressure underneath the northern sector of the edifice plummeted, which caused the strong blast recorded in the photographs.
6. Vegetation may influence stability through mechanical cohesion and removal of water via evapotranspiration.

2.2.1.1 Factor of Safety

To quantitatively assess the stability of a slope in engineering geology, a parameter F known as *factor of safety* is introduced. The factor of safety is the ratio between the resistive forces and gravity pull

$$\text{Factor of safety } F = \frac{\text{Resistance forces}}{\text{Gravity force parallel to slope}} \qquad (2.19)$$

A value $F > 1$ indicates stability, whereas $F < 1$ implies instability. Thus, the transition between stability to collapse may be envisaged mathematically as a decrease in the factor of safety to values below unity.

2.2.2 An Example: Layered Slope

A simple model of stability analysis consists in analyzing a homogeneous slope like in Fig. 2.4, including the possible presence of water at a certain depth. From Eq. 2.18, the resistive forces deriving from cohesion and friction can be written in the following way

$$F_{\text{res}} = CwL + (\sigma - P_W)wL \tan \phi. \qquad (2.20)$$

where the effect of water resulting in pore pressure P_W has been added. Water tends to destabilize the slope, because as evident from Eq. 2.20, it acts in the direction

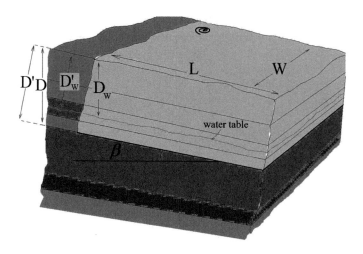

Fig. 2.4 Layered slope for the calculation of the factor of safety F

of reducing the contribution of the effective friction angle. The water table is assumed to lie at a constant depth. Because the normal pressure is (←Sect. 2.1)

$$\sigma = \rho g D \cos^2 \beta$$
$$P_W = \rho_W g (D - D_W) \cos^2 \beta \qquad (2.21)$$

it is found that

$$F_{\text{res}} = CwL + (\Delta \rho D + \rho_W D_W) gwL \cos^2 \beta \tan \phi \qquad (2.22)$$

whereas accounting also for the weight of pore water

$$F_p = [\rho D + \rho_W \chi (D - D_W)] gwL \sin \beta \cos \beta \qquad (2.23)$$

where $\chi < 1$ is the volume fraction of water for the case of 100% saturation. Considering that $\rho D \gg \rho_W \chi (D - D_W)$, this term can be neglected for simple estimates.

The factor of safety becomes so

$$F = \frac{F_{\text{res}}}{F_=} = \frac{\tan \phi}{\tan \beta} \left(\frac{\Delta \rho}{\rho} + \frac{\rho_W D_W}{\rho D} \right) + \frac{C}{D \rho g \sin \beta \cos \beta} \quad \text{if } D_W < D$$

$$F = \frac{F_{\text{res}}}{F_=} = \frac{\tan \phi}{\tan \beta} + \frac{C}{D \rho g \sin \beta \cos \beta} \quad \text{if } D_W \geq D \qquad (2.24)$$

If a tensile stress C_T contributes to stability along the surface of area WD, then (2.24) can be generalized to

$$F = \frac{F_{res}}{F_=} = \frac{\tan\phi}{\tan\beta}\left(\frac{\Delta\rho}{\rho} + \frac{\rho_W D_W}{\rho D}\right) + \frac{1}{\rho g \sin\beta \cos\beta}\left[\frac{C}{D} + \frac{C_T}{L}\right] \quad \text{if} \quad D_W < D$$

$$F = \frac{F_{res}}{F_=} = \frac{\tan\phi}{\tan\beta} + \frac{1}{\rho g \sin\beta \cos\beta}\left[\frac{C}{D} + \frac{C_T}{L}\right] \quad \text{if} \quad D_W \geq D$$

(2.25)

As a simple application, let us assume absence of water. Imposing $F > 1$ we find the condition of instability as

$$\tan\beta > \frac{1}{\rho g \cos^2\beta}\left[\frac{C}{D} + \frac{C_T}{L}\right] + \tan\phi. \tag{2.26}$$

The cohesive term (first term on the right-hand side) becomes very small for long and deep slabs, $\rho g D \gg C$ and $\rho g L \gg C_T$; thus, cohesion in soils is particularly important for a shallow landslide. In compact rocks, where cohesion and tensile stress typically reach values of tens of MPa, the cohesive term may become important also for large landslides.

It is also interesting to solve for D as a function of the angles

$$D > \frac{1}{\tan\beta - \tan\phi}\frac{C}{\rho g \cos^2\beta}. \tag{2.27}$$

where the contribution from the tensile strength has been neglected for simplicity. This equation shows that a minimum thickness is necessary for instability to occur. If the angle of dipping approaches the friction angle, the minimum thickness tends to infinity. Thus, landslides developing at angles close to the friction angle will be particularly large (Cruden and Krahn 1973).

If cohesion is more important than friction in stabilizing the slope, assuming $L \gg D$ one obtains the depth of detachment as

$$D = \frac{C}{\rho g \sin\beta \cos\beta}. \tag{2.28}$$

It is sometimes observed that the pressure of water is greater than that of the hydrostatic value. This may occur, in particular, when the soil permeability is low so that pressurized water cannot seep to zones of lower pressure. By inserting a pipe in the soil at the height of the water level, water will rise up in the pipe to a height η, called the piezometric height. Formally, another pressure term should be added to Eq. 2.21 accounting for the excess pressure. In practice, however, it is better to directly write the equations in terms of the piezometric height η previously measured on the terrain

$$P_W = \rho_W g \eta \tag{2.29}$$

and so the factor of safety becomes

$$F = \frac{\tan\phi}{\tan\beta}\left(1 - \frac{\rho_W \eta}{\rho D}\right) + \frac{1}{\rho g \sin\beta \cos\beta}\left[\frac{C}{D} + \frac{C_T}{L}\right] \tag{2.30}$$

Box 2.1 One Step Back: The Stress Tensor

The present Box is rather concise. A more thorough presentation can be found in Middleton and Wilcock (1994).

Some physical properties, like the temperature of a body, can be described by just one number. These quantities are called *scalars*. Other quantities, *vectors*, require three numbers. The velocity of a material point is an example of vector quantity. The existence of three components derives from the three-dimensionality of space. In Cartesian coordinates, the vector components are the projections of the vector along the directions in space x, y, and z. Vectors satisfy certain transformation rules between different reference systems. This is because a vector has an essence in itself, which is independent of the reference system used to represent it. Mathematically, this has the consequence that, although the components of a vector depend on the reference system, its components are linked to the specific condition that the vector magnitude (i.e., its length) should be the same in all reference systems.

Other physical quantities necessitate an extension of the concept of vector. The state of stress of an elastic body is described by a *tensor*, a mathematical object of nine components. Let us consider an infinitesimal cube like in Fig. 2.5. Let the six faces be perpendicular to the directions of the Cartesian coordinates. We label with the letter "x" the two faces perpendicular to the coordinate x, and similarly we do with the other faces. Of the six faces, in Fig. 2.5, we consider only the three facing the observer. These faces are denoted as "positive" because the directions x, y, and z of the coordinate

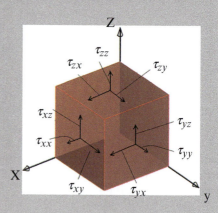

Fig. 2.5 For the definition of the stress components

(continued)

Box 2.1 (continued)

system cross the cube faces from the interior outwards. In other words, a positive face is a face whose outer normal points in the positive direction. On the surface of the positive "x" face we can draw a vector with its three Cartesian components along x, y, and z. We call these components $\tau_{xx}, \tau_{xy}, \tau_{xz}$, respectively. In this notation, the first index identifies the face; the second index denotes the direction of the vector component. Thus, τ_{xy} is the component y of the vector acting on the face x of the figure. Similarly, we introduce the other components $\tau_{yy}, \tau_{yx}, \tau_{yz}, \tau_{zz}, \tau_{zx}, \tau_{zy}$. We call tensor the mathematical object so defined.

So far, the definition of tensor has been purely formal, as no particular significance was attributed to the vectors of the form τ_{xy}. We now specify the equations to an important mathematical object used in continuum mechanics, the stress tensor. Suppose that the elementary cube shown in Fig. 2.5 resides within a larger volume of material subjected to an external stress, for example a portion of rock on which tectonic lateral thrust in addition to gravity is acting. The external stress field will act on the three faces of the elementary cube through three stress vectors on each of the front faces of the cube. In turn, each of these stress vectors can be decomposed into three components, for a total of nine vectors. For example, the stress vector acting on the x face is evidently

$$\vec{\tau}_x = \tau_{xx}\hat{i} + \tau_{xy}\hat{j} + \tau_{xz}\hat{k} \qquad (2.31)$$

where $\hat{i}, \hat{j}, \hat{k}$ are the versors (i.e., vectors of unit magnitude) pointing in the directions x, y, and z, respectively. Similar equations specify the stress vector on the other two frontal faces of Fig. 2.5. One might question why the other three faces of the cube (the negative ones) are not considered. This is because according to the standard geometrical construction of Fig. 2.5, the other three back faces belong to the front faces of another neighboring cube.

We can thus specify these nine numbers as the component of stress in a medium. This particular Cartesian tensor is denoted as the stress tensor, and provides the state of stress within the entire volume of the medium. The components $\tau_{xx}, \tau_{yy}, \tau_{zz}$ are called diagonal; the components τ_{ij} with $i \neq j$ are denoted as the off-diagonal terms. The denomination derives from the possibility to represent the component of the stress tensor in a matrix, i.e., a square table of the form

$$\begin{pmatrix} \tau_{xx} & \tau_{xy} & \tau_{xz} \\ \tau_{yx} & \tau_{yy} & \tau_{yz} \\ \tau_{zx} & \tau_{zy} & \tau_{zz} \end{pmatrix}. \qquad (2.32)$$

(continued)

Box 2.1 (continued)

A component of stress is positive when it is directed toward the positive axis of the reference system, and on a positive face. The components shown in Fig. 2.5 are positive. Likewise, the stress is positive if it points in the negative direction of a negative face. A component is negative if it points in the negative direction of a positive face, or in the positive direction of a negative face. As an example of negative component, think changing direction to the component τ_{xx} in Fig. 2.5. The opposite definition can also be found (and is adopted in the calculation of →Sect. 2.2.3 for convenience), where a positive component of the stress tensor is associated to the positive direction of a negative face.

An important property of the stress tensor is its symmetry, i.e., the equality between off-diagonal components with exchanged indices, namely: $\tau_{xy} = \tau_{yx}; \tau_{xz} = \tau_{zx}; \tau_{yz} = \tau_{zy}$. This condition results from stability considerations on the elementary volume. If for example were $\tau_{xy} \neq \tau_{yx}$, the torque directed along z acting on the elementary volume would be unbalanced, and the equations of dynamics would predict it to spin around the z axis. For very small volumes (the cube is infinitesimal) the spinning rate would tend to an infinite value. These symmetry conditions reduce the independent components of the stress tensor from nine to six. In two-dimensional problems, the independent stress tensor components are evidently three.

The diagonal components of the stress tensor are called the normal stresses, and are often denoted with the Greek symbol sigma: $\sigma_x, \sigma_y, \sigma_z$. The nondiagonal components of the stress tensor are called shear stresses. It is always possible to find a local reference system such that the shear stresses $\tau_{xy}, \tau_{xz}, \tau_{yz}$ vanish, and the state of stress is specified by the sole diagonal components. This procedure, also called diagonalization, is schematized in Fig. 2.6. The three diagonal stress vectors so identified ($\bar{\tau}_{xx}, \bar{\tau}_{yy}, \bar{\tau}_{zz}$ or also $\bar{\sigma}_x, \bar{\sigma}_y, \bar{\sigma}_z$) are called the principal stresses. In other words, diagonalization corresponds in finding the three perpendicular planes for which the shear stresses are zero. The recipe to find the principal stresses is sketched in the MathApp.

A final set of definitions will be useful for the following discussions. Once we have identified the magnitude and the direction of the principal stresses throughout each point of a stressed elastic body, it is possible to draw three families of lines parallel to the three principal stresses (only two will be necessary in two dimensions). These lines are called the stress trajectories, and give a clear-cut picture of the state of stress in the body.

How will a particular material respond to a certain stress field? It is obvious that a stiff material will deform less than a soft one for the same stress. Thus, to answer this question, further equations specifying the materials properties are needed. In particular, two kinds of materials are relevant

(continued)

Box 2.1 (continued)

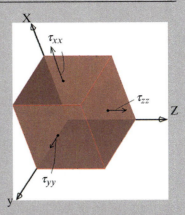

Fig. 2.6 Geometrical interpretation of the tensor diagonalization. At a certain point, a particular orientation of the local reference system reduces the tensor in a diagonal form

for the following discussion, which give different relationships between the stress and the deformation:

1. Some materials, such as the elastic ones, exhibit a relationship between the stress and the deformation. The state of stress in an elastic medium is specified in terms of the Navier equation, which is beyond the scope of the book (see, e.g., Middleton and Wilcock 1994).
2. For a different class of materials, the relevant relationship is between the stress and the deformation rate. These are the plastic and liquid substances. The resulting relations are the basis of fluid mechanics and will be explored in detail in Chaps. 3 and 4.

Box 2.2 Example of Stress Tensors in Stability Problem

Let us consider the case of an infinite outcrop of rock. Choosing the z direction parallel to the vertical, we seek for the expression of the stress tensor at a depth D under the surface. The zz component is evidently given as $\tau_{zz} = \rho g D$. This is the stress that would be measured by a pressure transducer oriented with the vertical face. The other two diagonal components xx and yy depend on the type of material. Elastic materials like hard intact rock follow the Hooke behavior ($\rightarrow$Sect. 5.1.1) for which $\tau_{xx} = \tau_{yy} = \frac{v}{1-v}\tau_{zz} = \frac{v}{1-v}\rho g D$. With the Poisson coefficient v typically of the order 0.25, the xx and yy components are about 30% smaller than the component zz. The stress tensor becomes so

(continued)

Box 2.2 (continued)

$$\begin{pmatrix} \frac{\nu}{1-\nu}\rho g D + \Sigma & 0 & 0 \\ 0 & \frac{\nu}{1-\nu}\rho g D & 0 \\ 0 & 0 & \rho g D \end{pmatrix} \quad (2.33)$$

where the shear terms vanish due to the symmetry. Further notice that independent stresses sum up linearly. Thus, if a constant tectonic stress of magnitude Σ is acting horizontally along the xx component, its contribution appears summed to the corresponding matrix term, as shown in (2.33).

A graphic way of illustrating the stress field within a stressed medium is by using the stress trajectories, (i.e., lines parallel at each point to the directions of the principal stresses, ←Box 2.1). Stress trajectories for the previous problem are drawn in Fig. 2.7 (exploiting the invariance along the horizontal direction, we can draw the stress trajectories in just two dimensions). In this particular case, they are simply a network of two families of straight lines, respectively, parallel and perpendicular to the ground. Another useful concept is the one of stress ellipsoid. At each point of the stressed medium, an ellipsoid can be drawn with axes equal to the magnitude of the three principal stresses and oriented as the principal stresses. In two dimensions, the ellipsoid becomes an ellipse. The stress ellipse is shown in Fig. 2.7 for the example above.

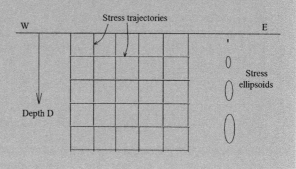

Fig. 2.7 Illustration of the stress trajectories and stress ellipses for the tensor 2.33 with $\Sigma = 0$

2.2.3 A Few Basics Concepts of Soil Mechanics and an Application to Slumps

In contrast to a layered slide like that of Fig. 2.4, in homogeneous soils there are no leading weakness layers. The surface of rupture thus develops following the internal and external forces, rather than on the preexisting geometry. The most likely surface of detachment is the one that gives the smallest factor of safety. The detachment surface that minimizes the factor of safety turns out to be sub-spherical, or spoon-like shaped. The corresponding landslide is thus a rotational

2.2.3.1 Mohr Circle

Consider a two-dimensional prism like in Fig. 2.8, making an angle β with respect to the horizontal. Both shear and normal stresses act on the prism from the lower "*B*" and left "*A*" faces of the figure. We wish to calculate the shear and normal stresses τ and σ acting on the upper tilted face denoted as "*S*." Consider the force

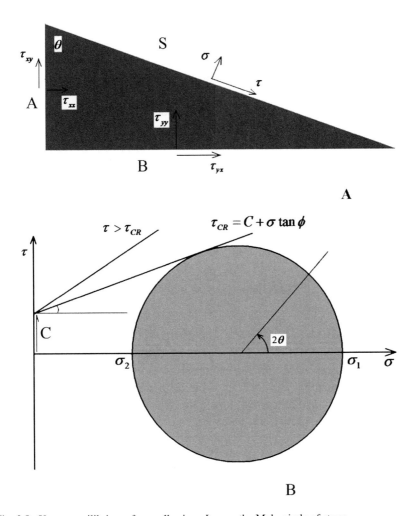

Fig. 2.8 *Upper*: equilibrium of a small prism. *Lower*: the Mohr circle of stress

on S directed to x, that we call S_x. The forces contributing to S_x are two shear forces and two normal forces acting on the faces A and B, for a total of four. Because the prism is in equilibrium, $S_x = 0$. Similarly, the four forces directed vertically on the faces A and B sum up to give the total force S_y acting on S and directed along y. To obtain the normal and shear stresses on S, the resultant forces S_x and S_y are projected along the directions normal and parallel to S. Dividing by the areas of the faces finally provides the stresses. The result for the stress on S, respectively, normal and shear reads

$$\sigma = \frac{1}{2}\tau_{xx}[1 + \cos(2\theta)] + \tau_{xy}\sin(2\theta) + \frac{1}{2}\tau_{yy}[1 - \cos(2\theta)]$$
$$\tau = \frac{1}{2}(\tau_{xx} - \tau_{yy})\sin(2\theta) - \tau_{xy}\cos(2\theta). \tag{2.34}$$

Evidently, for $\theta = \pi/2$ the face S becomes horizontal and $\sigma = \tau_{yy}$; $\tau = \tau_{xy}$. For $\theta = \pi/4$ (S at 45° with respect to the horizontal) it follows $\sigma = \frac{1}{2}(\tau_{xx} + \tau_{yy}) + \tau_{xy}$; $\tau = \frac{1}{2}(\tau_{xx} - \tau_{yy})$. For $\theta = 0$ (surface "P" vertical) $\sigma = \tau_{xx}$; $\tau = -\tau_{xy}$.

Consider now a situation where the stress tensor is in diagonal form so that there are no shear stresses acting on the A and B surfaces. Evidently, in this case the equations for σ and τ are similar to (2.39), but with the following replacements: $\tau_{xx} = \sigma_1$; $\tau_{yy} = \sigma_2$; $\tau_{xy} = 0$ where σ_1, σ_2 are the two principal stresses. Equation 2.34 becomes so

$$\sigma = \frac{1}{2}\sigma_1[1 + \cos(2\theta)] + \frac{1}{2}\sigma_2[1 - \cos(2\theta)]$$
$$\tau = \frac{1}{2}(\sigma_1 - \sigma_2)\sin(2\theta) \tag{2.35}$$

The first equation of (2.40) can also be rewritten as

$$\sigma = \frac{1}{2}(\sigma_1 + \sigma_2) + \frac{1}{2}(\sigma_1 - \sigma_2)\cos(2\theta) \tag{2.36}$$

and the shear stress τ can now be plotted as a function of σ with the angle θ varying parametrically. The locus is a circle of center $\frac{1}{2}(\sigma_1 + \sigma_2)$ and radius $\frac{1}{2}(\sigma_1 - \sigma_2)$, called the Mohr circle (Fig. 2.8b). The circle gives a pictorial view of the behavior of the normal and shear stresses with changing inclination angle of a plane with respect to the direction where the stress tensor is diagonal. For $\theta = 90°$ ($\sigma = \sigma_1$; $\tau = 0$) the S plane is horizontal. Likewise, when $\theta = 0°$ the S plane is vertical and the shear stress vanishes. For a generic value of θ the surface S is inclined and both the normal and shear stresses acquire a nonzero value. The maximum shear stress, reached on top of the circle, is found for $\theta = \pi/4 = 45°$.

2.2 Slope Stability

Fig. 2.9 (a) If a sample of frictional-cohesive rock or soil is subjected to constant confining pressure (*vertical arrows*) and compressive stress (*horizontal arrows*), it typically fails at an angle greater than 45°. If the confining pressure is increased (**b**), a point is reached where the material will not fail for the same compressive stress. The compressive stress must be increased for failure to occur (**c**)

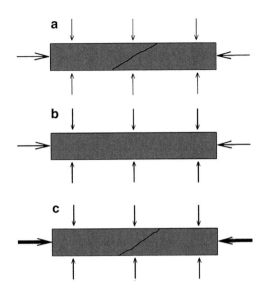

2.2.3.2 Failure Criterion for a Cohesive-Frictional Material

Mohr's circle allows also a graphic representation of the failure of cohesive-frictional soil or rock. Experiments show that cohesive-frictional materials fracture at angles θ slightly greater than 45° (Fig. 2.9).

The Navier–Coulomb criterion states that fracture occurs when the shear stress exceeds a critical value given as

$$\tau_{CR} = C + \sigma \tan \phi \tag{2.37}$$

where ϕ is the internal friction angle and C is the cohesion. Note that the normal pressure has a stabilizing effect, raising the critical shear stress required for failure. The geometrical locus of τ_{CR} as a function of σ is a straight line, which is also reported in Fig. 2.8b. If the line does not intersect the Mohr circle, the shear stress in the medium is not exceeded and the material does not break. The line tangent to the Mohr circle corresponds to the critical situation when rupture is about to occur. From the geometry of the figure, the critical angle for failure becomes

$$\theta_{CR} = \frac{\pi}{4} + \frac{\phi}{2} = 45° + \frac{\phi}{2}. \tag{2.38}$$

2.2.3.3 Slumps

Consider now a slope with a step like in Fig. 2.10. The two families of stress trajectories for this particular case are shown in Fig. 2.10a, b. Because the two

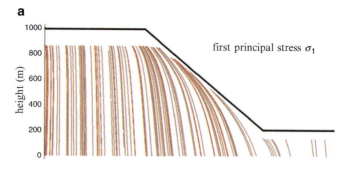

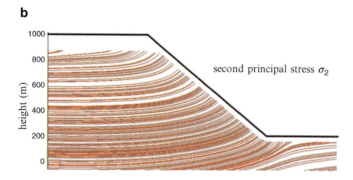

Fig. 2.10 Stress trajectories in an elastic medium with the properties of granite

principal stresses are mutually perpendicular at every point, the two families of stress trajectories are perpendicular too (this problem is two-dimensional so there is no third principal stress). Consider now the point of intersection of two stress trajectories as shown in Fig. 2.11. The failure is more likely to occur along a line inclined by $45° \pm \phi/2$ with respect to the directions of the two principal stresses. One of these lines is schematically shown in Fig. 2.11; it does assume a sub-circular form, so outlining the concave form of slumps.

Let us fix one point deep in the soil, for example one of the coordinates $x = 1,500$ m; $y = 200$ (the orientation of the global reference system is chosen with the x and y axes parallel and perpendicular to the base, respectively). The shear and the normal stresses are measured along the planes A, B, and C of Fig. 2.11. Because the step is just a small perturbation at such depth, a resulting small value for the shear stress will be measured in A. The plane A is nearly parallel to the principal stress τ_{xx}; thus, the normal stress recorded in A is close to τ_{yy}. The plane B of Fig. 2.11 measures the component τ_{xx} of the stress tensor. It has been shown ($\leftarrow$Box 2.2) that in elastic media the stress component τ_{xx} is proportional to τ_{yy}. In frictional soils a relationship of proportionality

2.2 Slope Stability

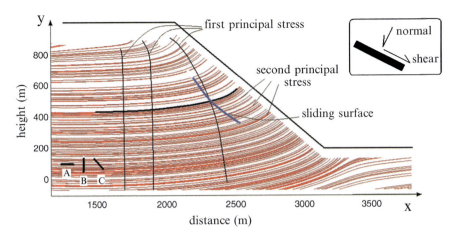

Fig. 2.11 The surface of detachment of a slump

$$\tau_{xx} = k\tau_{yy} \qquad (2.39)$$

also holds where k, called the coefficient of earth pressure, depends on the state of tensional-compression state of the medium. If the medium is compressed (corresponding to the so-called Rankine passive state) the coefficient is greater than unity. If the medium is expanding laterally (Rankine active state), the coefficient is lower than unity. If the medium is static there is an indeterminacy as to the actual value of the earth pressure coefficient, which falls somewhere between the active and the passive values. At failure, the Mohr circle is tangent to the line $\tau_{CR} = C + \sigma \tan\phi$. Thus, the radius of the circle is fixed, and so the ratio σ_2/σ_1 can be determined. This ratio corresponds to the earth pressure coefficient at failure. The situation in which the lateral stress is smaller than the vertical stress corresponds evidently to the active case, and is the one of more interest for landslides, because failure normally occurs during a tensional phase. The result is

$$\tau_{xx} = \frac{1-\sin\phi}{1+\sin\phi}\tau_{yy} + 2C\frac{\cos\phi}{1+\sin\phi}. \qquad (2.40)$$

Notice also that τ_{xx} may become negative for very small τ_{yy}. Remembering that $\tau_{yy} = \rho g D$ where D is the depth, this implies that the lateral stress close to the surface is negative for cohesive soil. The result is the formation of tension cracks in regions of soil instability that heal at sufficient depth (see also the opening figure of the chapter). Coming back to Fig. 2.11, note that the stress in the point C has a shear component in addition to the normal component.

Figure 2.13 shows a small slump in cohesive soil (top) and cracks in the soil of an unstable area (bottom).

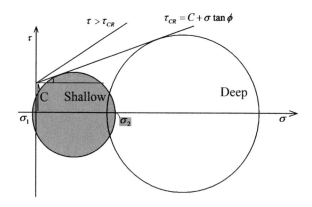

Fig. 2.12 The Mohr circle for a deep soil layer at failure is shown with the large circle. In a shallower layer at failure conditions, the circle shifts to the *left*, always remaining tangent to the Mohr–Coulomb line. A point is reached where the smaller of the principal stresses becomes negative (*small, gray circle*). This shows the existence of a state of tension at the surface. The finite value of cohesion C is necessary for the existence of a tensional state

2.2.3.4 Factor of Safety and a Simple Criterion for a Rotational Slide

In a first approximation, the shape of a rotational slide can be assumed as part of a circle of length L (Fig. 2.14). Because the driving force is the gravity acting on the center of mass which is displaced horizontally from the center of the circle, a torque develops of magnitude equal to

$$m = MgX - CLRW \tag{2.41}$$

where W is the width (perpendicular to the drawing), M is the mass, and X is the horizontal distance between the center of mass and the center of the circle. We have also assumed a purely cohesive material, absence of friction, and invariance along the direction perpendicular to the figure. Imposing that the torque >0, the condition of instability becomes

$$X \geq \frac{CLR}{(M/W)g}. \tag{2.42}$$

while the factor of safety is defined as the ratio between the moment of the resistance forces and the torque (Fig. 2.14).

$$F = \frac{CLR}{(M/W)gX}. \tag{2.43}$$

Note that because the geometrical quantities scale as

$$L \propto R; \ M/W \propto R^2; \ X \propto R, \tag{2.44}$$

2.2 Slope Stability

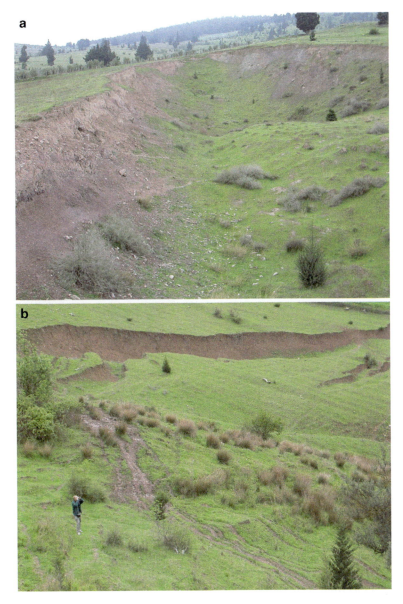

Fig. 2.13 (**a**) A small slump in soft terrain. (**b**) A new slump is often preceded by tension cracks. Both near Fatalak, northern Iran. See also the figure at the beginning of the chapter. Impulsive forces may result from earthquakes. Water seeping into the terrain may have diminished cohesion and friction and promoted instability, but the final trigger of this landslide was an earthquake on a previously unstable terrain

Fig. 2.14 The torque acting on a rotational slide

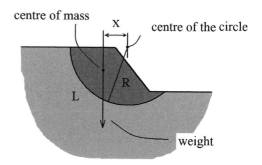

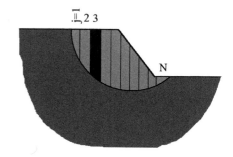

Fig. 2.15 According to the Fellenius method, the slump is divided into vertical slices and the factor of safety is obtained summing up the contribution of each slice to both gravity and resistance

it follows that $F \propto 1/R$, i.e., the factor of safety decreases with the radius of the slump. For this reason, the rotational slides require a minimum radius to develop.

For a soil with nonzero friction angle, the calculation is more complicated because the resistance is proportional to the local thickness of the slump. Several methods have been developed to calculate the factor of safety in this case. The rupture surface is initially conjectured. The landslide material is divided into segments separated by parallel, vertical sectors like in Fig. 2.15. Each sector, identified with a progressive number "j," contributes to a shear resistance in the form

$$\left(P_j \cos \beta_j - u\delta L_j\right) \tan \phi + C\delta L_j \qquad (2.45)$$

where P_j is the weight of each sector and δL_j is the arc length. The shear stress is evidently $P_j \sin \beta_j$, so that the factor of safety can be obtained summing up all the contributions from each sector

$$F = \frac{\sum\limits_{j=1}^{N} \left(P_j \cos \beta_j - u\delta L_j\right) \tan \phi + CL}{\sum\limits_{j=1}^{N} P_j \cos \beta_j} \qquad (2.46)$$

The weight P_j can be calculated from the geometry of the slice and the density of the material. Note that also soils with nonuniform properties may be considered in the calculations. This method, called the ordinary method of slices (or also the Fellenius method) is often used in analytical estimates of stability, although a computer will speed-up the calculations for complicate shapes and allow for numerous slices. It has some drawbacks, however. It does not consider properly the curvature of the sliding surface, and the Earth pressure force is not accounted for. This is reflected in characteristically low values of the factor of safety (of the order 10–15%) compared to those obtained with more advanced models. For this reason, methods like the modified Bishop method are currently used. In this method, the calculation of the factor of safety F is more involved, as it requires previous knowledge of F itself. With the use of modern computers, however, an iterative convergence of the solution can be easily attained. These methods are described at some length by Duncan (1996).

2.2.4 Other Factors Contributing to Instability

Numerous landslides are caused by earthquakes. The first documented suggestion is due to Dante Alighieri in the thirteenth century who mentions in the *Divine Comedy* the "tremuoto" (i.e., earthquake) as possible trigger for the Lavini di Marco landslide in Northern Italy (→Chap. 6). However, it is only with the seismic swarm of 1783 in Calabria (southern Italy) that the correlation between earthquakes and landslides began to be studied scientifically (Keefer 2002). All kinds of slides can be potentially triggered by earthquakes: approximate magnitudes necessary to produce a failure range from a modest $M = 4$ for rock falls to $M = 6$, $M = 6.5$ for rock and soil avalanches (Keefer 2002). Many devastating landslides have been triggered by earthquakes, such as the Nevados Huascaran rock and ice avalanche (→Chap. 6).

Correlations have been suggested between the area affected by landslides and the magnitude of the quake. For example, based on data fit a relationship of the form has been suggested (Keefer 2002)

$$\log_{10} A(\text{km}^2) = M - 3.46(\pm 0.47) \tag{2.47}$$

An earthquake of magnitude $M = 7$ can so potentially affect an area of about 10,000 km^2.

Physically, the action of earthquakes is to temporarily increase the ratio $\dfrac{\Delta\tau}{\sigma}$ between earthquake-induced shear stress and normal pressure.

In addition to increasing the shear stress, earthquakes may also liquefy water-saturated sands. During the shock, grains partially rest on the fluid rather than on other grains. When the contact between grains is lost, the pore water pressure tends to become equal to the total pressure. Hence, the soil is transformed into a dense liquid mixture of water and sand. This was the condition to blame when the Kensu debris flow (China) killed 200,000 people in 1920. During liquefaction, the ratio $\frac{\Delta\tau}{\sigma}$ can be estimated with the Seed–Idriss formula valid for depths $D < 12$ m

$$r = \frac{\Delta \tau}{\sigma} = 0.65(1 - 0.008 D)\frac{a_{max}}{g}\frac{P_N}{P_N - P_A} \qquad (2.48)$$

where a_{max} è is the maximum soil acceleration, P_N, P_A are the normal pressure without the effect of the quake, and the water pressure, respectively. Earthquakes of magnitudes 5, 6, 7, and 8 generate accelerations of 0.06, 0.15, 0.5, and 0.6 g, respectively, in the epicenter region. Thus, the ratio $\frac{\Delta \tau}{\sigma}$ may grow up to 0.5 for strong earthquakes. Far from the epicenter, the acceleration of the terrain during a quake decreases markedly. For example, at a distance of 10–100 km the acceleration is reduced by 53% and 13% of the epicenter value, respectively.

Vegetation may alter the stability of superficial soils, too. Firstly, roots may help stabilize the slope partly because they deprive the soil of water, and also as a consequence of their mechanical action similar to natural reinforcement shafts. However, vegetation may also have a negative impact on stability. Some species of bushes may increase instability by promoting the catchment of superficial water. In addition, trees add weight to the slope, an effect that may be significant for superficial soils. To deal with stability change, a factor ΔC is formally added in the equation for the factor of safety so that

$$F = \frac{\tan \phi}{\tan \beta} + \frac{1}{\rho g \cos \beta \sin \beta}\left[\frac{C + \Delta C}{D} + \frac{C_T}{L}\right] \qquad (2.49)$$

Values for ΔC have been collected for example by Sidle and Ochiai (2006). Maximum values may reach $\Delta C \approx 20$ kPa, but lower figures in the range of some kPa are more typical.

> **Box 2.3** External Link: Are Glacial Cirques the Remains of Ancient Landslide Scars?
>
> Glacial cirques are typical mountain landforms well-known to both geomorphologists and mountaineers. They have the shape of an upside-down helmet with steep wall, and typical width between 400 and 800 m. Cirques often punctuate the mountain environment at altitudes higher than the firn line (i.e., the level where the snow retreats during the thaw season). Cirques at lower altitude in the Alps are ascribed to a lower firn line corresponding to the coldest climate during the glaciations. Cirque size is not much dependent on the past history and glacial characteristics of the area. In a way, cirques represent a deviation from the fractal-like character of mountain altitudes, which is otherwise characterized by a power-law distribution of the topography spectral density.
>
> The standard model of cirque development considers equally important the periglacial and glacial processes. Periglacial processes commence when snow drifts into a hollow, a process termed nivation; freeze-thaw cycles are

(continued)

Box 2.3 (continued)

then responsible for headwall retreat until the hollow reaches a critical size. As the snow gathers in sufficient amount to form a small cirque glacier, the process of cirque excavation and wall retreat continues due to direct ice abrasion.

An alternative hypothesis for the formation of glacial cirques has been recently suggested (Turnbull and Davies 2006). The authors consider the measured erosion rates of glacial cirques as too low to explain the depth of most cirques (500 m in some cases). If cirques mostly grow due to glacier erosion, then it is reasonable as a working approximation to assume an erosion rate proportional to the shear stress at the base of the cirque glacier, and so

$$\frac{dy}{dt} = ky. \qquad (2.50)$$

which for a constant value of k gives an exponential increase. This parameter is estimated from erosion measurement as $\approx 3 \times 10^6 a^{-1}$. Thus, it takes over one million years to excavate the cirque, which is more than the duration of the Pleistocene glaciations. The alternative explanation suggested by Turnbull and Davies (2006) is that glacial cirques are in reality the scars of deep-seated landslides. In this way the problem of the slow growth in relationship to size is avoided. The authors cite as demonstrative example the Acheron rock avalanche in New Zealand. The mass failure, about 1,000 years old, has left a scar indistinguishable from those normally attributed to genuine glacial cirques. Indeed, if it had not been for the voluminous deposits proving the mass wasting nature of the Acheron, probably geomorphologists would consider it as a typical glacial cirque. Another indication in favor of this hypothesis is that cirques appear most often at the top of the slopes, exactly where the instability is supposed to occur.

One serious challenge to the model is the global orientation of cirques. Cirques in the Northern hemisphere are preferably directed to North, and in the southern hemisphere to South. This is easily explained in the standard model by the fact that both nivation and glacial erosions are faster for the slopes where the sun is weakest. The orientation effect is, conversely, difficult to reconcile with the landslide model. Another difficulty is that glacial cirques often appear in series (staircases), an uncommon feature in landslide scars. The lack of landslide deposits at the foot of most cirques is also problematic: it might be tentatively explained by removal by glaciers and river erosion (and perhaps mixing with morain deposits), but this seems partly an ad hoc explanation. Note also that owing to its high cohesion, rock rarely forms slumps, except perhaps for earthquake-induced failures. Nevertheless,

(continued)

Box 2.3 (continued)

although the hypothesis faces some objective difficulties to explain most glacial cirques, it is appealing in suggesting that at least some of the hollows previously attributed to cirques might in reality be the relicts of ancient catastrophes.

General references Chapter 2: Sidle and Ochiai (2006); Dikau et al. (1996); Turner and Schuster (1996), Duncan (1996), Dikau et al. (1996), Middleton and Wilcock (1994).

Chapter 3
Introduction to Fluid Mechanics

Everyone is familiar with matter in the fluid state such as air, water, or oil. The most obvious attribute of a fluid is its capability of flowing. Fluids can travel down slope like water in a river or crude oil flowing along pipes; they give rise to currents in the presence of a pressure gradient, such as wind created by pressure differences in the atmosphere. Fluid mechanics, the science of fluids, is crucial not only in science, but also in engineering problems like the flight of an aircraft and the lubrication of joints in an engine. Fluid mechanics is a complicated and varied subject where new experiments and theoretical calculations are constantly reported by scientists and engineers. This is testified by the numerous (in the number of some hundreds) original papers appearing monthly on specialized journals.

Fluid mechanics enters in many ways in Earth Sciences. Examples include atmospheric flows, convection in the Earth mantle, lava flow, the dam break problem, pyroclastic flows, or river mechanics. Even restricting the focus to landslides, fluid mechanical problems remain numerous.

1. *Rheological flows, such as mudflows and debris flows, can be often described as non-Newtonian fluids.*
2. *Subaqueous landslides and debris flows are strongly affected by the ambient sea water. Water exerts a resistance on the landslide but may also enhance mobility.*
3. *A mudflow in the ocean is a two-phase fluid flow where both mud and ambient water should be properly described by fluid mechanics methods.*
4. *Landslides falling onto a water reservoir can generate hazardous water waves. If the volume of water is small compared to the landslide, most of water may be displaced in a deadly splash.*
5. *Fluid dynamical models can also be a guide to understand dry granular materials in rapid flow.*
6. *Turbidity currents are solid suspensions in water, occurring mostly in the ocean. They represent a formidable problem of fluid dynamics of great interest in sedimentology and marine geology, with significant applications to petroleum exploration. Although not properly landslides themselves, they are related to submarine landslides.*
7. *Subaqueous landslides may generate tsunami waves. The description of wave propagation and run-up all along the coast is an important problem of geohazard prevention.*

The present chapter deals with some of the most important concepts of fluids and their basic applications to the simplest fluids, called Newtonian.

The image of a mountain torrent opening the chapter seems appropriate to symbolize the complexity of fluid flows we observe in nature.

3.1 Introduction

3.1.1 What Is a Fluid?

The typical stiffness exhibited by a solid body is a consequence of the stable lattice structure of the building block (atoms, molecules, or ions). Only applying external stresses greater than a threshold value will the solid break. In contrast, the lack of enduring bonds between molecules of water results in a deformable state that makes it acquire the shape of the container. Water is an example of fluid. Fluids comprise *liquids*, characterized by high density and small compressibility (e.g., water), and *gases* like air that have generally much lower density and high compressibility. From the microscopic viewpoint, molecules in a liquid are separated by distances of the order of their diameter and collide frequently. Compressing a liquid makes the electron molecules penetrate and also explains their low compressibility. In contrast, because the molecules of a gas are several molecular diameters apart, the compression of a gas just diminishes the empty space between molecules.

Some fluids, called *Newtonian*, tend to flow along an inclined plate for whichever value of the inclination angle. However, there are also fluids capable of maintaining a proper shape even when resting on a flat area, provided that the state of stress inside the fluid does not exceed a certain limit. These belong to the class of fluids called *non-Newtonian*. Familiar examples include certain paints, or simply flour mixed with water. Mud and mudflows may flow like fluids when subjected to intense stresses and maintain the rigidity typical of a solid when the external stress becomes low. Let us consider as an example the behavior of mud with increasing water content. At very low water contents, the material maintains a solid-like appearance and may fracture in response to an intense shear stress. Increasing the amount of water, the material responds to external shear stress by deforming without breaking, a behavior called ductile. The transition between the solid and plastic material occurs in correspondence of the *plastic limit*. Beyond the plastic limit, the material does not fracture, and flows if the external solicitation is higher than a property called the *shear strength* of the material. For higher water amount the material becomes more fluid and the shear strength vanishes.

To introduce fluid dynamics, a standard approach is followed here dealing especially with Newtonian fluids. The flow of a fluid parcel results from the combined movement of a myriad of molecules. It is impossible, even in principle, to know the trajectories of single molecules. However, a precise knowledge of the single-molecule dynamics is not even required for a description of fluid motion, inasmuch as average properties of the fluid can be used. Such average properties are ideally taken over a sufficiently large volume to include many molecules (else statistical fluctuations become large), but sufficiently small to avoid an appreciable variation of the macroscopic properties within the volume. For example, the water velocity in a river can be obtained by measuring the flow around a small volume in front of a gauge pipe.

3.1.1.1 Materials Properties of Fluids

For an understanding of fluid flow, the material properties of the fluid must be known. For Newtonian fluids there are two important properties: density, which quantifies the amount of matter per unit volume, and viscosity, which measures the resistance to flow. For non-Newtonian fluids additional materials properties may be needed.

3.1.1.2 Density and Pressure

The density of a fluid is defined as the ratio between the mass and the volume of a portion of the fluid

$$\text{Density } \rho = \frac{\text{Mass}}{\text{Volume}}. \tag{3.1}$$

Units for the density are kg m^{-3}. Values for water and air at a temperature of 20 K are 10^3 kg m^{-3} and 1.1 kg m^{-3}, respectively.

The pressure is defined as the force exerted by a fluid on the unit area

$$\text{Pressure} = \frac{\text{Force}}{\text{Area}} \tag{3.2}$$

and is measured in pascal = $\frac{\text{Newton}}{\text{m}^2}$.

Often the multiples of pressure are used: kPa (kilopascal) and MPa (megapascal).

3.2 Fluid Static

3.2.1 Isotropy of Pressure

Let us consider a certain amount of water at rest in a vessel. Water molecules move incessantly at high speed, but on the average there is no macroscopic movement of portions of liquid from one region of the vessel to another, i.e., the fluid is static. Let us consider an infinitesimal triangular prism within the fluid with basal angle α, as shown in Fig. 3.1. We assume that the pressure acquires different values p_x, p_y, p_z, p_s on the different faces of the prism as shown in the figure. The elementary volume is sufficiently small for the pressure p_s to be approximately constant on the upper surface. The condition of force equilibrium along x and z can be written, respectively, as

$$p_x \, dy \, dz = p_s \, ds \, dy \, \sin \alpha$$

$$p_z \, dy \, dx = p_s \, ds \, dy \, \cos \alpha + \frac{1}{2} \rho g \, dx \, dy \, dz \tag{3.3}$$

3.2 Fluid Static

Fig. 3.1 Defining the element to find the isotropy of pressure

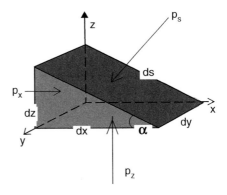

where the meaning of symbols is apparent from Fig. 3.1. From Eqs. 3.3 it is found $p_x = p_s$ and $p_z = p_s + \frac{1}{2}\rho g\, dz$; moreover, because dy can be arbitrarily small and the angle α is arbitrary, it follows that

$$p_x = p_y = p_s \equiv p \tag{3.4}$$

i.e., the pressure is independent of the direction of the surfaces. To summarize: pressure in a static fluid is *isotropic*.

3.2.2 Pressure Increase with Depth

The atmosphere exerts a pressure of about 1 bar (or 100 kPa) at the sea level. Pressure decreases with altitude as a consequence of the reduction in the total air content along the vertical column, but the difference in pressure becomes appreciable to the human body for altitude differences of 1,000 m or so. In a similar way, pressure under water increases with depth. However, because of the greater density of water, the pressure raise is noticeable already at a few meters depth.

The expression for the pressure increase as a function of depth can be easily worked out considering a cylinder containing a fluid at rest. The density is constant inside the volume, and the z-direction is chosen as the vertical one, parallel to the direction of the gravity field and pointing upward. Because the fluid inside the cylinder is at rest, the total forces along any radial direction and along the vertical must be zero. Equating the forces to zero along one radial direction merely shows that pressure depends only on the height z. Let us consider an infinitesimal height dz like in Fig. 3.2. The total force acting on the infinitesimal cylinder of height between z and $z + dz$ is

$$P(z)\, dx\, dy - P(z + dz)\, dx\, dy - \rho g\, dx\, dy\, dz = 0 \tag{3.5}$$

and in differential form

Fig. 3.2 Calculation of the pressure as a function of the depth

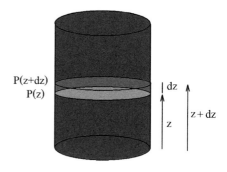

$$\frac{dP}{dz} = -\rho g. \tag{3.6}$$

If the density of the fluid is constant, the length dz needs not be infinitesimal. Water is, to a very good approximation, incompressible with static forces. Thus, the increase in pressure for a finite depth ΔH under water is given as

$$P(\Delta H) = P_{atm} + \rho g \Delta H \tag{3.7}$$

where P_{atm} is the atmospheric pressure.

Gases are much more compressible than liquids. Thus, to calculate air pressure at a height z, the relation (3.6) necessitates an integration

$$P(z) - P_0 = \int_0^z \rho(z) g \, dz. \tag{3.8}$$

Finally, note that pressure is completely independent of the shape of the containing vessel.

3.2.3 Pressure Exerted by a Suspension: Total Pressure and Water Pressure

We now consider the pressure at the bottom of a glass containing water and sand (Fig. 3.3). A distinction should be made between pore water pressure and total pressure. Pore water pressure is measured based on the level reached by water in a pipe connected to the base of the glass. If there is an increase in pore pressure, water rises vertically (Fig. 3.3); a metallic net keeps water clear from sand in the pipe. Total pressure, on the other hand, is measured based on the total load at the base of

Fig. 3.3 When sand is deposited at the bottom of the glass (**a**), pore pressure and total pressure differ. However, if sand is suspended, (**b**) the two pressures are both equal to $P = \langle\rho\rangle gD$ where $\langle\rho\rangle$ is the density of the suspension

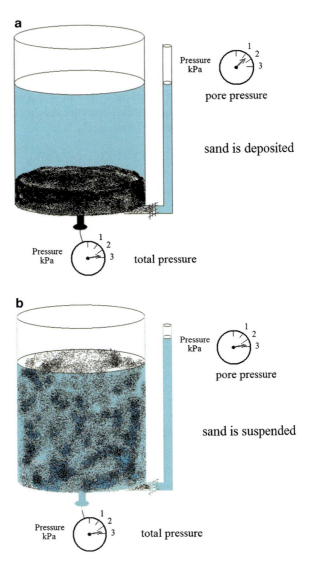

the cylinder. It is thus equal to the total weight of the material (water plus sand) divided by the area of the base of the cylinder.

When sand is deposited at the base, the measured water pressure is the one of a column of pure water (Fig. 3.3a). The total pressure is greater because the mixture contains sand, which is heavier than water. However, upon stirring, sand goes in suspension, a situation called fluidization (Fig. 3.3b). The total and pore pressures become both equal to $P = \langle\rho\rangle gD$ where $\langle\rho\rangle$ is the density of the water–sand suspension.

3.2.4 Force Exerted on a Dam by Water in a Reservoir

Water exerts enormous forces on artificial dams. The figure at the beginning of (→Chap. 10) and →Fig. 10.2 shows the remains of dams that collapsed under water lateral pressure. Let us idealize the dam wall with a rectangle of length a and height h and assume that the basin is filled up to this height. The force exerted by water of a strip of height dz and height z is from Eq. 3.8

$$dF = \rho g z a \, dz \qquad (3.9)$$

and the total force is

$$F = \int_0^h dF = \int_0^h \rho g z a \, dz = \frac{1}{2}\rho g a h^2. \qquad (3.10)$$

Thus, the force increases with the square of water height. For a 20 m high dam the force per unit length becomes of the order $F/a \approx 2 \times 10^6$ N/m

In the simple calculation considered here, the dam is idealized as a rectangle. Dams are usually bowl-shaped, with the concave side pointing valley-ward. This geometry is preferred to a straight wall because water exerts a compressive stress on the body of the dam, which is more tolerated than tensile stress.

3.3 Simple Treatment of Some Topics in Fluid Dynamics

In this section we consider some examples of fluid flows. Each example introduces one or two major concepts in a heuristic way. The purpose is to bring in key notions and help grasp some physical elements of the subject, without mathematical complexities. A more rigorous conceptual and mathematical framework of fluid dynamics is deferred to (→Sects. 3.5 and 3.6).

3.3.1 Fluid Flow (Key Concept: Velocity Field, Streamlines, Streamtubes)

The flow of a fluid can be described as a velocity field changing in space and time. Measuring the velocity at a particular position and instant of time implies considering an enormous number of fluid particles within the given volume. At a given time, the local velocity field of a fluid can be represented with a series of arrows representing the velocity vectors.

Fig. 3.4 Illustration of the velocity field, streamlines, and streamtubes

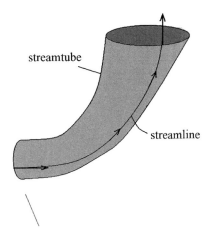

A *streamline* is a line tangent to the local fluid velocity vector (Fig. 3.4). Streamlines are useful as a pictorial description of fluid flow. In a three-dimensional flow, one also defines a streamtube as a tube-shaped volume whose boundaries are streamlines.

3.3.2 Fluid Flow in a Pipe with a Constriction (Key Concepts: Continuity, Incompressibility)

This case is illustrated in Fig. 3.5. The laminar flow in a pipe is called *Poiseuille flow*. A constant pressure gradient along the pipe ensures a constant fluid flow. The fluid has average velocity u_1 in the position 1 where the pipe has the largest diameter. The flux is defined as the mass of fluid passing through a certain section of the pipe in 1 s. In the position "1," the flux is given $J_1 = \rho_1 u_1 S_1$ where S_1 is the area of the section in position 1 and ρ_1 is the density in the same position. In position "2" at the constriction point, the flux is $J_2 = \rho_2 u_2 S_2$. Because fluid does not accumulate anywhere along the pipe, we must have $J_1 = J_2$. If the fluid is incompressible like water, then $\rho_1 = \rho_2$. Thus,

$$u_2 = \frac{S_1}{S_2} u_1 \qquad (3.11)$$

which shows that the velocity of the fluid increases at the constriction. This is the reason why one can increase the jet velocity of a garden hose by squeezing the nozzle.

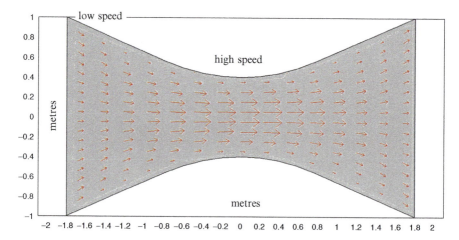

Fig. 3.5 Flow through a pipe with a constriction. The *arrow* length is proportional to the local velocity. 2D-FEM calculation at low Reynolds number

3.3.3 Lift Force on a Half-Cylinder (Key Concept: Energy Conservation and the Bernoulli Equation)

Fluids exhibit a certain resistance to flow. A perfect fluid is an ideal fluid with zero internal resistance (no viscosity). The flow of a perfect fluid is called potential flow because the velocity field can be calculated as the gradient of certain scalar functions, called potentials. Here an important result is anticipated for the case of perfect flow without rigorous demonstration: the Bernoulli equation.

Let us consider a streamtube and a parcel of perfect fluid flowing through it (Fig. 3.6). Because there is no dissipation, the total mechanical energy of the fluid is conserved during flow. At some point A along the streamtube, the kinetic and potential energy densities are given, respectively, as

$$\frac{\text{Kinetic energy in } A}{\text{Volume}} \equiv k_A = \frac{1}{2}\rho U_A^2$$
$$\frac{\text{Potential energy}}{\text{Volume}} \equiv p_A = \rho g H_A \quad (3.12)$$

Likewise, at another position B

$$\frac{\text{Kinetic energy}}{\text{Volume}} \equiv k_B = \frac{1}{2}\rho U_B^2$$
$$\frac{\text{Potential energy}}{\text{Volume}} \equiv p_B = \rho g H_B \quad (3.13)$$

3.3 Simple Treatment of Some Topics in Fluid Dynamics

Fig. 3.6 For the illustration of the Bernoulli equation

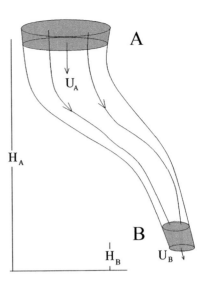

Because the energy is conserved, we must have

$$k_A + p_A = k_B + p_B + L(A \to B) \tag{3.14}$$

where $L(A \to B)$ is the work per unit volume associated to a transfer of the fluid from A to B. This contribution turns out to be the difference between the pressures in the two points

$$L(A \to B) = [P_B - P_A]. \tag{3.15}$$

To grasp the role of pressure in Eq. 3.18, consider that the kinetic energy carried by the molecules is made up of two contributions: the one resulting from the ordered macroscopic motion of the fluid, and the one due to the random movement of molecules. While the first gives the kinetic energy density k, the second result in the pressure. Hence, altogether

$$\frac{1}{2}\rho U_A^2 + \rho g H_A + P_A = \frac{1}{2}\rho U_B^2 + \rho g H_B + P_B \tag{3.16}$$

which is called the Bernoulli equation. Because the points A and B are arbitrary, this equation shows that for a perfect fluid the sum of the kinetic energy density, the potential energy density, and the pressure are constant along the streamtube.

Dividing by the density and gravity acceleration, Eq. 3.19 can also be expressed as

$$\frac{1}{2}\frac{U_A^2}{g} + H_A + \frac{P_A}{\rho g} = \frac{1}{2}\frac{U_B^2}{g} + H_B + \frac{P_B}{\rho g} \tag{3.17}$$

where all the terms have now the dimension of a height.

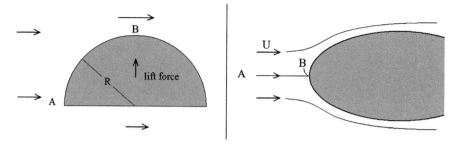

Fig. 3.7 *Left*: Geometry for the calculation of the lift force acting on a half-cylinder in the middle of a current of velocity U. The figure also shows the lift force. *Right*: For the calculation of the stagnation pressure

As an application of the Bernoulli equation, let us consider the following problem: a half-cylinder of radius R made of material of density ρ_0 is in the middle of a flow with a velocity U like in Fig. 3.7a. The experiments demonstrate that increasing U a point is reached, where the cylinder is lifted up against gravity. Among the other things, the problem is relevant for the lift of a pebble by a river flow. Calling A, B the points indicated in Fig. 3.7a, we have

$$\frac{1}{2}\rho U_A^2 + P_A = \frac{1}{2}\rho U_B^2 + \rho g R + P_B. \tag{3.18}$$

The velocity U_B can be calculated with the standard techniques used for potential flow (see, e.g., Julien 1998). In an approximate treatment followed here (a better calculation would require integrating the pressure difference based on the velocity field) we simply argue that because the fluid has to follow the outline of the cylinder's surface, the velocity in B must be greater than in A. The fluid has to take the same time flowing along the circumference of the cylinder as flowing between A and the right end of the cylinder, if the cylinder were not there. The average velocity along the cylinder is calculated taking the average between the velocities in A and B: $(U_A + U_B)/2$ and in this approximation we have

$$\frac{U_A + U_B}{2} = U_A \frac{\text{Length of semicircumference } A \to B \to C}{2R} = \frac{\pi}{2} U_A \tag{3.19}$$

from which it follows $U_B = (\pi - 1)U_A$. Substituting in Eq. 3.18 it is found

$$P_B - P_A = -\rho g R - \rho U_A^2 \left(\frac{(\pi - 1)^2 - 1}{2} \right) \tag{3.20}$$

3.3 Simple Treatment of Some Topics in Fluid Dynamics

Thus, the pressure difference between the top and the bottom of the cylinder is

$$\rho U_A^2 \left(\frac{(\pi-1)^2 - 1}{2} \right) \approx 3.579 \; \rho U_A^2/2 \quad (3.21)$$

(the exact value is $\frac{10}{3} \approx 3.333$ instead of $(\pi-1)^2 - 1 \approx 3.579$). In a rough approximation, the cylinder is lifted up if the pressure difference times the area of the cylinder base exceeds the weight of the cylinder, namely,

$$3.579 \frac{\rho U_A^2}{2} 2R > \frac{\pi R^2}{2} (\rho_0 - \rho) g. \quad (3.22)$$

Equation 3.25 yields a critical velocity

$$U_A > \sqrt{\frac{\pi/2}{(\pi-1)^2 - 1} R \frac{\rho_0 - \rho}{\rho} g}. \quad (3.23)$$

For reasons to be elucidated in ($\rightarrow$Chap. 9), the quantity

$$P_{\text{DYN}} = \frac{1}{2} \rho U^2 \quad (3.24)$$

is also called the dynamic pressure.

Consider now a body immersed in a fluid flowing at velocity U. The velocity of the fluid at the front point of the solid body must become zero. This point is called the stagnation point. Let us write the Bernoulli equations for the two points A and B in the Fig. 3.7b. A is far from the solid body and B is the stagnation point. Then using the stagnation condition $U_B = 0$, it is found from Eq. 3.16

$$\frac{1}{2} \rho U_A^2 + P_A = P_B \quad (3.25)$$

and calling $P_B - P_A = \Delta P$ it is found that at the stagnation point an excess stagnation pressure develops, having magnitude

$$\Delta P = P_{\text{DYN}} \quad (3.26)$$

3.3.4 Flow of a Plate on a Viscous Fluid (Key Concepts: No-Slip Condition, Viscosity, Newtonian Fluids)

Consider an experiment like the one sketched in Fig. 3.8, where a vessel is filled with a fluid like oil up to a certain height D. The upper cap of the vessel,

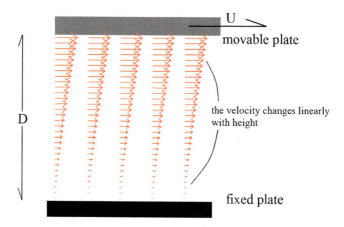

Fig. 3.8 For the illustration of the viscosity of a fluid

which is in contact with oil, can be moved horizontally. The fluid can be sheared by slowly moving the upper plate with a constant velocity U. Experiments show that if the velocity is very low, the flow pattern is laminar. The velocity decreases linearly with depth from a value U in contact with the upper plate to zero at the lower plate. Moreover, a constant shear force is needed to maintain a steady velocity. The shear force divided by the surface area of the plate gives the *shear stress* τ. The experiments show that provided that the velocity U is sufficiently low, for some fluids the shear stress is proportional to the ratio between U and D. In formulas

$$\frac{\text{Shear force}}{\text{Surface area of the plate}} = \mu \frac{U}{D} \qquad (3.27)$$

where the proportionality coefficient μ is called the *viscosity* and is measured in pascal per second (Pa s). To conform to common usage, we use here the same symbol μ as for the friction coefficient, as long as this will not cause ambiguity. The ratio $\dot{\gamma} = U/D$ is also called the *shear rate*. We can thus write $\tau = \mu \dot{\gamma}$. Fluids exhibiting this linear relationship between shear stress and shear rate are termed *Newtonian*.

The condition that the fluid velocity in contact with the solid bodies should be equal to the velocity of the body itself is an experimental condition that avoids discontinuous step in the velocity, which would else result in infinite shear rate at the boundary. This is also called the no-slip condition and is normally used as a boundary condition for fluid-solid object boundary in the calculations of fluid mechanics.

The viscosity of water at temperature of $10°C$ is $\mu_0 = 0.0013$ Pa s. The viscosity of oil is typically one order of magnitude greater, the one of air is more than one order of magnitude smaller.

3.3.5 Fluid Pattern Around a Cylinder (Key Concepts: Reynolds Number, Turbulence)

A fluid flowing with a velocity U past a cylinder generates particular flow patterns that depend not directly on the velocity, but on a dimensionless combination of velocity, density ρ, cylinder diameter D, and viscosity μ called the Reynolds number Re

$$Re = \frac{U\rho D}{\mu}. \tag{3.28}$$

In other words, the same flow pattern is reproduced doubling the velocity and halving the diameter of the cylinder, or changing consistently the density or the viscosity so as to maintain the same value of Re. At very low Re ($Re < 6$), the pattern is the one shown in the first of the Fig. 3.9. The flow is laminar and symmetrical from the front to the rear end (even though the pressure is greater at the front). The corresponding flow around a sphere is called the Stokes flow. The Stokes flow pattern for both cylinder and sphere can be calculated analytically, but this is more the exception than the rule in fluid mechanics. It is also found that in the Stokes regime the force transferred from the fluid to the solid body (called the drag force) is proportional to the velocity. A typical example where the Stokes regime is applied concerns the fall of small particles in water ($\rightarrow$Chap. 10).

At higher Re, a couple of permanent vortices form in the rear of the cylinders, that increases the drag against the vortex. For higher still Re, unsteady vortices form that cause the drag force to oscillate in time. At very high Re (last figure of Fig. 3.9) a large wake appears where the behavior of the fluid is unpredictable and variable in time (gray).

The reason for the change in the behavior is linked to a very important concept in fluid mechanics: turbulence. A good way to introduce turbulence is to refer to the early systematic investigations by the American engineer Osborne Reynolds. He injected a dye streak in the flow of water along a pipe (Fig. 3.10). At low velocities (Fig. 3.10a), the dye stretches along a straight line. This is the signature of a laminar flow with negligible vertical velocity. For increasing velocities (Fig. 3.10b) some eddies form, which partly displaces the dye. This is due to the onset of a marked vertical component in the fluid velocity field. At higher velocities (Fig. 3.10c), the dye tends to diffuse through the whole pipe. Turbulence is thus describable as a highly chaotic fluid flow, in which the fluid velocity varies irregularly in space and time. However, whereas fluctuations in the velocity can be very large, the average values are usually well defined. Turbulent flows are dissipative (much more than the laminar flow), nonlinear, and highly diffusive. They are dissipative because the high shear stresses developing in the fluid results in fast degradation of the fluid kinetic energy. The nonlinearity of the Navier–Stokes equations ($\rightarrow$Sect. 3.6.2) makes each single turbulence pattern unpredictable in the details. And finally, they are diffusive because the eddies displace the fluid and the dye similar to a diffusive process, much more efficiently than molecular diffusion in quiet or laminar fluid. Microscopically, turbulence has been described as a flow where eddies of sizes spanning several orders of magnitude develop.

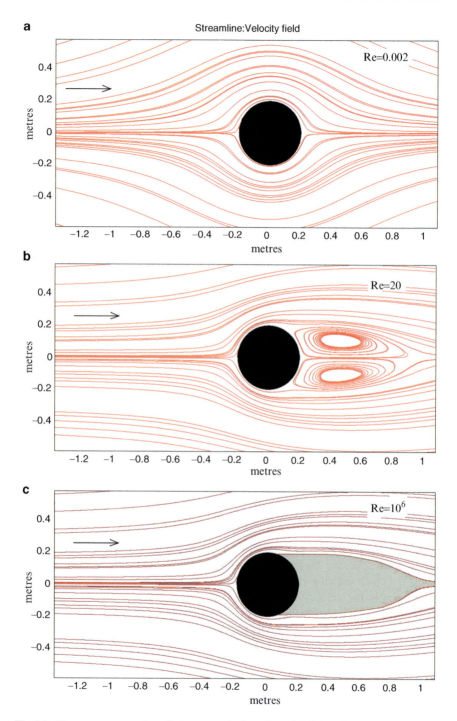

Fig. 3.9 Flow patterns around a cylinder, respectively, at $Re = 0.002$ (**a**), 20 (**b**), and 1,000,000 (**c**)

3.4 Microscopic Model of a Fluid and Mass Conservation   69

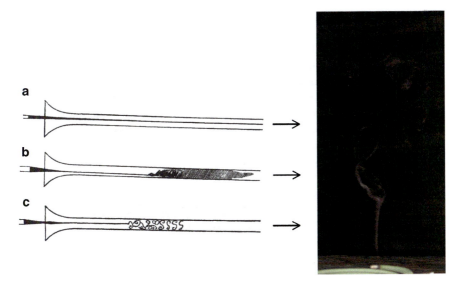

Fig. 3.10 *Left*: transition from laminar to turbulent flow in the original drawing by Reynolds. Water flows from *left* to *right* along a pipe, with a dye injected in the *center*. When the velocity is small (**a**), the dye stays at the axis of the pipe (a small molecular diffusion occurs, blurring a little the dye distribution; however, this effect is much less than the turbulent diffusion). (**b**) Situation where the velocity is about critical for turbulence to set in. (**c**) Fully turbulent flow. In modern terminology, (**a**), (**b**), and (**c**) are situations at low, critical, and supercritical Reynolds numbers. *Right*: a tiny smoke like the one from a cigarette acquires velocity as it streams upward, becoming turbulent after 10–20 cm

Turbulence is very common in fluids like air or water, for which the kinematic viscosity (i.e., the ratio between the viscosity and density) is of the order 10^{-6} and $10^{-6} s^{-2}$, respectively, and thus Reynolds numbers become high at relatively low velocities. Although easy to generate in low-viscosity fluids, the investigation of turbulence has not proven to be easy and in the last decades much experimental and theoretical-numerical research has been aimed at understanding turbulence. Turbulence affects the drag coefficients of objects moving in water, and it is thus important in understanding the dynamics of submarine landslides. Rheological flows moving at high speeds may also develop turbulence.

3.4 Microscopic Model of a Fluid and Mass Conservation

3.4.1 The Pressure in a Gas Is Due to the Impact of Molecules

The origin of pressure, viscosity, and the other properties of a fluid can be understood by considering the motion of its microscopic constituents. This has historically led to the kinetic theory, of great importance in the development of modern physics. Although in practice one seldom needs to introduce microscopic concepts in the study of fluids, it is interesting to appreciate the atomic origin of pressure and viscosity, also because similar concepts apply to granular flows.

When pouring water into a glass, a turbulent flow results, as one can easily notice with the aid of a dye. After a few seconds, however, water appears to be at rest. This does not mean that single molecules are motionless. Molecules are restlessly moving at a speed between one hundredth and one thousandth of the speed of light. However, their motion is disordered: each molecule moves randomly, constantly colliding against other molecules found along the path. In contrast, during the short time before motion is dissipated, molecular motion is partly ordered, as a very large number of molecules flow collectively along the same direction.

Let us consider the molecules of a gas contained in a cubical box of side lengths a, b, and c, respectively, parallel to the x, y, and z axes. One molecule of velocity components v_x, v_y, v_z will hit the face a, b, and c with frequencies

$$f_x = v_x/2a$$
$$f_y = v_y/2b \qquad (3.29)$$
$$f_z = v_z/2c$$

Considering a face perpendicular to x, the momentum transferred to this face upon one collision is

$$\Delta v_x = 2mv_x \qquad (3.30)$$

where m is the mass of one molecule. Thus, the force exerted by all the molecules is

$$F_x = \sum_{\text{all molecules}} \Delta v_x f_x = \frac{2m}{a} \sum_{\text{all molecules}} v_x^2 \qquad (3.31)$$

We obtain the pressure by dividing the force by the area of the face

$$P = F_x/(bc) = \sum_{\text{all molecules}} \Delta v_x f_x/bc = 2m/(abc) \sum_{\text{all molecules}} v_x^2$$
$$= 2m/V \sum_{\text{all molecules}} v_x^2 \qquad (3.32)$$

where $V = abc$ is the volume of the cube and $\rho = mn$ is the density of the gas. For statistical reasons, we must have that

$$\sum_{\text{all molecules}} v_x^2 = \sum_{\text{all molecules}} v_y^2 = \sum_{\text{all molecules}} v_z^2 = \frac{1}{3} \sum_{\text{all molecules}} v^2 \qquad (3.33)$$

where v is the total velocity of a molecule, and so

3.4 Microscopic Model of a Fluid and Mass Conservation

$$P = 2m/V \sum_{\text{all molecules}} v_x^2 = 2mN\langle v^2 \rangle/V = 2\rho\langle v^2 \rangle \quad (3.34)$$

where $\langle v^2 \rangle = \frac{1}{N} \sum_{\text{all molecules}} v^2$ is the average value of the squared velocity and N is the total number of molecules in the box. Equation 3.34 shows that pressure is proportional to the density and to the average square of the molecular velocity. In particular, the faster the molecules move, the higher the pressure they exert on the wall. Although we considered the case of a gas, where molecular collisions are rare, a similar conclusion holds for liquids, too.

3.4.2 Viscosity

Another property of real liquids is viscosity, intuitively identified with the resistance to flow (←Sect. 3.3.4). Intuitively, viscosity is related to the attraction between molecules. However, although viscosity is sensitive to the intermolecular forces, the origin of viscosity does not necessarily involve particle attraction. It may seem counterintuitive, but viscosity in an ideal gas does not derive from intermolecular attraction (which is missing by definition for an ideal gas) but from momentum exchange between molecules (Fig. 3.11).

An effective image suggested in the book by Guyon et al. (2001) is that of two trains traveling on parallel trails at different velocities. There is one worker standing on each train. Each of them shovels the same mass of gravel M to the other train at a constant rate, in such a way that the total amount of gravel on each train remains constant. Because one of the two trains is moving at higher speed, a certain quantity of momentum along the direction of flow is exchanged between the trains. Altogether, the slow train acquires a momentum along the direction of flow given by $M\Delta v$ where Δv is the velocity difference between the trains, and is thus slightly accelerated. The fast train loses the same amount and decelerates. Thus, although the two trains do not physically touch, they manage to exchange momentum along the traveling direction; their velocities become more equalized with time as if a shear force acted to limit the velocity difference between them.

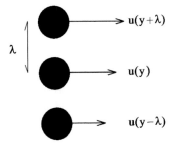

Fig. 3.11 Illustration of sheared gas for the discussion of viscosity

Let us now apply this model to gases. Consider a stationary shear flow as the one of Fig. 3.10. We can identify two parallel layers, located at heights $y + \lambda$ and $y - \lambda$ where λ is for the moment an unspecified length. The macroscopic velocities along x in the upper and lower layers are $u(y + \lambda)$, $u(y - \lambda)$, respectively. These velocities are in excess to the much greater random thermal velocity $\bar{u}$ of the molecules (in the train representation of the molecular velocity is symbolized by the shoveling rate). The flux of momentum (momentum exchanged per unit area and unit time) from the lower to the upper layer is

$$\vec{J}_\uparrow = \frac{1}{6} mu(y - \lambda) n\bar{u} \tag{3.35}$$

where n is the number of molecules per unit volume and the factor 1/6 has a geometrical origin. The flux from the upper to the lower level is

$$\vec{J}_\downarrow = \frac{1}{6} mu(y + \lambda) n\bar{u} \tag{3.36}$$

and the difference is

$$\vec{J}_\uparrow - \vec{J}_\downarrow = \frac{1}{6} mn\bar{u}[u(y - \lambda) - u(y + \lambda)] = -\frac{1}{3} mn\bar{u} \frac{u(y+\lambda) - u(y-\lambda)}{2\lambda}$$

$$= -\frac{1}{3} mn\bar{u} \frac{\partial u}{\partial y} = -\mu \frac{\partial u}{\partial y} \tag{3.37}$$

The difference $\vec{J}_\uparrow - \vec{J}_\downarrow$ has the dimension of a stress and can be identified with the shear stress. Comparing with the definition of viscosity provided in (←Sect. 3.3.4) and especially Eq. 3.27 giving the ratio between shear stress and shear rate $\mu U/D$, we can recognize the viscosity as the quantity

$$\mu = \frac{1}{3} mn\bar{u}\lambda \tag{3.38}$$

The length λ is the order of the path traveled by a molecule before encountering another molecule, what is called the *mean free path* of the molecule. From geometry, the mean free path can be found as

$$\lambda = \frac{1}{4\pi R^2 n} \tag{3.39}$$

where R is the radius of the molecule.

The two Eqs. 3.38 and 3.39 tell us that viscosity of a gas is related to the mass of the molecule, to its radius, and to its thermal velocity. In turn, the mean square velocity in an ideal gas is proportional to the temperature.

3.5 Conservation of Mass: The Continuity Equation

In a liquid, matters are more complicated because molecules are much closer and strongly interacting. When the liquid is subjected to an external shear, a molecule has to find its way through all the other molecules surrounding it. The physics can be described in terms of a potential barrier the molecule has to surmount. It is interesting to see that these concepts also apply to granular materials in state of strong agitation, which is one of the reasons for considering them here.

3.5 Conservation of Mass: The Continuity Equation

3.5.1 Flux

Consider a box filled with air. If a small window on the surface of the box is suddenly opened in vacuum, air flows out. The amount of air crossing the opening per unit time must be equal to the change of air content in the box. The current density (or flux) is defined as follows:

$$\text{Flux } j = \text{Air mass passed per unit time and unit area.} \tag{3.40}$$

The dimensions for j are

$$[j] = \frac{\text{kg}}{\text{m}^2 \text{ s}}. \tag{3.41}$$

An expression for the flux can be obtained considering that the amount of air passing through the opening during a time δt is $v \delta t S \rho$ where v is the exit velocity of air and S is the opening area. Dividing by the time and the area we find

$$j = \rho v \tag{3.42}$$

i.e., the flux is simply equal to the product between the velocity and the density. The mass of air δM flowing out of the opening during a certain time δt is so

$$jS = \frac{\delta M}{\delta t}. \tag{3.43}$$

If we visualize the box oriented with its axes along a Cartesian reference system, and imagining three openings are cut on each face pointing to the positive directions of x, y, and z, we can define the fluxes along each direction exactly as before

$$\begin{aligned} j_x &= \rho v_x \\ j_y &= \rho v_y \\ j_z &= \rho v_z \end{aligned} \tag{3.44}$$

and in compact vector notation

$$\vec{j} = \rho \vec{v}. \tag{3.45}$$

This equation defines in a general way the flux of a fluid. If a certain area is skewed with respect to fluid flow direction, in order to calculate the amount of liquid passed through the area it is necessary to take the scalar product between the flux and the versor normal to the surface. If a fishing net is oriented with an angle ϑ with respect to water velocity, the amount of water through the net (and so the probability of catching a fish) is given by the area of the net opening projected along the river velocity vector

$$\text{Amount of water passed through the net} = \rho v S \cos \vartheta \tag{3.46}$$

3.5.2 Continuity Equation in Cartesian Coordinates

The conservation principle of fluid mass leads us to the *continuity equation* for a fluid. Let us consider an elementary volume shaped as a parallelepiped, with its six faces parallel to a Cartesian reference system (Fig. 3.12). In contrast to the box considered in (←Sect. 3.5.1), these surfaces are not physical, but purely geometrical. The parallelepiped is immersed in the flow of a fluid, so that the local velocity vector of the fluid is not necessarily perpendicular to any of the faces. We assume the parallelepiped to be so small that fluid velocity and density are practically constant inside its volume. We wish to express the concept of matter continuity according to which the flow through the six surfaces must be equal to the total variation of the amount of fluid inside the box. Let us first consider the two opposite faces directed toward y. The net inflow from the surface "1" (situated at the left in Fig. 3.12) is

$$\rho v_y dx\, dz \tag{3.47}$$

and the outflow from the surface "2" on the right is

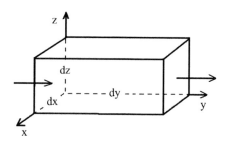

Fig. 3.12 For the demonstration of the continuity equation

3.5 Conservation of Mass: The Continuity Equation

$$\left[\rho v_y + \frac{\partial(\rho v_y)}{\partial y}dy\right]dx\,dz \tag{3.48}$$

(in case of need see the →MatApp for first-order Taylor expansion). Thus, the net outflow along the direction y is the difference between the two expressions above

$$\left[\rho v_y + \frac{\partial(\rho v_y)}{\partial y}dy\right]dx\,dz - \rho v_y dx\,dz = \frac{\partial(\rho v_y)}{\partial y}dy\,dx\,dz \tag{3.49}$$

The calculation for the other faces is similar. The net outflow along x and z are so, respectively, $\frac{\partial(\rho v_x)}{\partial x}dx\,dy\,dz$ and $\frac{\partial(\rho v_z)}{\partial z}dx\,dy\,dz$. The total outflow is the sum of these three terms

$$\text{Outflow} = \left[\frac{\partial(\rho v_x)}{\partial x} + \frac{\partial(\rho v_y)}{\partial y} + \frac{\partial(\rho v_z)}{\partial z}\right]dy\,dx\,dz \tag{3.50}$$

Because the amount of fluid is conserved, this quantity must be equal to the change of fluid mass per unit time in the whole volume

$$-\frac{\partial \rho}{\partial t}dx\,dy\,dz \tag{3.51}$$

and equating Eqs. 3.51 and 3.52, we find

$$\frac{\partial \rho}{\partial t} + \frac{\partial(\rho v_x)}{\partial x} + \frac{\partial(\rho v_y)}{\partial y} + \frac{\partial(\rho v_z)}{\partial z} = 0 \tag{3.52}$$

which is the *continuity equation* for a fluid. The notation using the symbol nabla (→MathApp) is also used

$$\frac{\partial \rho}{\partial t} + \nabla \cdot (\rho v) = 0 \tag{3.53}$$

If the density of the fluid is constant, Eq. 3.52 simplifies to

$$\frac{\partial v_x}{\partial x} + \frac{\partial v_y}{\partial y} + \frac{\partial v_z}{\partial z} = 0. \tag{3.54}$$

We will also use the following notation for the velocity components: $u \equiv v_x \equiv u_x;\ v \equiv v_y \equiv u_y;\ w \equiv v_z \equiv u_z$.

3.6 A More Rigorous Approach to Fluid Mechanics: Momentum and Navier–Stokes Equation

After the phenomenological introduction of the previous sections, the following part is more rigorous in the approach to fluid mechanics.

3.6.1 Lagrangian and Eulerian Viewpoints

When observing the motion of a fluid, two kinds of equivalent descriptions are possible. In the first description, the physical quantities (such as velocity or pressure) are measured from a fixed position in space. This description of fluid flow is called *Eulerian*.

Consider the flow pattern of water dropping from a faucet (Fig. 3.13). Immediately after opening, the fluid drops vertically pulled by gravity. After a transient initial time (times t_1 and t_2), the flow pattern becomes independent of time, or stationary (t_3 and t_4).

We can gauge water velocity at a given height from the faucet, for example with the camera "E" ("E" standing for "Eulerian"). The fact that the flow is stationary does not obviously mean that the fluid is always the same: the camera records new fluid molecules passing along; it is the flow pattern that does not change with time so that two snapshots taken at different times would be indistinguishable.

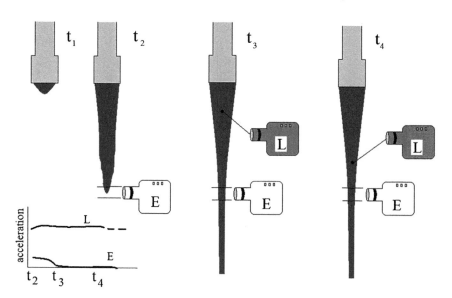

Fig. 3.13 Flow from a faucet

3.6 A More Rigorous Approach to Fluid Mechanics

A gauge located in a fixed position would always measure the same pressure and velocity. When at instant t_2 the water enters into the field of the camera, the Eulerian acceleration may be initially different from zero; however, as stationarity is reached, the Eulerian acceleration approaches zero.

Another possible description is based on the idea of following the fluid flow, so that the viewer is always observing the same fluid element. The description of fluid flow based on following the fluid trajectory is called *Lagrangian*. In Fig. 3.13 the "L" camera records the Lagrangian acceleration. Note that the Lagrangian acceleration corresponds to the one defined in mechanics for the moving point.

As a further example, consider some amount of clay carried by a water flow. Clay particles are very small, and so they are dragged along by water following faithfully the flow pattern. Let the spot of clay be in the position "1" at time t and in the position "2" at time $t + \Delta t$. The Eulerian acceleration is defined as the one always calculated in the same position, "1" in this case

$$\frac{\partial \vec{v}}{\partial t} = \lim_{\Delta t \to 0} \frac{\vec{v}(\text{time}=t+\Delta t; \text{position}="1") - \vec{v}(\text{time}=t; \text{position}="1")}{\Delta t} \quad (3.55)$$

where the symbol ∂ (partial derivative) is used in fluid mechanics to characterize the Eulerian derivative. The Lagrangian derivative is calculated by measuring the velocity at the same instants of time, but following the fluid

$$\frac{D\vec{v}}{Dt} = \lim_{\Delta t \to 0} \frac{\vec{v}(\text{time}=t+\Delta t; \text{position}="2") - \vec{v}(\text{time}=t; \text{position}="1")}{\Delta t}. \quad (3.56)$$

The relationship between the Eulerian and Lagrangian derivative can be specified in the following way. For a scalar quantity like the concentration T of a substance in the fluid or the temperature, the Lagrangian (which is mathematically a "total derivative") can be expressed based on expansion in terms of the partial derivatives

$$\frac{DT}{Dt} = \frac{\partial T}{\partial t} + \frac{\partial T}{\partial x}\frac{\partial x}{\partial t} + \frac{\partial T}{\partial y}\frac{\partial y}{\partial t} + \frac{\partial T}{\partial z}\frac{\partial z}{\partial t} = \frac{\partial T}{\partial t} + \vec{v} \cdot \nabla T. \quad (3.57)$$

This equation shows that the concentration changes either because of the local variation of the clay (first term on the last equation) or because of the advection of clay in the fluid (second term).

The same relationship can be applied to a vector quantity, for example, the velocity itself

$$\frac{D\vec{v}}{Dt} = \frac{\partial \vec{v}}{\partial t} + \vec{v} \cdot \nabla \vec{v} = \frac{\partial \vec{v}}{\partial t} + \left[v_x \frac{\partial \vec{v}}{\partial x} + v_y \frac{\partial \vec{v}}{\partial y} + v_z \frac{\partial \vec{v}}{\partial z} \right]. \quad (3.58)$$

giving the relationship between Lagrangian and Eulerian acceleration.

3.6.2 Momentum Equation

3.6.2.1 General

The principle of momentum conservation, which in mechanics results in the three Newton's laws of dynamics, states that in the presence of an external force acting on a particle the time derivative of momentum equals the force

$$\frac{d\vec{Q}}{dt} = \vec{F}. \qquad (3.59)$$

The momentum of a particle on mass m is given as $\vec{Q} = m\vec{v}$. If the mass is constant, (3.59) acquires the more familiar form

$$m\vec{a} = \vec{F} \qquad (3.60)$$

where $\vec{a} = d\vec{v}/dt$ is the acceleration. Thus, Eq. 3.59 shows that the trajectory of a particle can be calculated if the external force is known. As a corollary of Eq. 3.59, note that in the absence of external forces, the momentum of the point particle remains constant in time; if the mass is constant, the particle moves at constant velocity.

For every molecule of a fluid, collisions with other molecules and the movement in an external field like gravity always occur respecting Eq. 3.59. It has been discussed earlier that since the movement of each fluid particle is not computable, in fluid dynamics one considers the momentum conservation of a parcel of fluid, i.e., an ideal chunk of fluid ideally bounded by a streamtube comprising a large number of molecules, yet sufficiently small to be representative of the movement of the fluid in a particular point in space. In this section, the equations of momentum are applied to a parcel of fluid to find the fluid dynamical counterpart of Newton's law (3.59 and 3.60). For Newtonian fluids like water the result is the Navier–Stokes equation, a relationship at the foundation of fluid mechanics.

3.6.2.2 Listing the Forces

The momentum Eq. 3.59 is a vector equation that results equating the rate of momentum change of a liquid parcel to the forces acting on it. The forces we shall consider are the gravity force, the pressure force, and the internal friction of the fluid involving the viscosity. One commonly distinguishes between body forces and surface forces. Body forces are those acting on the bulk of the material; their magnitude is proportional to the mass of the fluid parcel. Surface forces, on the other hand, are transmitted through the parcel's surface. Gravity is a body force. Pressure and viscous forces are surface forces. For convenience,

3.6 A More Rigorous Approach to Fluid Mechanics

it is preferable to write the momentum conservation equation in terms of force per unit volume (N/m³).

3.6.3 Analysis of the Forces: The Momentum Equation

3.6.3.1 Gravity

The acceleration due to gravity can be considered constant on the Earth surface (←Box 1.1). The gravity force is the mass of a fluid parcel multiplied by gravity acceleration

$$d\vec{F}_{gravity} = \vec{g}dM = \vec{g}\frac{dM}{dV}dV \qquad (3.61)$$

where dM is the mass of the parcel, and dV its volume. From the above equation we find the force per unit volume

$$\vec{f}_{gravity} \equiv \frac{d\vec{F}_{gravity}}{dV} = \rho\vec{g} \qquad (3.62)$$

3.6.3.2 Pressure Force

Consider again an infinitesimal volume shaped as a parallelepiped (Fig. 3.14). Without loss of generality, we consider the edges parallel to the local Cartesian reference system. Let us assume that the pressure varies spatially within the volume. The pressure force acting along the y direction of the parallelepipid is of the form

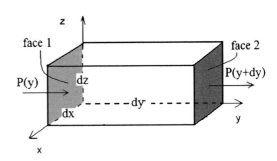

Fig. 3.14 For the calculation of the pressure force

$$dF_{pressure,y} = [P(y) - P(y+dy)]dx\,dz$$
$$= \left[P(y) - P(y) - \frac{\partial P}{\partial y}dy\right]dx\,dz \qquad (3.63)$$
$$= -\frac{\partial P}{\partial y}dx\,dy\,dz$$

where the Taylor expansion to first order is typically used in this kind of demonstrations (→MathApp) and thus the pressure force per unit volume becomes

$$f_{pressure,x} = \frac{dF_{pressure,x}}{dV} = \frac{dF_{pressure,x}}{dx\,dy\,dz} = -\frac{\partial P}{\partial x}. \qquad (3.64)$$

In a similar way the components of the pressure force along the x and z coordinates can be obtained. In a compact form we write

$$\vec{f}_{pressure} = -\frac{\partial P}{\partial \vec{x}} = -\nabla P \qquad (3.65)$$

of also for the generic coordinate i

$$f_{pressure,i} = -\frac{\partial P}{\partial x_i} \qquad (3.66)$$

The minus sign in (3.63)–(3.66) implies that the force pushes a fluid parcel from high to lower pressures.

3.6.3.3 Viscous Forces

In the calculation of the pressure force (Eq. 3.63), the contributing forces were perpendicular to the surfaces. In contrast, the viscous forces result also from the shear forces parallel to the surfaces of the elementary volume. Based on the definition of stress tensor and with reference to the Fig. 2.5, the total force along the direction x is written as the sum of six terms, one for each face (note that although the elementary volume in Fig. 2.5 represents a cube, the faces need not be equal in the present reasoning)

$$dF_{viscous,x} = \left(\tau_{yx} + \frac{\partial \tau_{yx}}{\partial y}dy\right)dx\,dz - \tau_{yx}dx\,dz + \left(\tau_{zx} + \frac{\partial \tau_{zx}}{\partial z}dz\right)dy\,dz$$
$$- \tau_{zx}dy\,dz + \left(\tau_{xx} + \frac{\partial \tau_{xx}}{\partial x}dx\right)dy\,dz - \tau_{xx}dy\,dz \qquad (3.67)$$
$$= \left[\frac{\partial \tau_{xx}}{\partial x} + \frac{\partial \tau_{yx}}{\partial y} + \frac{\partial \tau_{zx}}{\partial z}\right]dx\,dy\,dz$$

3.6 A More Rigorous Approach to Fluid Mechanics

and dividing by the volume dxdydz the force per unit volume becomes

$$f_{viscous,x} \equiv \frac{dF_{V,x}}{dV} = \frac{dF_{V,x}}{dx\,dy\,dz} = \left[\frac{\partial \tau_{xx}}{\partial x} + \frac{\partial \tau_{yx}}{\partial y} + \frac{\partial \tau_{zx}}{\partial z}\right] = \frac{\partial \tau_{jx}}{\partial x_j} \quad (3.68)$$

and similarly for y and z

$$f_{viscous,y} \equiv \frac{dF_{V,y}}{dV} = \frac{dF_{V,y}}{dx\,dy\,dz} = \left[\frac{\partial \tau_{xy}}{\partial x} + \frac{\partial \tau_{yy}}{\partial y} + \frac{\partial \tau_{zy}}{\partial z}\right] = \frac{\partial \tau_{jy}}{\partial x_j}$$

$$f_{viscous,z} \equiv \frac{dF_{V,z}}{dV} = \frac{dF_{V,z}}{dx\,dy\,dz} = \left[\frac{\partial \tau_{xz}}{\partial x} + \frac{\partial \tau_{yz}}{\partial y} + \frac{\partial \tau_{zz}}{\partial z}\right] = \frac{\partial \tau_{jz}}{\partial x_j} \quad (3.69)$$

Equations 3.68 and 3.69 adopt the Einstein notation according to which a summation over a certain index is intended whenever that index is repeated. For example: $A_j B_j = A_1 B_1 + A_2 B_2 + A_3 B_3$ where $\vec{A}$ and $\vec{B}$ are vectors. In Eq. 3.69 the repeated index is j in the last term on the right-hand side.

3.6.3.4 Final Expression for the Equation of Motion of the Fluid

Following Newton's law, the total force per unit volume along the coordinate i is equal to the rate of momentum change per unit volume given as

$$\text{Momentum change per unit volume along } i = \frac{dM}{dV}\frac{Du_i}{Dt} = \rho \frac{Du_i}{Dt} \quad (3.70)$$

and summing up the gravity force, the pressure force and the viscous force it is found

$$\boxed{\rho \frac{Du_i}{Dt} = \rho g_i - \frac{\partial P}{\partial x_i} + \frac{\partial \tau_{ji}}{\partial x_j}} \quad (3.71)$$

which is the momentum equation for the fluid. The inertia term on the left-hand side is written in terms of the Lagrangian derivative. In Eulerian form the left-hand side of the momentum equation should be written as (see Eq. 3.58)

$$\boxed{\rho \left[\frac{\partial u_i}{\partial t} + u_j \frac{\partial u_i}{\partial x_j}\right] = \rho g_i - \frac{\partial P}{\partial x_i} + \frac{\partial \tau_{ji}}{\partial x_j}} \quad (3.72)$$

In the Eqs. 3.71 and 3.72, the only explicit fluid property is the density. We know that some more properties must be contained in the stress tensor quantifying the resistance of fluid to flow.

3.6.4 Adding up the Rheological Properties: The Navier–Stokes Equation

One more tensor equation relating the stress tensor to other kinematic quantities is needed to close the momentum equation. These are called the *constitutive* or *rheological* equations.

3.6.4.1 The Simplest Rheological Equation: A Newtonian Fluid

The simplest case is the one of a Newtonian fluid. Water, air, and oil are examples of Newtonian fluids. Because air and water are the most important fluids in a human perspective, most fluid dynamics textbooks only deal with Newtonian fluids. For a Newtonian fluid the rheological equation is an extension of the equation seen in (←Sect. 3.3.4)

$$\tau_{ji} = \mu \left(\frac{\partial u_j}{\partial x_i} + \frac{\partial u_i}{\partial x_j} \right) + \lambda \delta_{ji} \frac{\partial u_k}{\partial x_k} \qquad (3.73)$$

where μ, λ are called the coefficients of viscosity and second viscosity, respectively. The second viscosity is related to the microscopic character of the fluid. For example, for a monoatomic gas, $\lambda = -(2/3)\mu$. Note, however, that the second viscosity is multiplied by the divergence of the fluid

$$\frac{\partial u_k}{\partial x_k} = \nabla \cdot \vec{u} = \frac{\partial u}{\partial x} + \frac{\partial v}{\partial y} + \frac{\partial w}{\partial z} = 0 \qquad (3.74)$$

which from the continuity equation must be equal to zero for an incompressible fluid. We are thus left with the rheological equation

$$\boxed{\tau_{ji} = \mu \left(\frac{\partial u_j}{\partial x_i} + \frac{\partial u_i}{\partial x_j} \right)} \qquad (3.75)$$

Equation 3.75 is of the form Eq. 3.27 (←Sect. 3.3.4). In the two-dimensional system introduced in (3.34), the only velocity derivative different from zero is

$$\partial u / \partial y$$

We can now use (3.75) to calculate the contribution of viscous forces to the momentum equation. Taking the derivative of the generic component of the stress tensor (3.75) we find

$$\frac{\partial \tau_{ji}}{\partial x_j} = \mu \frac{\partial}{\partial x_j} \left(\frac{\partial u_j}{\partial x_i} + \frac{\partial u_i}{\partial x_j} \right) = \mu \left(\frac{\partial^2 u_j}{\partial x_j \partial x_i} + \frac{\partial^2 u_i}{\partial x_j \partial x_j} \right) \qquad (3.76)$$

3.6 A More Rigorous Approach to Fluid Mechanics

where it is assumed that the viscosity is spatially independent. For the y component ($i = y$) it is found that

$$\frac{\partial \tau_{jy}}{\partial x_j} = \mu \frac{\partial}{\partial x_j}\left(\frac{\partial u_j}{\partial y} + \frac{\partial v}{\partial x_j}\right) = \mu \left(\frac{\partial^2 u_j}{\partial x_j \partial y} + \frac{\partial^2 v}{\partial x_j \partial x_j}\right)$$
$$= \mu \left(\frac{\partial^2 u}{\partial x \partial y} + \frac{\partial^2 v}{\partial y^2} + \frac{\partial^2 w}{\partial z \partial y} + \frac{\partial^2 v}{\partial x^2} + \frac{\partial^2 v}{\partial y^2} + \frac{\partial^2 v}{\partial z^2}\right). \tag{3.77}$$

Note that in the last parenthesis, the first, second, and third terms can be written as

$$\frac{\partial^2 u}{\partial x \partial y} + \frac{\partial^2 v}{\partial y^2} + \frac{\partial^2 w}{\partial z \partial y} = \frac{\partial}{\partial y}\left(\frac{\partial u}{\partial x} + \frac{\partial v}{\partial y} + \frac{\partial w}{\partial z}\right) = 0 \tag{3.78}$$

where the rule of interchanging the derivatives has been used. It is so found that

$$\frac{\partial \tau_{jy}}{\partial x_j} = \mu \left(\frac{\partial^2 v}{\partial x^2} + \frac{\partial^2 v}{\partial y^2} + \frac{\partial^2 v}{\partial z^2}\right) = \mu \nabla^2 v \tag{3.79}$$

where the Laplacian symbol is defined as $\nabla^2 v \equiv \frac{\partial^2 v}{\partial x^2} + \frac{\partial^2 v}{\partial y^2} + \frac{\partial^2 v}{\partial z^2}$. Similar calculations for the x and z coordinates give

$$\frac{\partial \tau_{jx}}{\partial x_j} = \mu \left(\frac{\partial^2 u}{\partial x^2} + \frac{\partial^2 u}{\partial y^2} + \frac{\partial^2 u}{\partial z^2}\right) = \mu \nabla^2 u$$
$$\frac{\partial \tau_{jz}}{\partial x_j} = \mu \left(\frac{\partial^2 w}{\partial x^2} + \frac{\partial^2 w}{\partial y^2} + \frac{\partial^2 w}{\partial z^2}\right) = \mu \nabla^2 u \tag{3.80}$$

and substituting into the momentum equation, we finally find the expression for a Newtonian fluid

$$\rho \left[\frac{\partial u_i}{\partial t} + u_j \frac{\partial u_i}{\partial x_j}\right] = \rho g_i - \frac{\partial P}{\partial x_i} + \mu \nabla^2 u_i \tag{3.81}$$

which is called the Navier–Stokes equation (hereafter also called NS).

This equation, together with continuity Eq. 3.54, rules the behavior of a Newtonian fluid. The equation is nonlinear due to the inertia term of the left-hand side. Usually the solution of the NS requires a computer. Analytical solutions have been found for particular cases (e.g., very high viscosity and simple geometry) but a general analytical solution is not available. The discoverer of a general three-dimensional solution will receive one of the Clay prizes, one million USD worth.

An alternative form of NS is fully vector

$$\rho \left[\frac{\partial \vec{u}}{\partial t} + \vec{u} \cdot \frac{\partial \vec{u}}{\partial \vec{x}}\right] = \rho \vec{g} - \nabla P + \mu \nabla^2 \vec{u} \tag{3.82}$$

Remembering that the scalar product on the left-hand side is between $\vec{u}$ and $\vec{x}$, and not between $\vec{u}$ and $\vec{u}$! Yet another form of the NS equations for a Newtonian fluid is often met in the literature

$$\frac{\partial}{\partial t}(\rho u_i) = -\frac{\partial}{\partial x_k}\left[P\delta_{ik} + \rho u_i u_k - \mu\left(\frac{\partial u_i}{\partial x_k} + \frac{\partial u_k}{\partial x_i}\right)\right] + \rho g_i \qquad (3.83)$$

3.7 Some Applications

3.7.1 Dimensionless Numbers in Fluid Dynamics

Let us consider the Navier–Stokes equation in the stationary limit, where the Eulerian time derivative is set to zero:

$$\rho \vec{u} \cdot \frac{\partial \vec{u}}{\partial \vec{x}} = \rho \vec{g} - \nabla P + \mu \nabla^2 \vec{u} \qquad (3.84)$$

that can be rewritten as

$$\vec{u} \cdot \frac{\partial \vec{u}}{\partial \vec{x}} + \frac{1}{\rho}\nabla P - \frac{1}{\rho}\mu \nabla^2 \vec{u} - \vec{g} = 0. \qquad (3.85)$$

Let us consider the experimental situation of a sphere immersed in uniform fluid flow. Calling R_0 the radius of the sphere, we expect the variations of pressure and velocity to take place within a volume of radius $x_0 \approx R_0$. Let u_0, g_0, ρ_0, P_0, μ_0 be the velocity, gravity acceleration, density, pressure, and viscosity far from the sphere. The first, second, third, and fourth term in (3.85) are, respectively, of the order

$$\frac{u_0^2}{x_0};\quad \frac{P_0}{x_0\rho_0};\quad \frac{\mu_0 u_0}{x_0^2 \rho_0};\quad g_0. \qquad (3.86)$$

Notice that the first term $\frac{u_0^2}{x_0}$ represents the characteristic value of the acceleration induced in the fluid by the presence of the sphere. Dividing all the terms of (3.86) by the characteristic acceleration u_0^2/x_0, the relative values of the four terms become, respectively, the following dimensionless ratios

$$1;\quad \frac{P_0}{u_0^2 \rho_0};\quad \frac{\mu_0}{x_0 u_0 \rho_0};\quad \frac{g_0 x_0}{u_0^2} \qquad (3.87)$$

We can also define the following dimensionless numbers

$$\frac{P_0}{u_0^2 \rho_0} = Eu = \frac{P_0}{2P_{\text{stag}}};\quad \frac{\mu_0}{x_0 u_0 \rho_0} = \frac{1}{Re}\quad \frac{g_0 x_0}{u_0^2} = \frac{1}{Fr^2} \qquad (3.88)$$

that can be identified with standard numbers of widespread usage in fluid mechanics.

3.7 Some Applications

1. *Eu* is called the Euler number. It is half of the ratio between pressure and the stagnation pressure.
2. The Reynolds number *Re* has been introduced in (←Eq. 3.28). It becomes now apparent that *Re* gives the ratio between inertial to viscous forces. When *Re* is very large, the corresponding term in the Navier–Stokes equation becomes negligible, which reduces (3.85) to

$$\vec{u} \cdot \frac{\partial \vec{u}}{\partial \vec{x}} - \vec{g} + \frac{1}{\rho}\nabla P = 0. \qquad (3.89)$$

This is the equation for a perfect fluid, from which the Bernoulli equation (←Sect. 3.3.3) can also be demonstrated. When *Re* is very small, the NS equation becomes

$$\nabla P = \mu \nabla^2 \vec{u} \qquad (3.90)$$

which represents the so-called creeping or Stokes motion (←Sect. 3.3.5).

3. The Froude number *Fr* is the ratio between inertial to gravitational forces. When the Froude number is very large the NS equation becomes

$$\vec{u} \cdot \frac{\partial \vec{u}}{\partial \vec{x}} + \frac{1}{\rho}\nabla P - \frac{1}{\rho}\mu \nabla^2 \vec{u} = 0. \qquad (3.91)$$

The Froude number is important in open-channel flows. When *Fr* is large, the inertia dominates over gravity. This usually happens for rapid flows, like for example in correspondence of rapids or waterfalls. Else, river flow is typically subcritical ($Fr < 1$).

The reason why these dimensionless numbers are significant in fluid mechanics is that the solution to the Navier–Stokes equation depends on the geometry of the flow and on the physical parameters only through these particular combinations. For example, the flow pattern around a full-size airfoil can be reproduced at a smaller laboratory scale provided that the Reynolds numbers are identical (the other dimensionless numbers are not important in this example). This requires higher velocities for the laboratory model compared to the full-scale airfoil, or greater densities, or lower viscosities.

3.7.2 Application to Open Flow of Infinite Width Channel

As an application of the use of the Navier–Stokes equation we consider a sheet of fluid down an incline with constant slope angle β (Fig. 3.15). The *x* axis is taken along the slope direction while the *y*-axis is perpendicular to the terrain; the *z*-axis is directed to the line of sight in the figure. The sheet extends to infinity in the *z* and *x*

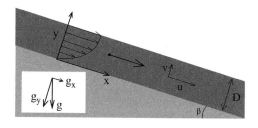

Fig. 3.15 For the illustration of the flow of a Newtonian fluid down an incline

directions, and has a finite depth D in the y direction. The Navier–Stokes equations along the x and y directions read

$$\rho\left[\frac{\partial u}{\partial t}+u\frac{\partial u}{\partial x}+v\frac{\partial u}{\partial y}\right]=\rho g \sin\beta+\mu\left[\frac{\partial^2 u}{\partial x^2}+\frac{\partial^2 u}{\partial y^2}\right]-\frac{\partial P}{\partial x}$$

$$\rho\left[\frac{\partial v}{\partial t}+u\frac{\partial v}{\partial x}+v\frac{\partial v}{\partial y}\right]=\rho g \cos\beta+\mu\left[\frac{\partial^2 v}{\partial x^2}+\frac{\partial^2 v}{\partial y^2}\right]-\frac{\partial P}{\partial y} \quad (3.92)$$

while the z-equation is irrelevant in this case. The full equations simplify as follows.

(1) Of the velocities, only u, the one along x, is different from zero. (2) Because we are interested in the stationary problem, u itself is independent of time. (3) By construction, the thickness of the debris flow does not change along the x coordinate, hence, u and P are independent of x.

With these substantial simplifications the NS equations yield

$$0 = \rho g \sin\beta + \mu\frac{\partial^2 u}{\partial y^2}$$

$$0 = \rho g \cos\beta - \frac{\partial P}{\partial y}. \quad (3.93)$$

The integration of the second of (3.93) gives the equation for the increase of the pressure with height seen earlier ($\leftarrow$Sect. 3.2)

$$P = P(D) + \rho g \cos\beta(D-y) \quad (3.94)$$

where $P(D)$ is the pressure at the upper limit of the fluid sheet, which coincides with the atmospheric pressure in subaerial conditions. The first of Eq. 3.93 is a second-order differential equation and thus requires a double integration. A first integration yields

$$\frac{\partial u}{\partial y} = \frac{\rho g \sin\beta}{\mu}(D-y) \quad (3.95)$$

where we have used the boundary condition that $\left|\frac{\partial u}{\partial y}\right|_{y=D}=0$

3.7 Some Applications

A second integration provides the final result

$$u(y) = \frac{\rho g \sin \beta}{\mu} \left(Dy - \frac{1}{2} y^2 \right) \quad (3.96)$$

where the no-slip condition $u(0) = 0$ has been used as boundary condition at the base of the flow. Equation 3.96 shows that the water velocity along x has a parabolic profile, reaching the maximum value at the top,

$$u(D) = \frac{\rho g \sin \beta}{2\mu} D^2 \quad (3.97)$$

The reader can try to apply the above equations to the case of water sheet of thicknesses $D = 1$ cm, 10 cm, and 1 m thick flow down a $1°$ slope. The values found for the velocity are really huge. What is going wrong with the calculation?

Let us define the Reynolds number for this problem as

$$Re = \frac{\rho u(D) D}{\mu}. \quad (3.98)$$

Experiments have demonstrated that this kind of open-channel flow remains laminar when the Reynolds number does not exceed a value of the order 1,000; for $Re > 2,000$ the flow is fully turbulent. When the flow is turbulent, most of the approximations that lead to Eq. 3.96 break down. The vertical velocity v becomes different from zero, and the derivatives are also nonvanishing.

When the flow in a channel is fully turbulent (and from the above examples it is clear that this happens in most cases of practical interest, at least dealing with clean water) it is customary to describe the flow based on semiempirical formulas of river mechanics and open-channel hydraulics, rather than attempting to describe the flow from the Navier–Stokes equations.

Chapter 4
Non-Newtonian Fluids, Mudflows, and Debris Flows: A Rheological Approach

People living in the lowlands around Mount Rainier, a stratovolcano of the Cascade mountains (USA), dwell on top of ancient landslide deposits. These deposits are due to mudflows descended from the top of the volcano in periods when the area was scarcely populated. However, Mount Rainier is still an active volcano; mudflows are certainly a current hazard. There are several other regions in the world at risk for volcanic mudflows, otherwise known as lahars. These flows are composed of a mixture of clay, silt, water, and coarser material like large blocks. They may be very fast and devastating. Wet mixtures of soils with clast size from clay to boulders are known with the generic name of debris flows. They are not necessarily volcanic but can derive from the failure of superficial soil in mountain environment, or even on flat areas in proximity of the sea.

The mechanical behavior of soil–water mixtures is scarcely known. It is certainly complex, as it involves both cohesion and friction. In addition, water may develop differential pressure in the pores of the granular material, thus complicating the dynamics even more.

As an initial step, this chapter starts addressing the physics of non-Newtonian fluids, a more complex class of fluids than the Newtonian fluids examined earlier. Rheology is introduced as the science describing the flow behavior of fluids subjected to stress. In a second step, water-rich mass flows like debris flows and mudflows are considered. As discussed in Chap. 1, debris flows and mudflows are also termed rheological flows.

The description of rheological flows based on fluid mechanics is a difficult task not only because the rheological properties are poorly known, but also because they may change radically and dramatically with time. For example, the shear strength of a lahar may change by orders of magnitude, as water seeps into the material or is expelled.

The figure below shows the lahar deposits of the volcano Cotopaxi, Ecuador. Unpublished photograph courtesy of Roberto Scandone and Lisetta Giacomelli.

4.1 Momentum Equations, Rheology, and Fluid Flow

In the previous sections dedicated to the laws of fluid mechanics, it has been shown that for a Newtonian fluid the relationship between the shear stress and the shear deformation is linear[1]

$$\tau_{ij} = \mu \left(\frac{\partial u_i}{\partial y_j} + \frac{\partial u_j}{\partial y_i} \right) \qquad (4.1)$$

where μ is the viscosity and τ_{ij} is the stress tensor. This equation is used in conjunction with the momentum Eq. 3.72 reproduced here

$$\rho \left[\frac{\partial u_i}{\partial t} + u_j \frac{\partial u_i}{\partial x_j} \right] = \rho g_i - \frac{\partial P}{\partial x_i} + \frac{\partial \tau_{ij}}{\partial x_j} \qquad (4.2)$$

to calculate the flow properties of the fluid like the velocity, pressure, and stress. Most of books on fluid dynamics deal exclusively with Newtonian fluids. This is because air and water, the most important fluids in the scientific and technological applications, are Newtonian.

However, rheological flows of interest for landslide studies seldom behave as Newtonian fluids. Thus, a more general relationship between stress and deformation rate should be used as a substitute to Eq. 4.1. Such a relation is called rheological. Rheology is the science aiming at measuring and elucidating the flow of the different materials in response to external stress. A rheological relation will normally involve several parameters, which need to be determined experimentally. The momentum equation together with the new rheological equation may then be used to determine the behavior of the non-Newtonian fluid.

4.2 Dirty Water: The Rheology of Dilute Suspensions

Let us consider a dilute suspension of solid particles in water, a problem tackled by Einstein in 1905 in relation to his famous paper on Brownian motion. The problem is as follows: how does the viscosity of the liquid change due to the presence of the suspension? Let us call C the volume fraction of solid particles (namely, the volume fraction occupied by the solid particles with respect to the total volume) and assume $C \ll 1$. We consider two points within the fluid separated by a distance δ. This length may be arbitrary, but the demonstration is shortened if we take δ as the average distance between the solid particles in the suspension. Thus, we can imagine the segment of length δ with the origin lying in the center of a suspended particle, and the other extreme at the center of the next particle. When a liquid contains a

---

[1] In the following we assume the fluid to be incompressible.

diluted suspension, the particles have the effect of excluding some regions of liquid, thus creating "holes" inside the fluid, through which the momentum is transmitted very rapidly.

The transfer of momentum is a diffusive process, exactly like the dispersal of an ink drop in water. The particles of ink, in moving away from the initial drop, collide with the molecules of water. Because of these collisions, they experience a kind of "drunk man" sequence of steps: they may move one step toward areas not yet filled with ink, or momentarily come back to the region of maximum ink concentration. However, on the average, they will more likely move away from the original position. It has been found that the mean square distance traveled by the front of the ink drop increases with the square root of time, or also $\tau \propto \text{distance}^2$.

Similarly, let us consider a volume of water with an excess of momentum along the direction x. This situation can be simply created by moving a plate along x, like in the experiments of (←Sect. 3.3.4). In full similarity with the diffusion of ink, momentum diffuses through the fluid too, because random collisions with fluid particles are a diffusive process. Because of the exclusion of the volume occupied by the solid particles, the diffusion time of momentum along the distance δ is of the order

$$\tau \approx \frac{(\delta - D)^2}{\mu_0/\rho} \tag{4.3}$$

where μ_0 ρ_0 are the viscosity and density of the *pure* liquid and D is the average diameter of a suspended particle. Note the dependence of the time on the distance squared, which is typical of diffusive processes. The diffusion time must also be equal to

$$\tau \approx \frac{\delta^2}{\mu/\rho} \tag{4.4}$$

where μ is the viscosity of the suspension. The slight density change between the pure liquid and the suspension is neglected. From geometrical considerations, one also finds that

$$D = \delta C. \tag{4.5}$$

Thus equating (4.3) and (4.4) we find

$$\tau \approx \frac{(\delta - D)^2}{\mu_0/\rho} = \frac{\delta^2(1 - C)^2}{\mu_0/\rho} = \frac{\delta^2}{\mu/\rho} \tag{4.6}$$

which yields Einstein's result

$$\mu \approx \frac{\mu_0}{(1 - C)^2} \approx \mu_0(1 + 2C) \tag{4.7}$$

where the last step is obtained by first-order Taylor expansion in C.

A more rigorous calculation would give

$$\mu = \mu_0(1 + 2.5C). \tag{4.8}$$

The above equations show that that the viscosity of a liquid increases with the addition of a solid suspension. It has been obtained by Batchelor and Green that to the second order in C

$$\mu \approx \mu_0(1 + 2.5C + 7.6C^2) \tag{4.9}$$

Einstein's approach is valid for very dilute suspensions ($C \ll 0.02$). One reason for such limitation is that for higher concentrations, hydrodynamic effect may become important. The range of validity of (4.9) is somehow broader ($C \ll 0.1$), because this equation accounts for effects depending on the second order in the concentration, like the Brownian motion of the particles, the stresses in the fluid flow, and the fact that flow lines are interrupted in correspondence of a particle. In general, the viscosity depends on the flow itself, which results in a very complicated problem. Some methods have been worked out for more concentrated suspensions, both as fitting formulas of the experimental data, and also as results of theoretical calculations. For applications to gravity mass flows one can refer to Coussot (1997).

A semiempirical formula that can be used also for higher concentration is the Krieger–Dougherty power law of the form (Coussot 1997)

$$\mu = \mu_0 \left(1 - \frac{C}{C_*}\right)^{-2.5C_*} \tag{4.10}$$

with $C_* = 0.6$. This equation does not contain information on the flow and should be considered as only indicative of the viscosity. The Krieger–Dougherty equation for the limit of small concentrations becomes equivalent to Einstein's results. Note that according to 4.10, the viscosity becomes infinite for $C \to C_*$.

4.3 Very Dirty Water: Rheology of Clay Slurries and Muds

4.3.1 Clay Mixtures

Let us suppose adding some clay to water in a bucket, and to measure the changes in the properties of the mixture as the percentage of clay increases. We have already used in the previous section the *concentration C*, defined as the volume fraction occupied by the solid material with respect to the total volume

$$\text{Concentration } C = \frac{V_S}{V} = n_N V_P \quad (4.11)$$

where V_S, V, V_P, n_N are, respectively, the volume occupied by the solid material, the total volume of the mixture, the average volume of a particle, and the number of particles per unit volume. The total density of the mixture is

$$\rho = \rho_F(1 - C) + \rho_C C \quad (4.12)$$

where ρ_C is the density of the clay mineral added into the slurry and ρ_F is now the density of the pure water fluid (differently from Chap. 3, in this chapter the symbol ρ is used for the density of the mixture).

4.3.2 Interaction Between Clay Particles

When a clay particle is immersed in water, it becomes surrounded by water molecules adsorbed at the surface. Bonding forces derive from hydration (incorporation into the crystal lattice), van der Waals attraction, and hydrogen bonding. At larger distances from the crystal, positively charged ions are attracted to the negatively charged surface of the clay particle. A *double layer* forms, where positive ions tend to be concentrated at the particle surface, and becomes less numerous as a function of the distance from the surface. A double layer can be very thick compared to the size of a clay crystal. For example, a crystal of montmorillonite is about 1 nm thick; the layer of absorbed water acquires a thickness of about 3 nm, while the double layer thickness is about 10–30 nm. Because of this enormous increase of the volume of influence, the interaction between clay crystals begins at much lower particle concentrations than expectable based only on the physical size of the crystals. Crystals interact if

$$C \gg \left(\frac{\text{Thickness of clay crystal}}{\text{Thickness of double layer}}\right)^3 \quad (4.13)$$

which gives a critical concentration of about 0.1% for montmorillonite. This is also the critical concentration below which the approximation of dilute concentration ($\leftarrow$Sect. 4.2) is reliable. A clay slurry becomes much stiffer when the concentration of clay particles is sufficient to form interconnected networks. Figure 4.1a shows crystals of kaolinite in water.

4.3.3 Rheology of Clay Mixtures and Other Fluids

To establish the rheology of a mixture, one can measure the relationship between the shear stress of the clay–water mixture and the shear rate. Tests are made with

4.3 Very Dirty Water: Rheology of Clay Slurries and Muds

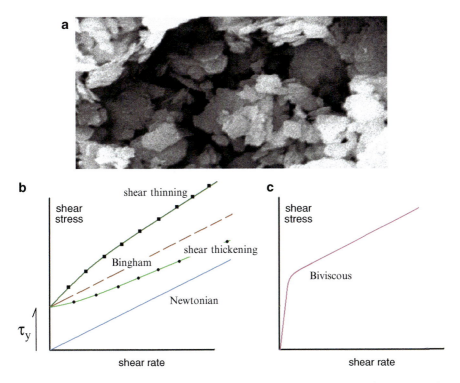

Fig. 4.1 (a) Electron photography of kaolinite crystals in water. Horizontal length approximately 10 μm. (Microphoto courtesy of Luca Medici.) (b) Idealized cases of rheological behavior. *Blue*: Newtonian fluid. *Brown*: Bingham fluid. *Dark green*: Herschel-Bulkley shear-thinning fluid. *Light green*: Herschel-Bulkley shear-thickening fluid. For the Newtonian and Bingham fluid, the viscosity is given by the slope of the straight line. The shear stress is given by the intercept with the ordinate axis. (c) Biviscous fluid characterized by a very high viscosity at small shear rate (initial slope of the curve) followed by more moderate viscosity at higher shear rates

a rheometer (for a simple introduction to rheological tests see for example Barnes et al. 1989). For the sake of illustration we can think of conducting the measurements with a Couette apparatus where the material to test is contained in the gap between two coaxial spinning cylinders. Commonly used rheometers are also the vane and the sphere rheometer.

The results of these measurements for different concentration of clay particles might look like the curves shown in Fig. 4.1b and c. For pure water a direct proportionality between τ (the shear stress) and γ (the shear rate) is found, the coefficient of proportionality being the viscosity (←Sect. 3.3.3). The shear rate is written explicitly as

$$\gamma = \frac{\partial u}{\partial y} \tag{4.14}$$

where y is the coordinate perpendicular to the two coaxial cylinders and u the fluid velocity parallel to the cylinders. As clay is added to the slurry, the viscosity increases. If the clay concentrations remain low, the rheology is Newtonian because even a very small stress produces a finite deformation rate. However, as a much more clay is added, the τ–γ curve changes in a more fundamental manner. For very small γ, τ does not go to zero but tends to a finite value τ_y of the stress (Fig. 4.1b), which indicates that a finite stress is necessary to deform the system. The stress τ_y needed to overcome the clay resistance at small shear rates is called the *yield stress* or also *shear strength* of the material (not to be confused with the *shear stress*, which is not a property of the material but a physical state, just like force or velocity). Microscopically, the existence of the shear strength is due to the long-range attraction of clay particles. To deform the system, breakage of some bounds is required.

Experiments show that the shear strength increases dramatically with clay content.

4.3.4 Bingham and Herschel-Bulkley

Figure 4.1b and c show some kinds of rheological behavior based on the analysis of the relationship between τ and γ. The functional form of the relationship between τ and γ can be written in some generality as a three-parameters fit

$$\tau = \tau_Y + K\gamma^n \qquad (4.15)$$

where K is called the consistency, and n is an exponent. In particular, for a Newtonian fluid $\tau_y = 0$, $K = \mu$ (the viscosity), and $n = 1$ and so we find again the usual relationship $\tau = \mu\gamma$.

The general form Eq. 4.15 for the rheology of clay–water system based on three parameters is called *Herschel-Bulkley* rheology. It represents well a behavior observed not only in clay–water mixtures, but also in cements, mud, polymers, or organic materials. If the exponent is greater than one ($n > 1$), the shape of the shear stress–shear rate curve is concave (Fig. 4.1). Materials with such behavior are called *shear-thickening* materials. Conversely, if $n < 1$ the curve appears convex and the material exhibits *shear-thinning* behavior. In both shear-thinning and shear-thickening fluids, the dependence of the shear stress is nonlinear in the shear rate. A typical shear-thickening fluid is wet sand fluidized in water.[2] At low shear rates, sand particles move freely within the liquid. At higher shear rates, however, particles begin to form arches. These are more stable structures which are capable of hindering the flow. In this situation, the viscosity increases with shear rate.

---

[2] "Fluidized" means that water circulates between grains, so that grains are not in contact, see (←Sect. 3.2).

4.3 Very Dirty Water: Rheology of Clay Slurries and Muds

In contrast, shear-thinning behavior is exhibited by certain paints. Paints must be easy to spread on a surface, and thus low viscosity is desired at high shear rates. However, the paint should not drip, which requires that at small shear rates the viscosity should be as high as possible.

The case with $n = 1$ is simpler to analyze. This is called the *Bingham* or *viscoplastic* rheology

$$\tau = \tau_Y + \mu_B \gamma \tag{4.16}$$

where the parameter μ_B (which substitutes K in Eq. 4.15) is called the *Bingham viscosity*. The name refers to the American engineer G. Bingham, the initiator of the science of rheology.

Note that the effective viscosity of the fluid goes like $\mu_{EFF} \propto \gamma^{n-1}$. Thus, in shear-thinning fluids the effective viscosity approaches zero for decreasing shear rates. In contrast, in shear-thickening fluids μ_{EFF} increases for small shear rates. Note that the Herschel-Bulkley formula is just a model that may be sufficient for many applications, but may become inadequate to describe the rheological behavior of some fluids for all shear rates. For example, colloidal suspensions exhibit a shear-thinning behavior followed by a strong increase of the effective viscosity for high stresses $\tau \approx 10^5$ Pa (Wagner and Brady 2009).

4.3.5 Shear Strength as a Function of the Solid Concentration

Figure 4.2 shows some experimental data for slurries composed of polystyrene beads, Sinard clay, and water slurry. Note that: (1) the slurry has finite shear strength. (2) The stress has a nonlinear behavior as a function of the shear rate.

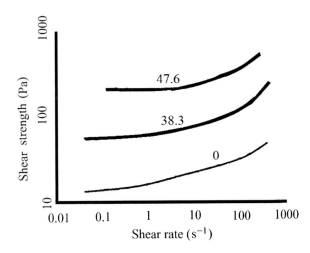

Fig. 4.2 Shear strength as a function of shear rate with increasing amount of large particles in the slurry. (From Coussot 1997, modified and simplified.) In the example shown, polystyrene beads are added to a Sinard clay–water slurry. The percentage of the added solid fraction is indicated

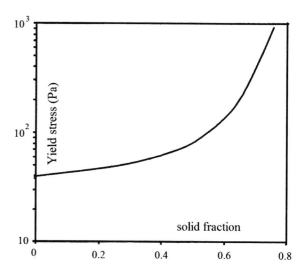

Fig. 4.3 Shear strength behavior of a slurry composed of water, clay, and polysized sand n as a function of the ratio of the solid fraction to the maximum possible solid fraction for sand alone (Figure modified and simplified from Coussot 1997)

(3) The shear strength increases enormously even with relatively small increase in the solid concentration.

The increase of the shear strength with solid concentration C is shown also in Fig. 4.3. Empirical fits of such curves fall in two categories: power law fits of the form

$$\tau_y \propto C^n \qquad (4.17)$$

or exponential fitting formulas

$$\tau_y \propto \exp(kC) \qquad (4.18)$$

where n and k are constants.

What is the microscopic origin of the shear strength and of the shear rate-dependence in a Herschel-Bulkley fluid? It has been discussed how the interaction between clay crystals has a very long range. Above a certain concentration, clay crystals start to develop networks in which single crystals form a quasi-crystalline phase. When an external shear stress is applied, some of the crystal–crystal bonds need to be broken for the slurry to move, which explains the existence of a finite shear strength.

To summarize, the rheological parameters of mixtures of solid particles in water vary dramatically with even a small increase of solid percentage. As a further example of the complications in dealing with rheological problems, it should be mentioned that the rheology may also depend on the previous stress history. For example, some substances exhibit *thixotropy*, whereby the viscosity decreases with time when subjected to a continuous stress. Thixotropic muds are used for example in oil drilling. The material flows efficiently in the interstices thus lubricating the drill head. During rest, the mud is capable to bear drilled rock chunks and to carry the stony load for long distances. For the same reason, also shear-thinning muds are used in drilling. Usually, a thixotropic substance also exhibits Bingham behavior, which indicates that thixotropy is linked to the formation of long bounds within the material. In state of rest, long-range forces can form resistant aggregates. During shearing, these aggregates begin to disassemble allowing the substance to become more fluid. If the fluid is put again at rest, aggregates form once more and the fluid goes back to the original value of viscosity. Among clay minerals, thixotropy is typical of bentonites.

4.3.6 Relationship Between Soil Properties and Fluid Dynamics Properties

The mechanical behavior of debris flow materials is somehow intermediate between the one of soils and that of fluids. It is thus useful to introduce some terminology used in soil mechanics. If water is added to dry soil, the material behaves like a solid material and begins to cracks if compressed. The soil remains initially solid, crumbly. Continuing to add water, it becomes easier to deform the soil without fracturing it. It is important for geotechnical engineers to devise straightforward operative criteria to describe the change in soil properties with added water. To this purpose, the Atterberg limits are widely used.

The *plastic limit* (PL) is defined as the minimum amount of water (percentage by weight) in correspondence of which it is possible to rework a small cylinder of soil to a diameter of 3 mm, without crumbling. The shear strength of a soil at the plastic limit is of about 0.1 MPa.

The *liquid limit* (LL) is defined as the amount of water in correspondence of which the cone of a penetrometer penetrates the sample by a length of 2 cm after a fall from a fixed height. The standard cone used for the measurements consists of a metal weight of 80 g with apex making an angle of 30°. In practice the liquid limit approximately corresponds to the minimum amount of water necessary for the soil to exhibit properties typical of liquids. The strength of the soil at the liquid limit is of the order 1 kPa.

The *plasticity index* is the difference $PI = LL - PL$. It is thus the amount of water necessary to diminish the strength of the sample by a factor 100, starting from the plastic limit. Usually, silts have a PI of about 5–10, while higher values of 30–45 are associated to clays. Highly reactive clays have PI of some hundreds.

The *liquidity index* of a soil is the quantity $LI = (W - PL)/(LL - PL)$ where W is the water content. The liquidity index at the liquit limit is by definition equal to one. Greater values indicate higher water content. Note, however, that the liquidity index indicates the amount of water in relationship to its effects in softening the soil. Thus, adding the same amount of water to a highly reactive clay ($PI = 400$) and to silt ($PI = 20$) will determine a much greater increase of the liquidity index in the silt than in the clay.

Locat and Demers (1988) have indicated an operative relationship to relate the liquidity index to the rheology of the material. The best fit for the viscosity and shear strength of slurries devoid of salt is based on a large number of measurements on different natural muds and is given as

$$\mu_B(\text{Pa s}) \approx \left[\frac{9.27}{LI}\right]^{3.3}$$
$$\tau_y(\text{Pa}) = \left[\frac{5.81}{LI}\right]^{4.55}$$
(4.19)

It is reasonable to expect an association between the shear strength and the Bingham viscosity, as both are affected by the attraction forces between the clay crystals. From the previous equations, one obtains this relationship in the form (Locat et al. 1997)

$$\mu_B(\text{Pa s}) \approx 5.2 \times 10^{-4} \left[\tau_y(\text{Pa})\right]^{1.12}.$$
(4.20)

where the Bingham model is used as a fitting model. Thus, according to these data, two mud samples with yield strength of 10 Pa and 1 kPa have a viscosity of about 0.008 and 1.3 Pa s, respectively.

4.4 Behavior of a Mudflow Described by Bingham Rheology: One-Dimensional System

As a first example of non-Newtonian fluid flow pattern, let us consider the stationary flow of a mud layer of thickness D in laminar flow along an inclined plane of infinite length (Fig. 4.4). Similar to the case for the Newtonian fluid (←Sect. 3.7.2), the velocity $u(y)$ parallel to the inclined plane depends on the height y from the bed. The momentum equations are rewritten here as

$$\frac{\partial u}{\partial t} + u\frac{\partial u}{\partial x} + v\frac{\partial u}{\partial y} = \frac{1}{\rho}\left(-\frac{\partial P}{\partial x} + \frac{\partial \tau_{xy}}{\partial y}\right) + g\sin\beta$$
$$0 = -\frac{1}{\rho}\frac{\partial P}{\partial y} - g\cos\beta$$
(4.21)

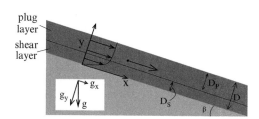

Fig. 4.4 Analysis of the Bingham fluid along a slope

4.4 Behavior of a Mudflow Described by Bingham Rheology

We further require invariance along the x-direction. Thus, the time and x-derivatives must be set to zero. In addition, we assume laminar flow, and so $v=0$. Similar to the analysis in ($\leftarrow$Sect. 3.7.2), the momentum equations simplify very much. Calling $\tau \equiv \tau_{xy}$ for simplicity, the simplified equations become

$$\begin{aligned} \frac{\partial \tau}{\partial y} &= -\rho g \sin \beta \\ \frac{\partial P}{\partial y} &= -\rho g \cos \beta \end{aligned} \quad (4.22)$$

from which it is found by integration

$$\begin{aligned} \tau &= -\rho g \sin \beta y + A \\ P &= -\rho g \cos \beta y + B \end{aligned} \quad (4.23)$$

where A and B are integration constants that can be found imposing appropriate boundary conditions: at the upper surface of the mud the shear stress is zero and the pressure equals the atmospheric pressure p_{atm}.[3] Thus

$$\begin{aligned} \tau &= \rho g \sin \beta (D - y) \\ P &= P_{\text{atm}} + \rho g \cos \beta (D - y) \end{aligned} \quad (4.24)$$

We now use the rheological equation for a Bingham fluid Eq. 4.16 and find

$$\begin{aligned} \mu \frac{\partial u}{\partial y} &= \tau - \tau_y = \left[\rho g \sin \beta (D - y) - \tau_y\right] \quad (\text{if } \tau > \tau_y) \\ \frac{\partial u}{\partial y} &= 0 \quad (\text{if } \tau < \tau_y) \end{aligned} \quad (4.25)$$

The first of the two equations (4.25) can now be integrated from the base up to the level where the right-hand side becomes zero. This defines a layer known as the *shear layer* whose thickness D_S can be found from Eq. 4.25 imposing $\tau - \tau_y = \left[\rho g \sin \beta (D - y) - \tau_y\right] = 0$:

$$D_S = D - \frac{\tau_y}{\rho g \sin \beta}. \quad (4.26)$$

For $y < D_S$, the velocity increases with height. For $y > D_S$, the weight of the overlying material is not sufficient to shear the material, as also obvious from

---

[3] There is a finite shear stress at the top due to the drag effect of air (or water for submarine mudflows). We neglect them here, but their role will be examined later for the case of submarine mudflows.

the second of the equations (4.25). As a consequence, the material moves rigidly like a plug. The thickness of this *plug layer* is

$$D_P = D - D_S = \frac{\tau_y}{\rho g \sin \beta} \quad (4.27)$$

The geometry of the Bingham flow is shown schematically in Fig. 4.4. To find the velocity profile, the first of Eq. 4.25 is integrated to find

$$u(y) = \frac{\rho g \sin \beta}{\mu} \left[D_S y - \frac{y^2}{2} \right] \quad \text{(if } y < D_S\text{)}$$
$$u(y) = u_{max} \quad \text{(if } y \geqslant D_S\text{)} \quad (4.28)$$

where

$$u_{max} = \frac{\tau_y}{\mu} \left[D - \frac{3}{2} \frac{\tau_y}{\rho g \sin \beta} \right] \quad (4.29)$$

is the maximum velocity reached by the Bingham fluid, coincident with the velocity of the plug layer. The parabolic profile of the velocity followed by the constant velocity in the plug layer is shown in Fig. 4.5.

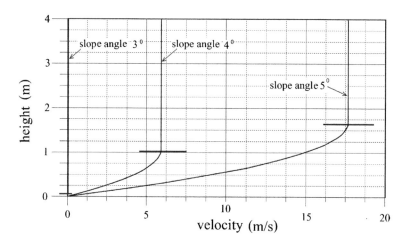

Fig. 4.5 Velocity profiles of a Bingham fluid based on Eq. 4.28. The shear strength is 3 kPa, the Bingham viscosity is 100 Pa s, and the density is 1,500 kg/m^3. Calculations are reported for a thickness of 4 m and slope angles of 3–5°. The boundary between the plug and the shear layers is indicated with a horizontal gray line. Note how the plug layer becomes thinner with increasing slope angle. For the slope of 3°, the plug layer becomes almost as thick as the whole debris flow. In this case, the debris flow is about to stop because the shear stress at the base becomes very close to the shear strength

4.5 Flow of a Bingham Fluid in a Channel

Equation 4.27 is relevant in the sedimentological analysis of debris flow deposits. When solved as a function of the shear strength, it yields

$$\tau_y = \frac{\text{Measured thickness of deposit}}{\rho g \sin \beta} \qquad (4.30)$$

thus allowing for an estimate of the shear strength based on the thickness of the deposit.

Finally, we report without demonstration the expression for the velocity profiles for a Herschel-Bulkley fluid

$$u(y) = \left(\frac{\rho g \sin \beta}{K}\right)^{1/n} \frac{1}{1+1/n} \left[D_S^{1+1/n} - (D_S - y)^{1+1/n}\right] \quad \begin{array}{l}(\text{if } y < D_S)\\ (\text{if } y \geqslant D_S)\end{array} \qquad (4.31)$$
$$u(y) = u_{max}$$

where

$$u_{max} = \frac{1}{1+1/n} \left(\frac{\rho g \sin \beta}{K}\right)^{1/n} D_S^{1+1/n} \qquad (4.32)$$

is the velocity of the plug layer.

4.5 Flow of a Bingham Fluid in a Channel

Often a debris flow travels along a valley, the bed of a river or other natural and artificial conduits. It is thus useful to examine the flow of a non-Newtonian fluid along a channel. A first approach to the problem is to search for the solution in idealized cases, where a regular channel geometry and simple rheological models are used. In the following only some basic results are presented.

4.5.1 Calculation for a Cylindrical Channel

The flow of a non-Newtonian fluid in more than one dimension can be complicated even for simple geometries of the channel. In the present section, the solution for the flow in a cylindrical channel of radius R is presented (e.g., Coussot 1997). We make use of cylindrical coordinates ($\rightarrow$MatApp). As in the previous one-dimensional case, many of the velocity components of the stress tensor are equal to zero. (1) Only the velocity component u_z along z (that we call $\equiv u$ for simplicity) is different from zero. (2) Because we are interested in the stationary problem, u itself is independent of time. (3) Of the nondiagonal components of the stress tensor only the r–z and z–r are different from zero. (4) The diagonal components of the stress tensor are considered equal to the hydrostatic pressure.

Using the momentum equations in cylindrical coordinates, the three following equations are found (see, e.g., Coussot 1997)

$$\frac{\partial p}{\partial r} = \rho g \cos \beta \cos \phi$$
$$\frac{\partial p}{\partial \phi} = -r\rho g \cos \beta \sin \phi \qquad (4.33)$$
$$\frac{\partial \tau}{\partial r} = -\frac{\tau}{r} - \rho g \sin \beta \cos \phi$$

and upon integration it is found

$$p = \rho g r \cos \beta \cos \phi$$
$$\tau = -\frac{1}{2} r \rho g \sin \beta \cos \phi \qquad (4.34)$$

The velocity for a Herschel-Bulkley fluid is reported here without demonstration (Coussot 1997)

$$u(r) = \left(\frac{\rho g \sin \beta}{2K}\right)^{1/n} \frac{1}{1+1/n} \left[(R-r_0)^{1+1/n} - (r-r_0)^{1+1/n}\right] \quad \text{when} \quad R>r>r_0$$
$$u(r) = u(r_0) = \left(\frac{\rho g \sin \beta}{2K}\right)^{1/n} \frac{1}{1+1/n} (R-r_0)^{1+1/n} \quad \text{when} \quad r<r_0 \qquad (4.35)$$

where the radius of the plug "layer" in the middle is

$$r_0 = \frac{2\tau_y}{\rho g \sin \beta} \qquad (4.36)$$

The material moves rigidly in the middle of the channel, for $r<r_0$ (Fig. 4.6a).

4.5.2 *Triangular Channel*

Rheological flows often stream along river beds and local depressions geometrically similar to a triangular channel shape. Theoretical calculations, experiments, and field observations show that a Bingham fluid flowing in a triangular channel forms three "dead regions" where the shear stress is too low to shear the material. The three dead regions are located at the bottom at the two sides of the channel (Fig. 4.6b). The side dead regions form where the material is very thin and the

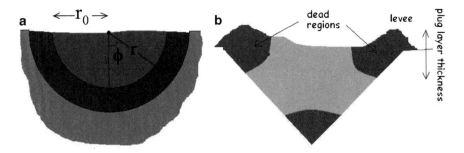

Fig. 4.6 The "dead regions" appearing in Bingham fluid flowing in a triangular channel

shearing effect of gravity is insufficient. Once the fluid at the sides begins to stagnate, it captures more material. The central bottom region forms because the fluid may stick to two valley walls.

4.6 Rheological Flows: General Properties

4.6.1 Introduction

Rheological flows have been previously defined as gravity-driven mass flows whose dynamical behavior is akin to the one of a non-Newtonian fluid. Thus, after having examined some rheological models of complex fluids, we consider the application to rheological flows observed in nature. Rheological flows form naturally as a mixture of water, clay, silt, and often coarser material such as gravel and boulders. They may also contain trees and human artifact. The distinctive characteristic of rheological flows is thus the presence of water and a cohesive matrix. Common types of rheological flows are *debris flows*, *mudflows*, and *hyperconcentrated flows*. In the present definition, debris flows are rheological flows of variable composition, which includes larger clasts. This generality of debris flows makes them nearly synonymous of rheological flows. Mudflows are prevalently formed of fine materials, while hyperconcentrated flows contain a large amount of water. These denominations are of widespread usage and will be adopted here.

Figure 4.7 gives us the opportunity to introduce some terminology of debris flows morphology. This particular case, occurred in the mountains of western Norway (Fjærland), originated from the failure of a moraine damming a glacial lake. The Suppellhe glacier is visible in the position A. Normally, the glacier feeds the valley visible in the foreground in the autumn with water and ice fall (white arrow). However, in 2004 things went differently. Already in early spring, water began to pull against the moraine damming the lake southward (A), determining its sudden collapse on May 8, 2004. The moraine material mixed with water travelled at high speed along the valley is visible in the figure (B). Approximately starting

Fig. 4.7 A debris flow in western Norway. The debris flow was caused by the failure of a moraine damming a subglacial lake in *A* (From Breien et al. (2008), reproduced with permission. Author: Andres Elverhøi)

from the point in C and then in D and E, it began to erode the soft soil along its path. Incidentally, these erodible deposits are partly due to previous historical debris flows. The debris flow opened up in a fan at the valley opening (F-G), and suddenly came to stop.

In this example, a large amount of water necessary for mobilizing the debris flow was provided by a lake. The incidence of this kind of event, also called GLOF (Glacial Lake Outburst Flood), will probably increase as a consequence of global warming.

4.6.2 Geological Materials of Rheological Flows

4.6.2.1 General

After rocks are exposed to geomorphologic agents (weathering, water, ice, wind) they are reduced in size. Usually, the efficiency of breakdown depends on the composition of the primary rock. Quartz, which is resistant to abrasion, will be the main constituent of advanced process of weathering, like in desert or beach sands. Clay materials derive from the weathering of feldspars. Feldspars are much more sensitive to weathering than quartz. As a consequence, size reduction is more pronounced. The small size of clay particles also implies a very large surface to volume ratio, and thus a high chemical reactivity. Clastic deposits may also be formed from the fragmentation of lavas during volcanic eruptions.

The processes driving particle erosion also determine the final grain size of a clastic deposit. In some cases, such as wind deposits (sand dunes, loess), particles may be sorted according to size or specific weight. In other cases, more various grain distribution results. Thus, glacial till is composed of unsorted material of grain sizes ranging from clay to boulders. This is because the till-depositing glacier has indistinctly abraded, grinded, and transported rocky debris irrespective of its size. Also taluses depositing at the flank of mountain cliffs are composed of fragments of different sizes, but the variability is more limited than for glacial till.

An overview of Earth materials important for gravity mass flows can be found for example in Selby (1993).

4.6.2.2 List of Debris Flow Source Materials

The geomorphic processes mentioned above produce the following source materials for debris flows.

- Soil. A large portion of land is covered by soil, a superficial layer of fundamental importance for plant, animal, and human life. Cutting a cross section of a soil would normally show a series of different layers merging gradually one into the other. Normally, the lower part at the contact with the bedrock is followed by a series of more recent horizons where the clay content increases due to the weathering processes at the surface. The total thickness of a soil may generally reach a few meters, but strong variations are possible depending on climate, weathering efficiency, recent geological history of the area, or local slope angle.
- Colluvium and talus are made up of undifferentiated rocky fragments. Common at the flank of mountain flanks, they often contribute as the source of debris flows. A talus, considered more in detail in ($\rightarrow$Chap.7), is a heap of loose rocky material originated from the periodic fall or rocks, and is devoid of fine material. The difference between colluvium and talus is that the former is also partly composed of soil.

- Pyroclastic materials are the source of a special class of debris flows, called lahars. These contain a high percentage of fines.
- Glacial materials, like the one of Fig. 4.7, are characterized by coarser materials like gravels and boulders, and may also contain silt.
- Loess is a wind deposit capable of maintaining very sharp angles if dry, but lacking cohesion if wet.
- Residual soils, distinctive of tropical regions.
- Quick clays typical of high-latitude areas in proximity of the sea level may give sudden origin to surprisingly fluid slides.
- Fluvioglacial deposits may be common in heavily glacialized areas and create large debris flows with variable granulometry.
- Artificial deposits (such as coal) and tiling dams.
- Some debris avalanches, originally composed of rock, may transform themselves in debris flows following water and soil entrainment.

4.6.3 Structure of a Debris Flow Chute and Deposit

The example in Fig. 4.7 allows us to identify the three main elements of the geometry of a debris flow chute and deposit: the source area, the debris flow track, and the deposition area. The most important distinctive characteristics are the following.

1. A source region of the debris flow material. The soil initially contributing to mass failure may have been destabilized by erosion or by failure of soft material. In correspondence of the original surface one can often observe a scraped smooth surface with exposed bedrock.
2. Debris flows frequently start traveling along a natural chute such as a torrent bed, a river, or a channel. Usually in the steepest part of the descent, the debris flow tends to erode rather than deposit.
3. In the middle part of its path, the debris flow usually increases in width due to entrained material and also because of widening of the valley. The debris flow may also form levees with height decreasing down slope. The dilution with water may also determine greater mobility and enhanced speed.
4. As the slope angle decreases, the debris flow slows down, initiating deposition.
5. When the slope angle decreases further, or in correspondence of a slope break, the debris flow often suddenly widens in a fan, and stops suddenly.

4.6.4 Examples of Rheological Flows

Here some examples of debris flows are presented, without pretence of full coverage.

4.6 Rheological Flows: General Properties

4.6.4.1 Small Nonchanneled and Superficial Debris Flows

An example of superficial, nonchanneled debris flow in mountain environment is shown in Fig. 4.8. The thin cover of colluvium was probably made unstable by water percolating between the bedrock (dipping at the same angle as the mountain flank) and the loose soil material. The figure and insets show some interesting features common to many debris flows: the erosion zone, the deposition zone, the tree scoured by the debris flow. Such kind of debris flow commences from a tiny region on the top. Flowing downward, the material affects the neighboring soil in a cascade process, so that the scar acquires the form of a triangle. This shape is also common in certain snow avalanches ($\rightarrow$Sect. 10.4). Once it has acquired momentum, in the deposition zone the debris flow may acquire a tongue-like shape.

4.6.4.2 Small Channeled Debris Flows

In the presence of a natural channel, a superficial debris flow may preferentially follow the main stream pathway. Figure 4.9 shows an example from the Alps. The next Fig. 4.10 shows three debris flows in Leirdalen, Norway, composed of blocks and soil (Shekesby and Matthews 2002). Levees, clearly noticeable in the photographs, border the sides of the debris flows. Levees are interpreted as the dead regions forming at the sides, where shear stress is small ($\leftarrow$Sect. 4.5, Johnson 1970). The average blocks size decreases downslope, as large blocks are transported only in the initial path.

4.6.4.3 Deeper Debris Flows Deriving from the Failure of Thick Unconsolidated Deposits

Figure 4.11 shows the scar of a debris flow initially started as a failure of fluvio-glacial deposit. Running into a mountain river, the debris flow was strongly diluted, reaching distances of several kilometers along the torrent (Crosta 2001).

4.6.4.4 Lahars

Lahars (a Javanese word) are debris flows composed of volcaniclastic materials. The ash from explosive volcanoes or degraded blocky lavas may combine with water to form a debris flow of consistency comparable to wet concrete. The water may derive from precipitation, from the melting of snow or ice following an eruption, or rain set off by the rising plume produced by the eruption itself. Often a triggering mechanism is provided by the tremors of the volcano. Lahars may form during the volcanic eruptions (primary lahars) or much later as the result of floods or torrential rains (secondary lahars). Owing to the enormous volumes of material erupted by some volcanoes, lahars can reach volumes of the order 1 km^3, far larger

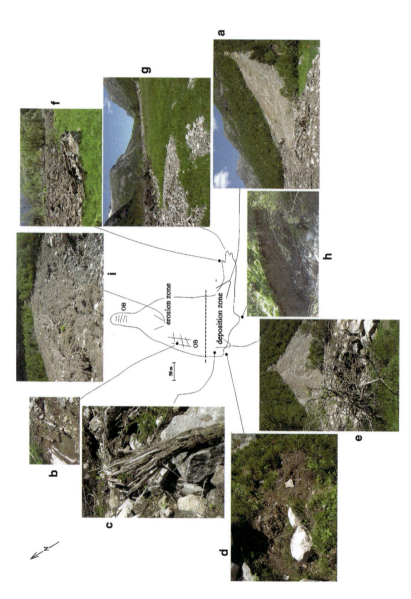

Fig. 4.8 A superficial debris flow in mountain environment probably triggered by water infiltration between the colluvial layer and the bedrock outcrop. The scheme of the scar and deposit is shown in the middle of the figure. *OB* outcrop of bedrock. (**a**) General view. (**b**) Detail of the outcropping rock. (**c**) Snapped tree trunk. (**d**) Boulders cast by the debris flow on the ground. (**e**) A lateral levee. (**f, g**) Details of the flow in the deposition zone. (**h**) The impact of the front with the river levee produced a scar. The event took place in May 2006. Lyngsetfjellet, western Norway. (**i**) The erosion zone

4.6 Rheological Flows: General Properties

Fig. 4.9 A debris flow in the Alps (Selvetta-Rodolo, Valtellina, Italy, July 2008; original photo courtesy of Giovanni Crosta)

than any other rheological flows. In addition to the primary material, lahars may entrain boulders and other material along the path, aided by the high density and shear strength. Figure 4.12 shows the lahar deposits of two volcanoes in Ecuador (see also the figure at the beginning of the chapter).

4.6.4.5 Quick Clay Slides

Quick clay slides are examples of lateral spread in which soil liquefies completely. In Scandinavia and Canada, terrains now on land were submerged during the last ice age due to isostatic sink caused by the weight of the thick ice sheet. A large amount of glacial clays was deposited in the sea, where clay crystals were partially supported by salt. After isostatic uplift following deglaciation, fresh water runoff removed part of the salt in some areas covered by glaciomarine clays. Lacking the support, the structure is unstable and may collapse like a deck of cards. Sometimes little stress is necessary to provoke instability. In 1978, a quick clay landslide occurred in Rissa (Trondheim, Norway) when part of the terrain of a farm was overloaded by a modest weight. A local collapse caused a retrogressive slide that progressively affected larger areas. A total of 330,000 m^2 of land was swept in the sea together with the load of soil, houses, and cars. The clay, solid just seconds before the first failure took place, was transformed into a fluid of zero shear strength.

Fig. 4.10 (a) Three channeled debris flows (length of each about 300–400 m). Notice the evident levees. (b) The final bulge of the debris flow in the middle appears made up of clasts of comparable size. (c) Panoramic view of the debris flow. The clasts composing the levees decrease in diameter downslope. In the proximal crown close to the headwall, the average clast diameter in the levees is about 0.5 m, while in the lower part they are about 0.2 m across. Levees look very fresh, almost resembling artificial rural walls. Leridalen (Jotunheimen), Norway

4.6 Rheological Flows: General Properties

Fig. 4.11 A debris flow from thick unconsolidated deposit. The scar of the Sesa landslide, Valcamonica, Italian Alps

Even on an essentially horizontal terrain, the liquefied soil ran swiftly toward the sea, causing also a small tsunami. In high-latitude areas, the quick clay slides cause much concern due to their rapidity and destructive potential and mapping of the hazardous areas are being completed (Fig. 4.13).

Box 4.1 Brief Case Study: The Osceola Lahar

Salient data
Name: Osceola
Mount involved: Mount Rainier
Location: Mount Rainier National Park, Washington (USA)
Coordinates: 46 51' 10"; 121 45' 37" (*summit of M. Rainier*)
Volume: (Mm^3): 1,000–3,000
Runout: 73 km.
Year: 5,000 *YBP* (5,040 *C*14).
Kind: lahar
Material involved: andesitic lava partially weathered to clay
Cause: unknown
Fahrböschung: 0.055
Special characteristics: it is one of the lahars with the longest runout and the largest volume.

(continued)

Fig. 4.12 Lahar deposits of the volcano Cotopaxi, Ecuador (*top*), and Tungurahua, Ecuador (*bottom*). The deposits of Tungurahua appear stratified. Deposits devoid of boulders can be ascribed to lahars with lower shear strength (Unpublished photographs courtesy of Roberto Scandone and Lisetta Giacomelli)

4.6 Rheological Flows: General Properties

Fig. 4.13 Hvittingfoss (Norway). The area is known as potential area for quick clay slides. The house has been evacuated (Original photograph courtesy May-Britt. Sæter)

Box 4.1 (continued)

Mount Rainier, in the Cascade Range of the Rocky Mountains, is a typical stratovolcano composed of successive layers of andesitic lava hydrothermally weathered to soft clay. The peak is covered with snow that as side effect provides additional water reservoir for mass failures. With a periodicity of some hundreds years, lahars descend from the summit of M. Rainier to distances of the order of 100 km. There are several preferred pathways for mudflows (Fig. 4.14), which tend to follow the beds of important rivers. Many local communities and settlements have grown on the deposits of ancient lahars.

The 5,000 years old Osceola mudflow has been the largest lahar from M. Rainier in the last 10,000 years (Fig. 4.15). It started as a debris avalanche, leaving a 600 m high scar, and subsequently changed to a debris flow. It moved initially northward, then changed direction westward, following the course of the White river. Important lahars in the area are also the Electron mudflow (N-NW, 500 years BP) and a constellation of smaller debris flows that flooded along many different routes branching from M. Ranier (some of these are indicated in the map of Fig. 4.14).

(continued)

Box 4.1 (continued)

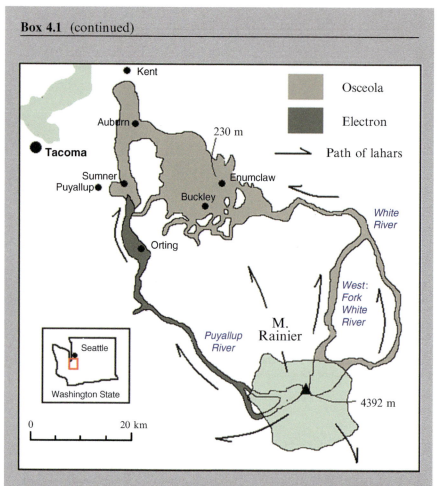

Fig. 4.14 Schematic Map of the Mount Rainier showing the Osceola and Electron lahars and other important lahar paths (*Source*: USGS web site, modified)

The making of mudflows from M. Rainier is probably not over. New flows could occur any time. Compared to the past, there is now the essential difference of a densely populated area. Although a reasonable action would be at least to avoid further settle in the area, the district is attractive and the population is steadily increasing. It becomes thus vital to identify with some accuracy the dynamical characteristics of future mudflows and their path. Estimates of the peak velocity of the Osceola mudflow are of at least 20–25 m/s, while the average velocity from the top of the volcano to the lowland could be in the range of 6 m/s (Scott and Vallance 1995). This would give some hours of warning time to the population living in the lowlands.

Other references: Crandell (1971); Scott and Vallance (1995).

(continued)

Box 4.1 (continued)

Fig. 4.15 A section of the Osceola Mudflow deposits, Mount Rainier National Park (Washington). On the left, the whitish traces of till almost completely covered by the mudflow deposits. People now dwell on the deposits of the ancient mudflows (Figure 13, U.S. Geological Survey Bulletin 1288, reproduced is accord with the USGS statute. Photographer: I.D. Crandell)

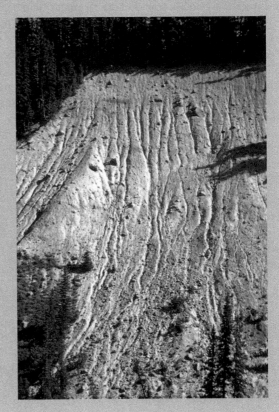

4.7 Debris Flows: Dynamics

4.7.1 Velocity

4.7.1.1 Measurement of the Velocity

Methods used for measuring the velocity of rivers are inapplicable for debris flows, as the solid fraction is too high. In principle, the average velocity can be directly obtained either by measuring the timing between two stations or by direct observation at one station based on radar, video analysis, or with ultrasonic sensors. This is possible when the debris flow occurs regularly in sites where the necessary equipment can be installed. In other cases, different methods have been devised to estimate the velocity from the observation of the deposits.

The most used method consists in measuring the superelevation (Fig. 4.16). A debris flow traveling along a watercourse curves in correspondence of a bend, rising on one of the bend sides and descending on the other, much like a bobsleigh. The height difference of the free surface between the two opposite sides of the channel is the superelevation. The radii of curvature R_1, R_2 for the two opposite sides of the channel differ (Fig. 4.16) but as a starting approximation we consider this difference to be small, and so $R_1 \approx R_2 \equiv R_C$. Let W be the watercourse width and H the superelevation (Fig. 4.16). From the fluid momentum equations ($\leftarrow$Eq. 3.81) along the direction of the bend (indicated with an arrow in Fig. 4.16), it follows that the horizontal pressure gradient must be equal to the centrifugal acceleration

$$\frac{\partial P}{\partial x} \cos \beta = \rho \frac{U^2}{R_C} \qquad (4.37)$$

where R_C is the radius of curvature. Because $\frac{\partial P}{\partial x} \approx \rho g \frac{H}{W}$, the velocity can be obtained as

$$U \approx \sqrt{\frac{1}{k} g \cos \beta \frac{H}{W} R_C} \qquad (4.38)$$

where the factor k in the denominator of the square root accounts for the viscosity and vertical sorting, and is essentially a semiempirical parameter. Values for k range between 1 and 5. The superelevation H can often be measured from the traces left by the debris flows, or by the height of the levees. A limitation of this equation is that it assumes subcritical conditions, i.e., the Froude number ($\leftarrow$Sect. 3.7) should be less than one.

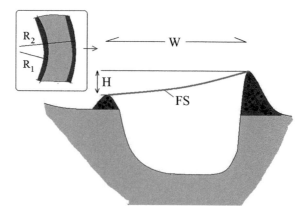

Fig. 4.16 Geometry of superelevation (H) for the calculation of the debris flow velocity. "FS" indicates the free surface of the debris flow

4.7 Debris Flows: Dynamics

In a more refined calculation, the difference in the radii of curvature along the two sides of the channel can be accounted for. From the momentum equation in polar coordinates one finds so

$$\frac{\partial P}{\partial r}\cos\beta = \rho\frac{U^2}{r} \qquad (4.39)$$

and the integration gives

$$U \approx \sqrt{g\cos\beta\frac{H}{\ln(R_2/R_1)}} = \sqrt{g\cos\beta\frac{H}{\ln(1+W/R_1)}} \qquad (4.40)$$

from which Eq. 4.38 is recovered in the limit $|R_1 - R_2| \ll R_1$ expanding the logarithm to second order, $\ln(1+W/R_1) \approx W/R_1$.

Table 4.1 reports velocities measured for some debris flows.

4.7.1.2 Hydraulic Jumps

Hydraulic jumps are frequently observed in fast-moving debris flows (Fig. 4.17). An hydraulic jump may appear as step, in correspondence of which the thickness and velocity of the debris flow change. The figure shows the basic elements for the analysis of the hydraulic jump. The fluid moves with two different speeds V_1 and V_2 before and after the jump, respectively. A classical calculation for water shows how the jump is related to the speed. The force exerted by the fluid against the surfaces

Table 4.1 Some velocities measured for debris flows

Debris flow	Density (g cm^{-3})	Solid fraction (by weight)	Estimated viscosity (Pa s)	Surface velocity (m/s)	Reference
Wrightwood Canyon, California, 1941.	2.4	79–85	210–600	1.2–4.4	Sharp and Nobles (1953).
Tenmile Range, Colorado.	2.53	91	3,000	2.5	Curry (1966).
Wrightwood Canyon, California, 1.62–2.131973.	1.62–2.13	59–86	10–6,000	0.6–3.8	Morton e Campbell (1974).
Mt. Thomas, New Zealand.	2.09	78	210–810	2.5–5	Pierson (1980).
Jiang-Jia ravine, Cina.	1.9–2.3	89	170 (field) 1.55 (laboratory)	3.19–13.1	Li et al. (1983).
Mount St. Helens	–	–	–	>30	Major et al. (2005)
Acquabona (Dolomites)	–	–	–	6–8	Galgaro et al. (2005)
Ghizar (Pakistan)	–	–	–	27	Nash et al. (1985)

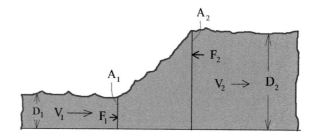

Fig. 4.17 Analysis of the hydraulic jump. The debris flow is moving from left to right

A_1, A_2 is $F_1 = \frac{1}{2}\rho g D_1^2$; $F_2 = \frac{1}{2}\rho g D_2^2$ where D_1, D_2 are the thickness of the debris flow before and after the jump, respectively. The difference in the force must be equal to the momentum carried by the fluid per unit time

$$\text{Momentum carried per unit time} = \frac{\delta M}{\delta t}(V_1 - V_2) = V_1 D_1 W \rho (V_1 - V_2) \quad (4.41)$$

and so

$$F_1 - F_2 = \frac{1}{2}\rho g W D_1^2 - \frac{1}{2}\rho g W D_2^2 = V_1 \rho W D_1 (V_1 - V_2) \quad (4.42)$$

and

$$\frac{1}{2}g D_1^2 - \frac{1}{2}g D_2^2 = V_1 D_1 (V_1 - V_2) \quad (4.43)$$

Calculating V_2 from the equation of continuity $V_2 = V_1(D_1/D_2)$, D_2 can be eliminated after simple manipulations to gain

$$\frac{1}{2}g(D_1 + D_2) = \frac{D_1}{D_2}V_1^2 \quad (4.44)$$

so that a second-order equation in D_1 can be found, whose solution is

$$D_2 = \frac{1}{2}\left[-D_1 + \sqrt{D_1^2 + \frac{8 D_1 V_1^2}{g}}\right] \quad (4.45)$$

which shows a relationship between the height of the jump and the velocity. Note that a necessary condition for the onset of the hydraulic jump is $D_2 > D_1$, and using Eq. 4.45 it is found that

$$\frac{V_1}{\sqrt{D_1 g}} \equiv Fr > 1 \quad (4.46)$$

The dimensionless ratio (4.46) is the Froude number, i.e., the ratio between the inertial and gravity forces (←Sect. 3.7). Hence, Eq. 4.46 shows that the hydraulic jump is due to the transition between critical conditions (inertia prevails over gravity) to subcritical (gravity force is greater than inertia). A debris flow may show variable Froude numbers along its path, and both subcritical and critical conditions have been observed. A hydraulic jump may also travel as a bore (in Fig. 4.17 a bore would move from the right to the left).

4.7.2 Dynamical Description of a Debris Flow

4.7.2.1 Depth Integration of the Momentum Equations

We now address the origin of the forces acting on a debris flow from a simplified but physically transparent viewpoint. This can be obtained depth-integrating the momentum equation from the basal level to the maximum height D.

4.7.2.2 General Equation of Motion

Let us consider a debris flow traveling along a slope and assume laminar flow. The momentum equations for the component of the velocity along slope in Lagrangian form are rewritten here

$$\frac{Du}{Dt} = \frac{\partial u}{\partial t} + u\frac{\partial u}{\partial x} + v\frac{\partial u}{\partial y} = -\frac{1}{\rho}\frac{\partial p}{\partial x} + g\sin\beta + \frac{1}{\rho}\frac{\partial \tau}{\partial y} \qquad (4.47)$$

$$p = \rho g(D - y)\cos\beta$$

We consider a portion of the debris flow mass limited by parallel planes.

4.7.2.3 Gravity

The integration of the gravity term on the right-hand side of Eq. 4.47 results in

$$\int_0^D g\sin\beta \, dy = g\sin\beta D \qquad (4.48)$$

4.7.2.4 Shear Stresses

The integration of the shear term gives

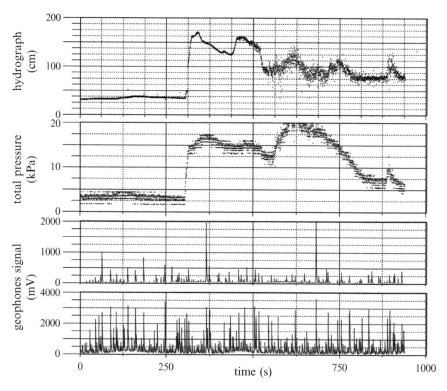

Fig. 4.18 Data for the debris flows in Acquabona creek, a basin stretching for 1.5 km in the Dolomites (Northern Italy) taken on August 25, 1998. The four panels show the hydrograph, the total pressure, and the measurements of geophones as a function of time. The bursts visible especially in the hydrograph data are correlated to increase of the velocity (Original data courtesy of A. Simoni)

$$\frac{1}{\rho}\int_0^D \frac{\partial \tau}{\partial y}dy = \frac{1}{\rho}\left(\tau_{\text{top}} - \tau_{\text{bottom}}\right) \approx \frac{1}{\rho}\left(\tau_{\text{DRAG}} - \tau_y - 2\mu_B \frac{U}{D}\right) \quad (4.49)$$

where U is the velocity of the plug layer. The physical interpretation of this result is that no internal stress can modify the acceleration of the system center of mass. Note also that the stress at the bottom is the sum of the yield and the viscous stresses, where a Bingham model has been assumed.

4.7.2.5 Friction with the External Medium: Skin Friction

The drag shear stress may be important for subaqueous debris flows. For this reason it will be examined thoroughly in relation to subaqueous landslides (→Chap. 9). Here we limit ourselves in giving a general expression

4.7 Debris Flows: Dynamics

$$\frac{\tau_{DRAG}}{\rho} = -\frac{1}{2}\frac{\rho_F}{\rho}C_{SF}U^2 \tag{4.50}$$

where ρ_F is the ambient fluid density and C_{SF} is the skin friction drag coefficient (→Chap. 9). Coefficients vary according to the smoothness of the debris flow surface. A value of the order 0.003 is a good estimate based on experiments with flat plates.

4.7.2.6 Earth Pressure Force

This term derives from the pressure term in Eq. 4.47. If the debris flow has not uniform thickness, a lateral force arises pushing from the thick to the thin parts of the debris flow. From (4.47)

$$-\frac{1}{\rho}\int_0^D \frac{\partial p}{\partial x}dy = -\frac{\rho g \cos \beta}{\rho}\int_0^D \frac{\partial (D-y)}{\partial x}dy = -g\cos\beta D\frac{\partial D}{\partial x}. \tag{4.51}$$

The origin of the earth pressure (or lateral pressure) term can be understood by this simple argument. Consider for simplicity the debris resting on a flat area. At two positions x and $x + \delta x$, the horizontal forces through a vertical slice are, respectively,

$$F(x) = \int_0^{D(x)} \rho g(D(x) - y)dy = \frac{\rho g D^2(x)}{2}$$

$$F(x + \delta x) = \int_0^{D(x+\delta x)} \rho g(D(x + \delta x) - y)dy = \frac{\rho g D^2(x + \delta x)}{2} \tag{4.52}$$

If the height of the material decreases with x, then the difference $F(x + \delta x) - F(x) \approx -\rho g D(x) \delta D$. Calling τ_b the shear stress at the base, then $\rho g D(x) \delta D = -\tau_b \delta x$, and so $\tau_b = -\rho g D(x) \frac{\partial D}{\partial x}$. Notice that in the presence of a granular component, the earth pressure form should be written as

$$-g\cos\beta D k_{ACT/PASS}\frac{\partial D}{\partial x} \tag{4.53}$$

where $k_{ACT/PASS}$ is a lateral earth pressure coefficient (←Sect. 2.2.3).

It is also interesting to show the role of the earth pressure force in shaping a heap of Bingham material resting horizontally. From (4.47) putting all the time derivatives and the velocities to zero, it is obtained that

$$0 = g\cos\beta \frac{\partial D}{\partial x} + \frac{\tau_y}{\rho D} \tag{4.54}$$

from which

$$\frac{\partial D^2}{\partial x} = -\frac{2\tau_y}{\rho g} \quad (4.55)$$

and so

$$D(x) = \sqrt{D_0 - \frac{2\tau_y}{\rho g}x} \quad (4.56)$$

where D_0 is the maximum height of the deposit. The Bingham fluid thus acquires a radial distribution with radius

$$R = \frac{\rho g D_0}{2\tau_y} \quad (4.57)$$

Concerning the left-hand side of (4.47), we assume for simplicity the debris flow thickness D to be constant and so $\int_0^D (dU/dt)dy \approx DdU/dt$. Notice also that the derivative symbol d is used instead of D as this might generate confusion.

Summarizing, the acceleration of an element of debris flow becomes

$$\frac{dU}{dt} \approx g\sin\beta - g\cos\beta \frac{\partial D}{\partial x} - \frac{1}{2}C_{SF}\frac{\rho_L}{\rho D}U^2 - \frac{1}{\rho D}\left(\tau_y + 2\mu_B \frac{U}{D}\right) \quad (4.58)$$

4.7.2.7 Generalized Rheological Equation: Accounting for a Granular Component

Debris flows are often composed of grains of different sizes such as sand, gravel, or boulders. The simplest way to account for friction is to extend the simple Bingham (or Herschel-Bulkley) behavior with a Coulomb frictional component. A simple model is the Johnson's Coulomb-viscous rheological model (Johnson 1970) obtained by summing a Coulomb frictional term to the stress of a pure Bingham model including also pore water pressure p_u

$$\tau = \tau_y + [\rho g(D-y) - p_u]\cos\beta\tan\phi + \mu_B \frac{\partial u}{\partial y}. \quad (4.59)$$

Iverson (2005) showed the importance of pore pressure advection and diffusion in the mechanical behavior of debris flows. In a series of experiments, it was shown that the basal pore pressure amounts to about one third of the total pore pressure, which determines a decrease of the effective friction coefficient. In the presence of pore pressure at the bed p_{BED}, the equation of motion of the debris flow (4.58) assumes the form (e.g., Iverson 1997a, b)

$$\frac{dU}{dt} \approx g\sin\beta - gk_{\text{ACT/PASS}}\left(\cos\beta\frac{\partial D}{\partial x} - \frac{1}{\rho g}\frac{\partial p_{\text{BED}}}{\partial x}\right) - g\left(\cos\beta - \frac{p_{\text{BED}}}{\rho g D}\right)\tan\phi$$
$$-\frac{1}{2}C_{\text{SF}}\frac{\rho_L}{\rho D}U^2 - 2\mu_B \frac{U}{\rho D^2}$$
(4.60)

where the excess pore pressure may change in space and time due to diffusion through the pores and exchange of material from different zones of the debris flow (Iverson and Denlinger 2001).

4.7.3 Impact Force of a Debris Flow Against a Barrier

A debris flow colliding against a barrier or a wall communicates an impact force. Let us visualize the impact process as the action of a machinegun shooting against a wall. Each bullet carries a momentum mU, where m is the mass and U is the velocity. The momentum transferred to the wall by each bullet is $P = mU$ if the bullet is absorbed by the wall, and $P = 2mU$ if the bullets bounce elastically ($\leftarrow$3.5.1). Thus, if n bullets per unit time are shot against the wall, the force mediated in time is $F = GmU/\Delta t = GmUf$ where Δt is the interval between successive bullets, f is the impact frequency, and the coefficient G is 1 if bullets are absorbed, and 2 if they bounce elastically. The mass per unit time hitting the wall is thus the product of the frequency times the mass of each bullet, namely, mf. The principle is the same if the bullets are now substituted by a continuum medium. Calling D the height and W the width of the wall, the projectile mass per unit time becomes $mf = \rho DWU$ and so the force against the wall is

$$F \approx G\rho U^2 DW \qquad (4.61)$$

which is also applicable to the case of the debris flow. As expected, this dynamical force term is of the form of a stagnation-pressure form of a fluid ($\leftarrow$Sects. 3.3.3 and $\rightarrow$9.6.2). A second contribution derives from the hydrostatic load earth pressure ($\leftarrow$Sect. 4.7.1). The total impact pressure exerted by a debris flow on an obstacle becomes so

$$P_{\text{IMP}} = \frac{F}{WD} \approx \frac{1}{2}\theta\rho U^2 + \frac{1}{2}K\rho g D\cos\beta. \qquad (4.62)$$

where θ and K are constants. Based on the previous analysis, the coefficient $\theta = 2G$ should acquire values between 2 and 4. In practice, it is more often obtained empirically. Values are found to range between 1.2 and nearly 40 (data collection in Hübl et al. 2009). Values for the earth pressure coefficient K are also determined empirically. They typically range between 4 and 10, but higher values have been suggested. In addition, shear stresses should also be considered, as the debris flow is cohesive, viscous, and frictional. Typical maximum values measured for the impact stress of a debris flow against an obstacle range between a fraction of MPa to some MPa.

4.7.4 Quasi-Periodicity

4.7.4.1 Intermittent Debris Flows and Surge Dynamics

Some debris flows do not come in just one single occurrence, but rather exhibit quasi-periodical surges. A classic example is the one of the Jiang-jia ravine in China, where calm periods of 20–30 min are interrupted by bursts of mudflow waves advancing on top of the previously deposited material. The slurry is water-rich and is primarily composed of fines. Another example of surging debris flow is the Acquabona creek in the Italian Dolomites (Fig. 4.18).

In contrast to the Jiang-jia ravine, the material of Acquabona is composed of more assorted material, and gravel is common (Galgaro et al. 2005).

The surging behavior has been explained based on the presence of different classes of material sizes, as shown in the conceptual model of Fig. 4.19. In a debris flow, the base travels slower than the upper layers (←Sect. 4.4). Thus, the material on top is transported faster. When reaching the front, the material on the top is dumped to the front. If the material is homogeneous, the process causes remixing with relatively little change of the dynamics. However, if the debris flow is composed of clasts of different sizes, both vertical and horizontal segregation may take place. Often large blocks have a tendency to reside at the top generating

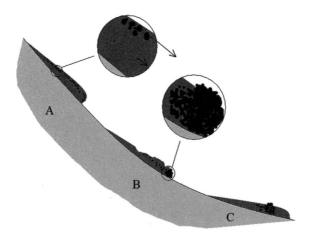

Fig. 4.19 Illustration of the origin of the intermittent behavior of some boulder-rich debris flows. (**a**) the higher speed of the top of the debris flow where large blocks are present creates a boulder-rich front; (**b**) the boulder-rich front, being more frictional than cohesive, determines a greater resistance; the material behind pushes against the natural boulder plug; (**c**) the plug resistance is overcome and the flow continues. The process may occur several times

4.7 Debris Flows: Dynamics

an inversely graded sequence (→Chap. 5). This may occur because of the Brazil-nut effect if blocks are in contact (→Sect. 5.5.2). If the debris flow is volumetrically dominated by the matrix, blocks will not necessarily settle since a viscous-cohesive matrix is capable of carrying large boulders of radius R less than a critical radius

$$R_{\text{critical}} \approx K \frac{\tau_y}{\Delta \rho g} \qquad (4.63)$$

where K is a dimensionless constant of the order 3–4 and $\Delta \rho$ is the difference between the block and the debris flow density. Thus, meter-size blocks can be carried by a highly cohesive debris flow. Large blocks conveyed to the front from the top of the debris flow will form a plug of coarse material. The debris flow material behind, enriched in fines, travels with an acceleration given by Eq. 4.58. The high frictional component in the front may force the frontal plug to stop. The debris flow lagging behind will increase in thickness, until the pushing force against the coarse plug causes the latter to start again. This geometrical division of coarse material leading the flow is frequently observed in some debris flows.

An analysis of surging debris flows can be found in Zanuttigh and Lamberti (2007).

4.7.5 Theoretical and Semiempirical Formulas to Predict the Velocity

Numerous formulas have been adopted to predict the velocity of a debris flow. The simplest approach is to consider the debris flow as a Newtonian fluid (←Sect. 3.7.2) from which

$$U \approx \frac{\rho g \sin \beta}{2\mu} D^2 \qquad (4.64)$$

that assumes a wide channel. The formula has been used for example by Curry (1966).

For non-Newtonian fluids the velocity is given by Eq. 4.29. Because the rheology is usually poorly known and moreover the debris flow may dramatically change its properties in space and time depending on the water amount and changes in the composition, hydraulic approaches for predicting the velocity are also used.

One approach is based on the Chezy and Manning equations for open channel flow. Let us consider a portion of the channel, modeled as rectangular shape (Fig. 4.20). If the river flow is stationary, the gravity force must equate the drag resistance (→Sect. 9.4.3) with the river walls

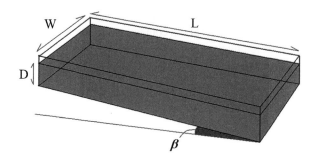

Fig. 4.20 For the derivation of the Chezy equation

$$Mg \sin \beta = \frac{1}{2} \rho_F C_{SF} U^2 [2DL + WL] \quad (4.65)$$

where the mass of debris flow element is evidently $M = LWD\rho_F$ and C_{SF} is the skin friction coefficient ($\rightarrow$Sect. 9.4.3). It is so found

$$U = \sqrt{\frac{2g}{C_S}} \sqrt{\frac{DW}{2D+W}} \sqrt{\sin \beta}. \quad (4.66)$$

The ratio $\frac{DW}{2D+W} = R$ between the cross section of the river and the wetted perimeter is called the hydraulic radius. For shallow flows ($D \gg W$) Chezy's equation becomes

$$U = C\sqrt{R \sin \beta} \quad (4.67)$$

where $C = \sqrt{\frac{2g}{C_S}}$ is Chezy constant, which according to Hürlimann et al. (2003) is of the order of 3.3–9.1 m$^{1/2}$s^{-1} for debris flows. Note that this estimate does not contain information on the rheological properties. However, we know that the rheology is very important in determining the velocity as obvious considering that rivers have a much greater Chezy constant (70–80 m$^{1/2}$s^{-1}). The Froude number of a debris flow following Chezy's equation is

$$Fr = C\sqrt{\frac{\sin \beta}{g}} = \sqrt{\frac{2 \sin \beta}{C_S}} \quad (4.68)$$

and thus in the Chezy model depends directly on the angle of slope.

Other empirical relations are described by Manning's hydraulic relation

$$U = \frac{D^{2/3}(\sin \beta)^{1/2}}{n} \quad (4.69)$$

where n is Manning's constant dependent on the rheology and volume concentration of the slurry (Cui et al. 2005). This relation has been applied to debris flows

4.7 Debris Flows: Dynamics

with $n \approx 0.1$ s m$^{-1/3}$. The use of a single coefficient is very reductive, however, as it is well-known because the velocity depends dramatically on composition and water content. Notice that Chezy's and Manning's hydraulic equations can be generalized to

$$U = kR^a(\sin \beta)^b \tag{4.70}$$

where k, a, and b are constants and R is the hydraulic radius. Manning's equation for turbulent river flow is obtained by setting $a = 2/3$, $b = 1/2$, and $k = 1/n$. Chezy equation corresponds to $a = 1/2$ and $b = 1/2$.

Semiempirical predictive relations for debris flows velocity have been derived either from field observations or based on theoretical considerations.

Just as examples: Pengcheng (1992) describes the turbulent debris flows of the the Jiangja ravine in China in terms of a Chezy-like formula $(gRS)^{0.5}$,

$$U = 25.38 \left(\frac{d}{R}\right)^{0.127} \left(\frac{\mu}{\rho g \sqrt{gR^3}}\right)^{0.0576} (gRS)^{0.5} \tag{4.71}$$

where d is the average block diameter. Notice that the small exponent of the viscosity indicates that the dissipation at high Reynolds numbers occurs more because of turbulence rather than viscous-cohesive forces. When turbulence is less important, the dependence on the viscosity increases. The same author reposts for the slower Majing river debris flows (China)

$$U = 2.77 \left(\frac{d}{R}\right)^{-0.737} \left(\frac{\mu}{\mu_A}\right)^{0.42} (RS)^{0.5} \tag{4.72}$$

where μ_A is water reference viscosity.

Another approach consists in comparing the velocity of the debris flow and that of water in the same watercourse before the debris flow took place (or between successive debris flows), like for example in this formula reported by Pengcheng (1992) for the Hunshiu ravine

$$U = \left(\frac{\rho_A}{\rho}\right)^{0.4} \left(\frac{\mu_A}{\mu}\right)^{0.1} U_A \tag{4.73}$$

where $U = \rho_A, U_A$ are water density and velocity in the absence of a debris flow.

Other authors have tried to relate the area inundated by a debris flow to its volume. For lahars, for example, a fit to existing data gives a law of the form

$$\text{Area inundated} = 200 \, V^{2/3} \tag{4.74}$$

where V is the volume of the debris flow. The above relation shows that lahar deposits scale geometrically, whereas the Bingham model would predict an exponent equal to 4/5. It is interesting that a similar relation has been found for granular avalanches in the Alps, with a coefficient equal to 6.2 instead of 400.

General references Chapter 4: Johnson (1970); Lorenzini and Mazza (2004); Jakob and Hungr (2005).

Chapter 5
A Short Introduction to the Physics of Granular Media

Granular materials are ubiquitous in many natural and industrial processes. The extraction, storage, grinding, and separation of ores are made with rock in the granular state. Granular media are also common in agricultural, food, and pharmaceutical industry. Practical problems frequently occur when dealing with granular media. Arches form with granular flows through pipes, to the point that the flow is often interrupted. Segregation takes place between granular materials made up of grains of different size.

Despite the practical importance, several aspects of granular behavior remain elusive. In the Earth sciences, granular media come about in a variety of situations: from the formation of sedimentological structures to landslides.

Fluid mechanics can be formulated precisely in terms of the momentum, continuity, and rheological equations. The situation is different for granular materials, where a corresponding general equation for the dynamics of granular media does not exist. Different models have been suggested for static and dynamical granular materials, but these models focus on specific physical situations. Devised to grasp only a partial aspect of the physics, such models cannot provide a coherent and thorough picture of granular materials. This chapter exclusively deals with some of the most important concepts of granular physics needed in the study of granular avalanches.

The figure below shows the result of a simple experiment on segregation. A container has been filled with sand (small grains in the figure) and pebbles (large grains). Upon shaking, most pebbles appear at the top.

5.1 Introduction to Granular Materials

5.1.1 Solid Mechanics: Hooke's Law, Poisson Coefficients, Elasticity

When a rock bar is compressed like in Fig. 5.1, it responds by deforming along the direction of the force. If the deformation does not exceed a certain limit, typically a small percent of the total length of the specimen, the deformation is reversible and elastic, i.e., if the force is removed the material goes back to the original length. Experiments show that in the elastic limit the force is proportional to relative deformation $\frac{\Delta L}{L}$ in accordance with Hooke's law

$$F = EA\frac{\Delta L}{L} \quad (5.1)$$

where A is the cross area of the specimen, ΔL is the deformation, L is the total length of the specimen along the direction of deformation, and E is a material parameter called the Young's modulus. Typical values for rock are in the range of hundreds of GPa. Defining as usual the stress σ as the normal force per unit area, Hooke's law can also be written as

$$\sigma = \frac{F}{A} = E\frac{\Delta L}{L} \quad (5.2)$$

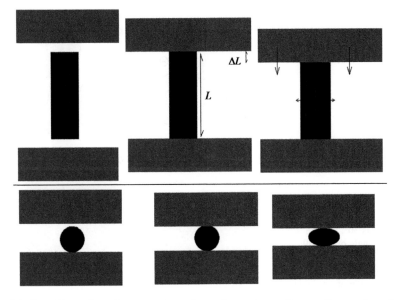

Fig. 5.1 *Top*: illustration of compression of a rock specimen. *Bottom*: the specimen has a spherical shape

The same equation holds when the rock is pulled instead of compressed; in this case the deformation ΔL is negative.

The compression or stretching of the specimen is accompanied by an expansion or contraction along the perpendicular direction. To quantify the amount of expansion ΔH from the original length H, it is necessary to introduce a second material parameter called the Poisson ratio σ_P; the experiment shows that

$$\frac{\Delta H}{H} = -\sigma_P \frac{\Delta L}{L} \qquad (5.3)$$

(sometimes the Poisson parameter is denoted as σ; to avoid confusion with the stress, the notation σ_P is preferred). A deformation greater than the elastic limit is plastic and irreversible.

5.1.2 Angle of Repose

A granular medium gently poured on a table forms a heap making a characteristic *angle of repose* with the horizontal. Although this angle depends on the internal friction angle of the material, the geometry of the grains and their size distribution also play a role. Figure 5.2a and b show that in an ideal ensemble of cylinders of identical radius, the angle necessary to displace one cylinder from its equilibrium position can be predicted by simple geometrical considerations to be 30°. This angle is termed the *pivoting angle*, and is conceptually different from the

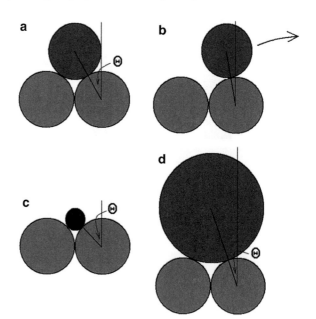

Fig. 5.2 (a, b) The pivoting angle of three identical cylinders is 30°. (c) The pivoting angle of a small cylinder on large ones is higher than 30°. (d) The pivoting angle of a cylinder on smaller cylinders is smaller than 30°

5.1 Introduction to Granular Materials

angle of repose. The angle of repose is a property of the medium, and can be defined based on the granular heap. The pivoting angle, on the other hand, can also be defined for a small number of building blocks. The two angles are, however, related. For a group of four spheres, three at the base and the fourth on the top, the pivoting angle reduces to 19°, 46°, but increases if the base is composed of four spheres. If the cylinders or the spheres differ in their size, the pivoting angle may change a great deal. The Fig. 5.2c shows some values for cylindrical shapes. When the cylinder on top has smaller radius, the pivoting angle increases considerably; when the cylinder on top has greater radius, the pivoting angle decreases (Fig. 5.2d), and for very different ratios of the radii it may reach the theoretical value of zero degrees. The differences in shape play an important role as well, as elliptically shaped grains have higher pivoting angle. One formula to predict pivoting angles for spherical grains is

$$\Theta = K \left(\frac{D_{MOV}}{D_{REST}} \right)^{-\beta} \tag{5.4}$$

where D_{MOV} is the diameter of the movable grain, D_{REST} is the one of the spheres at the base, K and β are positive constants. Thus, the pivoting angle decreases with the power law of the ratio between diameters.

The angle of repose results from a combination of the pivoting angle and of the internal friction angle of the material. Values for crushed rock are typically around 38 for millimeter-sized grains and 42° for centimeter-sized grains, while rounded grains may exhibit angle of repose around only 30°. The extreme case of very flat ellipses (tending to just flat blocks) illustrates the case in which there is no shape effect and the angle of repose is purely frictional. In this case, very high values may be observed.

5.1.3 Force Between Grains

Similar to the bar in (5.11), a spherical grain compressed between two parallel plates deforms elastically and reversibly if the compressive force does not exceed the plastic limit (Fig. 5.1, bottom). The strong resistance to compression typical of rocks can be described in terms of elastic energy accumulated by the grain. It was shown long ago by Hertz that the compressive energy acquired by a compressed elastic sphere is given by

$$E = \frac{1}{2} k \frac{E}{1 - \sigma_P^2} \sqrt{R} \delta^{5/2} \tag{5.5}$$

where δ is the deformation of the sphere, E is Young modulus, σ_P is Poisson coefficient, and $k = (4\sqrt{2}/15)$ is a geometrical constant. The coefficient ½ in (5.5) has been maintained for analogy with the Hooke's law for the deformation of elastic media. The derivative of the energy with respect to the deformation gives the repulsive force

$$F = -\frac{\partial E}{\partial \delta} = -\frac{5}{4}k\frac{E}{1-\sigma_P^2}\sqrt{R}\delta^{3/2} \qquad (5.6)$$

which shows that the force is not of the Hooke type (Hooke's law gives a force which is *linear* in δ). The reason for the 3/2 exponent of δ in Eq. 5.6 is connected to the finite size of grains.

We now consider not just a static force, but an elastic central collision between two grains with opposite velocities V and $-V$. In Hertz's model, the time needed for the collision to take place (partly taken by penetration up to a certain depth and then recoil) is

$$\tau \approx 2.94 \left(\frac{m(1-\sigma_P^2)}{Ek}\right)^{2/5} V^{-1/5} \qquad (5.7)$$

and the maximum deformation is

$$h = \left(\frac{m(1-\sigma_P^2)}{Ek}\right)^{2/5} V^{4/5} \qquad (5.8)$$

For example, the maximum deformation during the collision of two spheres of aluminium of 1 g at a speed of 2 m/s calculated with Eq. 5.8 is about one tenth of mm; the duration of the collision is about 4×10^{-5} s. For energetic collisions, plastic rather than elastic behavior may result, which dissipates energy. For higher energies, the particles may break.

5.2 Static of Granular Materials

5.2.1 Pressures Inside a Container Filled with Granular Material

A performance shown sometimes by magicians consists in filling up a glass cylinder with water. The cylinder should be long with a small cross area, so that the little amount of liquid involved makes the performance more impressive. The cylinder is open both at the top and at the bottom. When the cylinder is lifted in vertical position, the water develops a big pressure at the base, capable of breaking a sturdy container. The destructive power carried by a small amount of water appears amazing but is easily explained as the pressure at the base of a cylinder filled with a liquid of density ρ is hydrostatic ($\leftarrow$Sect. 3.2) $P = \rho g H$ and thus

5.2 Static of Granular Materials

independent of the area of the base of the cylinder. Now imagine performing the same experiment with granular material instead of water. Surprisingly, the pressure measured at the base of the cylinder turns out to be much smaller, and does not raise much by increasing the length of the cylinder (Fig. 5.3).

In contrast to liquids, the stress inside solid bodies and granular media is not isotropic. Let us consider a cylinder slice of infinitesimal height dh (Fig. 5.4). Because the medium is static, the sum of the forces acting on the slice is zero.

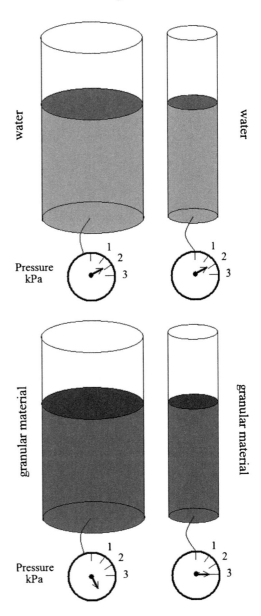

Fig. 5.3 The pressure at the base of these two cylinders filled with the same level of water is the same (*top*). If the cylinders are filled with sand, a greater pressure develops at the base of the cylinder with larger cross section (*bottom*)

Fig. 5.4 A cylinder filled with granular material and the pressure exerted by the granular material

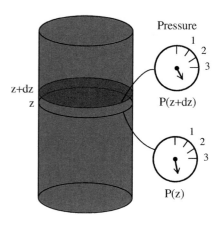

The first contribution to the vertical component of the force is the weight of the slice

$$dW = \rho g S dh \qquad (5.9)$$

where ρ is the density of the granular medium (accounting also for the voids) and S is the surface of the cylinder base. Similar to water, the granular medium exerts a certain lateral pressure perpendicular to the walls of the cylinder. However, because the granular medium is frictional, the lateral pressure implies a certain stress parallel to the wall of the cylinder, and so directed vertically. Let us indicate with P the *vertical* pressure within the granular medium at a depth h (with the terminology adopted in Chap. 2 it is the component τ_{zz} of the stress where the z-axis points vertically). The reference system is oriented downward so that $h = 0$ corresponds to the top of the cylinder. The horizontal stress is given as the product PK, where K is the earth-pressure constant (←Chap. 2). The total vertical force exerted on the lateral surface of the slice then becomes

$$dF_w = PK\mu\Pi dh \qquad (5.10)$$

where Π is the slice perimeter and μ is the friction coefficient of the granular medium.

A certain pressure also acts at the top and at the bottom of the slice due to the weight of the overlying material. This gives an additional force

$$dF_S = [P(\text{top}) - P(\text{bottom})]S = dP\,S. \qquad (5.11)$$

The sum of the elementary forces becomes

$$dW - dF_S - dF_W = \rho g S dh - S dP - PK\mu\Pi dh = 0 \qquad (5.12)$$

5.2 Static of Granular Materials

where because of the orientation chosen, the sign is positive when the force is directed downward. From (5.12) a differential equation is found

$$\frac{dP}{dh} = \rho g - \frac{K\mu\Pi}{S}P \qquad (5.13)$$

which has a solution

$$P(h) = \frac{\rho g S}{K\mu\Pi}\left[1 - \exp\left(-\frac{K\mu\Pi}{S}h\right)\right] + P_0 \exp\left(-\frac{K\mu\Pi}{S}h\right) \qquad (5.14)$$

where the pressure P_0 at height $h = 0$ is introduced as a boundary condition. In particular, for a cylinder of radius R we have $\Pi = 2\pi R$; $S = \pi R^2$ and so

$$P(h) = \frac{\rho g R}{2K\mu}\left[1 - \exp\left(-\frac{2K\mu}{R}h\right)\right] + P_0 \exp\left(-\frac{2K\mu}{R}h\right) \qquad (5.15)$$

This equation shows that:

- The pressure depends on the radius of the cylinder
- It is approximately hydrostatic down to small depths compared to the cylinder radius
- Deeper than a radius length, the pressure tends to a finite asymptotic value given by

$$P_\infty = \frac{\rho g R}{2K\mu}. \qquad (5.16)$$

This is also the maximum pressure attainable in the granular medium. Figure 5.5 shows the behavior of the vertical pressure as a function of depth calculated with Eq. 5.15.

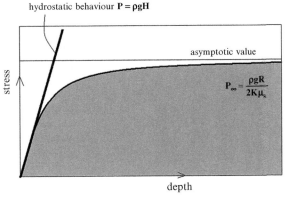

Fig. 5.5 The pressure in the cylinder filled with granular material as a function of the depth deviates from the hydrostatic value, as part of the weight is supported by the friction at the walls of the cylinder

5.2.2 Force Chains

A plastic material in a state of stress changes the polarization of light travelling through it, a phenomenon called photoelasticity. A photograph taken with a polarizer will thus reveal the state of stress in the material, a technique largely used in the industry.

This method has been significant also in the study of granular materials. In the earliest experiments, researchers used an array of long transparent cylinders with the axis parallel to the line of sight. A force was applied vertically and the stresses within the system were studied by examining optically the photoelastic response. Because the dimension along the cylinder axes does not play any role, this experiment essentially represents a two-dimensional granular medium. In Fig. 5.6, the zones with high compressive force materialize with a dense black tone. It appears that forces and stresses are not distributed uniformly like in a liquid and regularly like in a homogeneous solid, but occur in *force chains*. Force chains appear to diverge from the source of the force P much like the sparks depart from a high-voltage electrode through a dielectric medium. These chains also form in the absence of the weight P, as a consequence of the self-weight of the material. What is the source of these large concentrations of stress, and why is the stress nonuniform through the medium? In contrast to an ideal homogeneous solid body, in a compressed granular medium the geometry is inhomogeneous due to small differences in the positions and shape of the grains. This is true also for systems with apparently identical building blocks of regular shape. In the example of

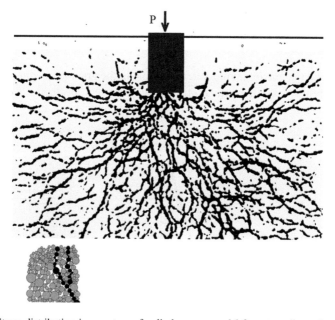

Fig. 5.6 Stress distribution in a system of cylinders as a model for a two-dimensional granular material. The stress is revealed with photoelastic technique. (Picture taken from Dantu (1968).) In the panel on the right, a schematic illustration of force chains in a granular medium. Two such chains are identified with a *black color*

5.2 Static of Granular Materials

Fig. 5.6, there are always small imperceptible differences in the position and radius of the cylinders. This implies that some of the cylinders merely rest on their neighbours, without supporting the cylinders above. These cylinders are the ones outside the force chains; touching them we would notice they are loose. In contrast, cylinders belonging to force chains appear strongly bound to their neighbors. Force chains form also in three-dimensional systems such as spheres, beads, or irregular grains.

Notice that in Fig. 5.6 the chains terminate abruptly at the internal surface of the container, both at the base and on the lateral surface. Thus, the force must be strongly inhomogeneous at the boundary with the container. This can be demonstrated and measured with a very simple technique. It consists in placing some carbon paper along the internal surfaces of the container; a high force is so visualized by black dots on the paper. From the diameter of the dot, previously calibrated, one can infer the value of the force (see e.g. Mueth et al. 1998). These data show that there are numerous contacts between particles and the walls with small transmitted force, but also few contacts with very large forces. The figure shows that a realistic fit of the distribution of forces $P(f)$ associated with force chains is exponential

$$P(f) \propto \exp(-f/f_0) \qquad (5.17)$$

where f, f_0 are, respectively, the force acting on grains, and a reference force related to the external pressure and that can be obtained by experimental data. The exponential distribution results from the statistical nature of chain force formation, and can also be explained theoretically.

More recent results show that for external pressures exceeding some MPas, the distribution function of the stress shifts from an exponential function to a Gaussian (Makse et al. 2000)

$$P_1(f) = \langle f \rangle^{-1} \exp[-f/\langle f \rangle] \to P_2(f) = \langle f \rangle^{-1} \exp\left[-(f/\langle f \rangle - 1)^2\right]. \qquad (5.18)$$

Force chains probably play an important role in landslide fragmentation. Thus, a question relevant for landslide physics is whether force chains also rise temporarily when the granular material is in motion (Fig. 5.7).

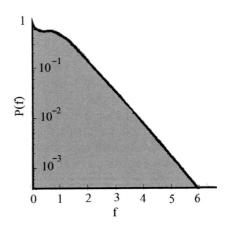

Fig. 5.7 Force distribution $P(f)$ at the internal surface of a confining vessel measured with the carbon paper technique. (From Mueth et al. (1998), modified and simplified.) The line represents an average of the forces recorded against the top piston, the bottom piston, and walls. The forces are in relative units

5.3 Grain Collisions

5.3.1 Grain-Wall Collisions

5.3.1.1 Coefficient of Restitution

Suppose dropping a rubber ball on the floor from to height of 1 m. Because of the energy dissipation at the impact, the ball bounces back a lower height, say 80 cm. We define the *normal coefficient of restitution* $\varepsilon_\perp$ as the ratio of the velocity after the impact U' to the one before, U

$$U' = -\varepsilon_\perp U \tag{5.19}$$

where the minus sign accounts for the fact that the velocity changes sign after impact and $\varepsilon_\perp$ is the normal coefficient of restitution. In this particular example,

$$\text{Coefficient of restitution } \varepsilon_\perp = \left|\frac{\text{Velocity after impact}}{\text{Velocity before impact}}\right| \\ = \sqrt{\frac{0.8 \text{ m}}{1 \text{ m}}} = 0.894 \tag{5.20}$$

where air resistance is neglected. We distinguish between *elastic* and *inelastic* collisions. When hitting the floor, the rubber ball transforms its kinetic energy into deformation energy (←Sect. 5.1.3), a process which will normally implies energy dissipation due to irreversible deformation, heat generation, and sound waves. The collision of rocky materials dissipates energy more efficiently than rubber balls. An *elastic* collision is an ideal process in which the kinetic energy is not dissipated. Although a perfectly elastic collision is not possible, the definition turns out to be useful.

5.3.1.2 Grain-Wall Collisions

Let us consider a sphere colliding against a wall, and neglect for the moment the gravity force and air resistance. The sphere collides for a very short time transferring momentum, energy, and angular momentum to the wall. If the sphere is not spinning, the motion occurs on the plane perpendicular to the table. We can thus consider the problem only in two dimensions. Figure 5.8 shows a possible result of the collision. If the angle of incidence is less than a threshold value to be determined experimentally, the grain rolls without sliding at the contact with the wall. By increasing the angle of incidence beyond this threshold value, a situation is reached where sliding occurs.

Both the total momentum and the angular momentum must be conserved throughout the collision, whereas the energy usually is not. Denoting with primes

5.3 Grain Collisions

Fig. 5.8 Collision of a sphere against a plane. If the collision occurs at a large angle of incidence, there will be a slip between the sphere and the wall, while no slippage occurs when the angle of incidence is small

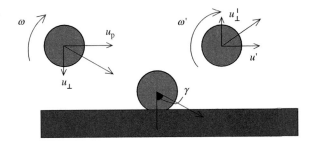

the quantities after impact, a first equation gives the velocity of the sphere perpendicular to the table after collision in term of the velocity before

$$u'_\perp = -\varepsilon_\perp u_\perp \qquad (5.21)$$

where $\varepsilon_\perp$ is the perpendicular coefficient of restitution.

In a similar way, we define the parallel coefficient of restitution $\varepsilon_=$ as the ratio between the velocities parallel to the wall before and after the collision. Here one needs to be careful because the parallel velocities should be taken relative to the point of impact, and not to the centre of mass of the particle. There is a difference between the two velocities, for the reason that the sphere is spinning around its axis with an angular velocity ω. Choosing the angular velocity as positive when the spin is clockwise in the figure, the velocities at the point of collision before $u_{C,=}$ and after the collision are, respectively, $u_{C,=} = u_= - R\omega$ and $u'_{C,=} = u'_= - R\omega'$. The parallel coefficient of restitution is introduced in the following way $u'_{C,=} = \varepsilon_= u_{C,=}$, which gives

$$u'_= - R\omega' = \varepsilon_=(u_= - R\omega). \qquad (5.22)$$

The experiments show that the parallel coefficient of restitution $\varepsilon_=$ depends on the angle of incidence γ, defined as the angle between the velocity vector before the impact and the normal to the surface ($\gamma = \left|\tan^{-1}(u_\perp/u_=)\right|$).

According to the model by Walton, a sphere can either roll or slide depending on the angle of incidence in relation to the physical characteristics of the surface. The parallel coefficient of restitution in the Walton model acquires the form

$$\varepsilon_= = \begin{cases} -1 - \tan\phi\left(1 + \dfrac{mR^2}{I}\right)(1+\varepsilon_\perp)\dfrac{u_\perp}{u_=} & (\gamma \geq \gamma_0;\ \text{slippage}) \\ \beta_0 & (\gamma < \gamma_0;\ \text{no slippage}) \end{cases} \qquad (5.23)$$

where γ_0 is the critical angle for slippage (it should be remembered that $\dfrac{u_\perp}{u_=}$ is negative). The first line of (5.23) refers to the case at large angle of incidence ($\gamma \geq \gamma_0$, "grazing collisions"), where the collision occurs with a certain slippage and no rolling. The second line concerns head-on collisions, in which the sphere rolls without slippage and the coefficient of restitution becomes equal to a constant β_0. The experimental data show a reasonable agreement with the Walton model. More complete models, not considered here, have been introduced.

We now assume lack of slippage. The conservation of angular momentum upon impact can be expressed as

$$I(\omega' - \omega) = -mR(u'_= - u_=) \tag{5.24}$$

where I is the moment of inertia: $I = \frac{2}{5}mR^2$ for a sphere and $I = \frac{1}{2}mR^2$ for a cylinder. Using (5.21) and (5.22), the equations for the postimpact velocity and angular velocity are

$$u'_= = \frac{2}{3}[(1 - \varepsilon_=/2)u_= + R\omega(1 + \varepsilon_=)/2]$$
$$u'_\perp = -\varepsilon_\perp u_\perp \tag{5.25}$$
$$\omega' = \frac{2}{3}[(1 + \varepsilon_=)u_=/R + \omega(1/2 - \varepsilon_=)]$$

In three dimensions, the collision equations must be generalized to account for all the possible geometrical situations. For example, the sphere may spin around an axis which is not parallel to the wall. This case is not considered here.

5.3.2 Grain–Grain Collisions

5.3.2.1 Elastic and Inelastic Collisions Between Spheres

For collisions between spheres a normal coefficient of restitution is defined as

$$\text{Coefficient of restitution } \varepsilon_\perp = \frac{\text{Velocity after impact}}{\text{Velocity before impact}}. \tag{5.26}$$

Let us consider the frontal collision of two nonspinning spheres of masses M_1, M_2 along a line. We call U_1, U_2 the velocities before the collision of the spheres 1 and 2, respectively, and U^*_1 U^*_2 the corresponding velocities after the collision (in the following the starred index will always refer to the postcollision variable). If the velocities before collision are known, the task is to determine them after the collision has taken place. For an elastic collision the solution can be easily found by imposing both the conservation of momentum and of kinetic energy

$$M_1 U_1 + M_2 U_2 = M_1 U^*_1 + M_2 U^*_2$$
$$M_1 U_1^2 + M_2 U_2^2 = M_1 U^{*2}_1 + M_2 U^{*2}_2 \tag{5.27}$$

5.3 Grain Collisions

From which it follows

$$U_1^* = \frac{M_1 - M_2}{M_1 + M_2} U_1 + \frac{2M_2}{M_1 + M_2} U_2$$
$$U_2^* = -\frac{M_1 - M_2}{M_1 + M_2} U_2 + \frac{2M_2}{M_1 + M_2} U_1 \tag{5.28}$$

To treat inelastic collisions, we should use the definition of relative velocity between the spheres before and after the impact

$$U_{12} = U_1 - U_2$$
$$U^*{}_{12} = U^*{}_1 - U^*{}_2 \tag{5.29}$$

and so the coefficient of restitution can be properly defined as

$$U^*{}_{12} = -\varepsilon_\perp U_{12}. \tag{5.30}$$

The condition of momentum conservation together with Eq. 5.30 yields

$$M_1 U_1 + M_2 U_2 = M_1 U_1^* + M_2 U_2^*$$
$$U_1^* - U_2^* = -\varepsilon_\perp (U_1 - U_2). \tag{5.31}$$

The result can be written as

$$U_1^* = U_1 - \frac{\mu}{M_1}(1 + \varepsilon_\perp) U_{12}$$
$$U_2^* = U_2 + \frac{\mu}{M_2}(1 + \varepsilon_\perp) U_{12} \tag{5.32}$$

where $\mu = \frac{M_1 M_2}{M_1 + M_2}$. If the two particles have identical mass then

$$U_1^* = \frac{1 - \varepsilon_\perp}{2} U_1 + \frac{1 + \varepsilon_\perp}{2} U_2$$
$$U_2^* = \frac{1 + \varepsilon_\perp}{2} U_1 + \frac{1 - \varepsilon_\perp}{2} U_2. \tag{5.33}$$

Because the total momentum is conserved, the velocity of the centre of mass

$$U_{CM} = \frac{M_1 U_1 + M_2 U_2}{M_1 + M_2} \tag{5.34}$$

is conserved during the collision.

As an example, let us consider two particles of identical mass moving at 12 and 10 m/s like in Fig. 5.9a. The relative velocity is thus 2 m/s. An observer moving

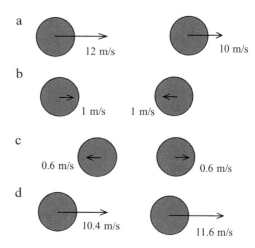

Fig. 5.9 Impact between particles in the laboratory system and in the centre of mass system

with the centre of mass at a velocity of $U_{CM} = \frac{1}{2}U_1 + \frac{1}{2}U_2 = 11$ m/s would thus observe a collision like in Fig. 5.9b, where each particle has a velocity of magnitude 1 m/s and opposite direction. Assuming a coefficient of restitution of 0.6, the magnitude of the velocity of each particle after collision is 0.6, and is directed opposite to the direction of impact like in Fig. 5.9c. Summing again the velocity of 11 m/s to retransform to the laboratory system, the final configuration of the velocity after impact appears like in Fig. 5.9d.

The energy lost in the collision is

$$\Delta E = \frac{1}{2\mu}(1 - \varepsilon_\perp^2)(U_1 - U_2)^2 \tag{5.35}$$

5.3.2.2 Parallel Coefficient of Restitution for Spheres

The parallel coefficient of restitution can be defined similar to the impact of a sphere against a wall. This coefficient is found to depend much on the impact angle and on velocity. Following a common practice, here parallel coefficients of restitution will be regarded as constant.

Box 5.1 One Step Forward: The Normal Coefficient of Restitution from First Principles

In many applications, the normal coefficient of restitution is considered constant and independent of both the velocity and the radius of the particles. In reality, the coefficient varies markedly with the velocity of the impacting particles, and we can appreciate the reason for this variation by applying collision theory. The colliding particles deform until a point is reached where

(continued)

Box 5.1 (continued)

the relative velocity of the centers has decreased to zero, after which they rebound, much like a loaded spring that is released.

If the Hertz theory is adopted to describe microscopically the rebound, the equation of motion of the sphere would result in equating the inertia to the force provided by Eq. 5.5

$$M\ddot{\delta} = -\frac{5}{4}k\frac{E}{1-\sigma_P^2}\sqrt{R}\delta^{3/2} \tag{5.36}$$

However, the rebound velocity calculated with Eq. 5.36 would be the same as the incoming velocity of the sphere. To account for the energy dissipation, two possibilities can be introduced: (1) velocity-dependent forces give the so-called viscoelastic model, and (2) the loading and subsequent unloading of the sphere deformation have different magnitudes.

Concerning model 1, let us consider two spheres colliding centrally. The forces on each particle are written as the sum of an elastic force and a dissipative force

$$F_{tot} = F_{elastic} + F_{dissipative} = \rho\delta^{3/2} + \frac{3}{2}A\rho\delta^{1/2}\dot{\delta} \tag{5.37}$$

where $\rho = \frac{2E\sqrt{R_{12}}}{3(1-v^2)}$ while A is a materials parameter. The equation of motion for the two particles is obtained by writing

$$\mu\ddot{\delta} = F_{tot} \tag{5.38}$$

with $\mu = m_1 m_2 /(m_1 + m_2)$. One can so calculate the velocity after impact, and obtain the coefficient of restitution as the ratio of the final to the initial velocity. If this way, in the limit of low dissipation the coefficient acquires the form $\varepsilon_\perp = 1 - Ku_\perp^{1/5}$ where K is a constant. Brilliantov and Poschel (2000, 2003) report to second order in the $u_\perp^{1/5}$ the following expression $\varepsilon_\perp = 1 - C_1 A\kappa^{2/5} u_\perp^{1/5} + C_2 A^2 \kappa^{4/3} u_\perp^{2/5}$ where $\kappa = (3/2)^{5/2}(\rho/m_{12})$ and $C_1 = 1.15344$; $C_2 = 0.79826$.

The above equations show that the coefficient of restitution approaches one for collisions with very small velocities. However, it decreases for very small velocities (for simplicity only the first three terms in the velocity expansion has been included in the solution). The fact that the coefficient of restitution decreases as a function of the velocity is verified experimentally.

Model 2 is called plastic, because it corresponds to an irreversible deformation of the sphere interior. In this model, the maximum indentation δ^* is given as (Johnson 1987)

(continued)

> **Box 5.1** (continued)
>
> $$\frac{1}{2}mv^2 = \int_0^{\delta^*} F d\delta \qquad (5.39)$$
>
> where the force F is the sum of a Hertzian and Hookean nondissipative term linear in the deformation (Eq. 5.1). The velocity v' after collision is calculated making use of the so-called compliance function P' that gives the repulsive energy of the two spheres
>
> $$\frac{1}{2}mv'^2 = \int_0^{\delta^*} P' d\delta' \qquad (5.40)$$
>
> From a linear model for the compliance function, Johnson has found the coefficient of restitution as $\varepsilon \propto v^{-1/4}$, which fits fairly well the behavior of metals at moderate speeds (<10 m/s).
>
> Measured values of the coefficient of restitution for rocks can range between 0.2 and 0.8. Empirical formulas have been suggested to deal with the decrease of the coefficient of restitution with the velocity for rocks ($\rightarrow$Sect. 8.4).

5.4 Dynamics of Granular Materials: Avalanching

5.4.1 General

Consider a thick layer of granular material resting on a rough table. By progressively tilting the table, grains start to flow when the angle of repose is reached. At sloping angles slightly greater than the angle of repose, the motion affects only the upper layers, while the material underneath remains at rest. We refer to this situation as superficial avalanching (Fig. 5.10a). In this figure, the white particles are in relative motion, whereas the black particles indicate lack of relative movement. The experiments and theory show that the velocity of grains parallel to slope increases almost linearly (see the scheme at the bottom of the figure) whereas the density slightly decreases (e.g., Ancey 2006). The collapse of sand toward the bottom resets the slope angle to a new smaller value.

At sloping angle much greater than the angle of repose, the whole material takes part to the avalanche (Fig. 5.10b). The flow velocity profile and density distribution also appear to be different and nonlinear as a function of the height. Studies also show that the highest frequency of grain–grain collision is reached at intermediate heights and that the fluctuation of the velocity ("granular temperature," $\rightarrow$Sect. 5.4.4) first increases, and then decreases with height.

5.4 Dynamics of Granular Materials: Avalanching

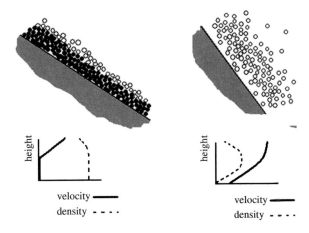

Fig. 5.10 Two flow regimes: superficial avalanching (*left*) characteristic of granular material tilted slightly above the angle of repose on a rough terrain. Avalanching affecting the whole granular medium (*right*) when the slope angle is much greater than the angle of repose

5.4.2 Dynamics of Granular Materials at High Shear Rate: Granular Gases and Granular Temperature

The molecules of a gas experience frequent elastic collisions with other molecules and with the walls of the container. A granular material in a state of agitation (also called a granular gas) is to some extent similar to a gas, but with the important difference that collisions are inelastic. This can be easily noticed shaking a box filled with sand. If shaking ceases, sand promptly settles to the bottom.

The temperature T of an ideal monoatomic gas in equilibrium is proportional to the mean square root of the particle velocity according to the equation

$$\text{Energy per molecule} = \frac{E}{N} = \frac{3}{2}k_B T = \frac{1}{2}m\left[\langle v^2 \rangle - \langle \vec{v} \rangle^2\right] \tag{5.41}$$

where k_B is Boltzmann's equation, $\langle v^2 \rangle = \frac{1}{N}\sum_{j=1}^{N} v_j^2$ is the mean square velocity, $\langle \vec{v} \rangle$ is the average velocity, and N is the total number of particles. Similarly, in the physics of granular materials one defines the granular temperature, without the Boltzmann constant

$$\text{Energy per grain} = \frac{E}{N} = \frac{3}{2}T = \frac{1}{2}m\left[\langle v^2 \rangle - \langle \vec{v} \rangle^2\right]. \tag{5.42}$$

The grain velocity is the vector sum of the average velocity and the fluctuating velocity. The granular temperature is referred to a reference system moving with the average granular velocity, for which evidently $\langle \vec{v} \rangle = 0$, so that the definition of granular temperature simplifies to yield

$$\text{Energy per grain} = \frac{E}{N} = \frac{3}{2}T = \frac{1}{2}m\langle v^2 \rangle. \tag{5.43}$$

Evidently, the granular temperature so defined has nothing to do with the thermodynamical temperature, even though dissipation of (granular) temperature will also cause a small increase in (thermodynamical) temperature. The concept of granular temperature, even if useful, has not the same generality and validity as in thermodynamics (see caveats at the end of the present section). Because of the strong dissipation, the granular temperature in an excited granular material decreases in time if energy is not provided externally. In the simplest view, the temperature decay is provided by Haff's equation.

5.4.3 Haff's Equation

Consider a homogeneous granular system composed of identical spherical grains of mass m and radius R. The coefficient of restitution ε is constant and independent of the velocity. At a certain time t, the system has granular temperature T. Because we assume that the system does not receive energy externally, the granular temperature must decay with time.

Similar to the molecules of a gas, the mean free path is defined as the average distance traveled by one grain between two consecutive collisions with another grain (←Eq. 3.39 is reported here)

$$\lambda = \frac{1}{4\pi n R^2} \tag{5.44}$$

where n is the number of grains per unit volume. The time between two successive collision experienced by a grain is on the average

$$\tau = \lambda/v = \frac{1}{4\pi n v R^2}. \tag{5.45}$$

where v is the average velocity. One such collision causes an energy loss of

$$\Delta E = \frac{1}{2}m\left(v'^2 - v^2\right) = -\frac{1}{2}mv^2\left(1 - \varepsilon^2\right) = -\frac{3}{2}T\left(1 - \varepsilon^2\right) \tag{5.46}$$

where $v' = \varepsilon v$ is the average velocity after a collision.

Calculating explicitly the energy derivative

$$\frac{dE}{dt} = \text{(Number of grains)} \times \frac{\text{Energy loss in an impact}}{\text{Time between impacts}} = N\frac{\Delta E}{\tau} \tag{5.47}$$

it is found that

$$\frac{dT}{dt} = -C(1 - \varepsilon^2)nR^2 T^{3/2} \tag{5.48}$$

5.4 Dynamics of Granular Materials: Avalanching

where $C = 4\pi(3/m)^{1/2}$ is a constant. The integration of (5.48) gives

$$T(t) = \frac{T_0}{(1+t/\tau_0)^2} \qquad (5.49)$$

where

$$\tau_0 = \frac{1}{CnR^2(1-\varepsilon^2)\sqrt{T_0}} \qquad (5.50)$$

gives the characteristic time for granular temperature dissipation. The characteristic time is less than 1 s for initial velocities of some m/s and with reasonable coefficient of restitution and cm-size particle radius.

A more realistic temperature decay has been suggested by Schwager and Poschel (see Brilliantov and Poschel 2003) based on the fact that the coefficient of restitution decreases with impact speed. Their equation reads

$$T(t) = \frac{T_0}{(1+t/\tau'_0)^{5/3}} \qquad (5.51)$$

where τ'_0 is a new time constant. This equation gives a faster cooling than predicted by Haff's equation, for the reason that faster grains lose energy at greater rate.

5.4.3.1 Caveats with the Concept of Granular Temperature

The most important difference between thermodynamical temperature for a gas and a granular medium is that, as seen earlier, the collisions in a granular material are highly dissipative. There are other important discrepancies between the two systems, which make the concept of granular temperature less useful than the temperature used in thermodynamics. The relationship between temperature and pressure of a gas (called the equation of state) is simple. A corresponding relationship cannot be easily worked out from first principles for a granular medium. The temperature can be defined for a system in equilibrium, which is often not the case for granular system. Moreover, temperatures are normally much easier to measure in a real gas than in a granular gas. Notwithstanding these important caveats, the concept of granular temperature may still be useful.

5.4.4 Fluid Dynamical Model of a Granular Flow

Based on observation especially on the Elm landslide, Hsü (1978) stressed that the geometry of rock avalanche deposits resembles the ones from "mudflows, lava flows, and glaciers" (→Chap. 6). It has also been noticed that rock avalanches often tend to spread in sheets like a viscous fluid.

In a fluid-dynamical model for rock avalanches, single grains are envisaged as molecules in a fluid, exactly like the granular gas considered earlier. The Navier–Stokes equations are written for this "fluid", with viscosity calculated from simple kinetic arguments (analogous to the expression derived in ←Sect. 3.4.2).

The Navier–Stokes equation for the granular material along the direction parallel to bedrock is (see, e.g., the formulation by Hwang and Hutter 1995) and (←Chap. 3)

$$\frac{\partial(\rho u)}{\partial t} + \frac{\partial(u u_k)}{\partial x_k} = -\frac{\partial P}{\partial x} + \frac{\partial}{\partial x_j}\left[\mu\left(\frac{\partial u}{\partial x_j} + \frac{\partial u_j}{\partial x}\right)\right] + \rho g \sin\beta \qquad (5.52)$$

where ρ is the density, g is gravity acceleration, β is the slope angle. The Coulomb friction is neglected in this formulation. The viscosity μ follows from kinetic arguments as (Haff 1983, ←Sect. 3.4.2)

$$\mu = qR^2 \rho \frac{\bar{v}}{s} \qquad (5.53)$$

where R is the radius of the particles, $\bar{v}$ is the velocity fluctuation, s is the average distance between the surfaces of two neighboring particles, and q is a dimensionless constant close to unity.

Because granular media are highly dissipative, energy dissipation must also be accounted for. The energy equation, that in fluid mechanics is seldom used, becomes of fundamental importance for a highly dissipative granular medium

$$\frac{\partial}{\partial t}\left[\frac{1}{2}\rho\bar{v}^2\right] \frac{\partial}{\partial t}\left[\frac{1}{2}\rho u^2\right] = \rho g(u\sin\beta + v\cos\beta) - \frac{\partial}{\partial x_k}\left[u_k\left(P + \frac{1}{2}\rho\bar{v}^2 + \frac{1}{2}\rho u^2\right)\right]$$
$$+ \frac{\partial}{\partial x_k}\left[u_j\mu\left(\frac{\partial u_k}{\partial x_j} + \frac{\partial u_j}{\partial x_k}\right)\right]\frac{\partial}{\partial x_k}\left[K\frac{\partial}{\partial x_k}\left(\frac{1}{2}\rho\bar{v}^2\right)\right]$$
$$- 2\gamma\rho\left(1 - \varepsilon^2\right)\frac{\bar{v}^3}{s}$$
$$(5.54)$$

The various terms in the equation have the following meaning. The left hand side gives the time derivative of the total kinetic energy density of the granular medium, which comprises the macroscopic velocity u and the fluctuating velocity $\bar{v}$. The granular temperature is due exclusively to contributions from $\bar{v}$. The change in the kinetic energy density is due to the fall in the gravity field (first term on the right hand side), advection (i.e., transport of energy due to particles flux, second term), viscous dissipation (third term), and thermal conductivity (fourth term; $K = 4r\langle R^2\rangle\frac{\bar{v}}{s}$ is the coefficient of thermal conductance, ε is the coefficient of restitution). These terms are present, with important differences, also in a fluid. The last term of (5.54) is peculiar to a granular gas. It gives energy dissipation in the Haff's model (Eq. 5.48) rewritten in terms of the fluctuating velocity instead of granular temperature.

5.5 Dispersive Stresses and the Brazil Nuts Effect

An equation of state (i.e., a relation between pressure and density) is also introduced for the granular material in the form suggested by Haff (1983) based on kinetic theory (←Eq. 3.34)

$$P = 2tR\rho \frac{\bar{v}^2}{s}. \tag{5.55a}$$

While, for example, Hwang and Hutter (1995) favor a dispersive-like form of the pressure (→Sect. 5.5.1). Fluid-dynamical models of granular media allow simulating a variety of situations like shaken granular media or rapid granular flow, but become problematic if the dissipation is so high that the granular material settles.

5.5 Dispersive Stresses and the Brazil Nuts Effect

5.5.1 Dispersive Pressure

In a series of experiments, Bagnold (1959) has studied the behavior of granular material contained in the annular space between two rotating cylinders. In order to eliminate the effect of gravity and obtain neutrally buoyant grains, he added the right amount of water to a glycerin–alcohol mixture and chose beads as granular medium of the same density of the fluid. Beads had diameter of 1.3 mm. By measuring the normal and shear stresses of the medium in conjunction with the shear rate (that can be determined from the relative rotational speed of the cylinders), Bagnold found that the collision between agitated grains produces a form of pressure which he called *dispersive pressure*.

Let us neglect the viscous stress between particles and the liquid, and consider two bead layers between the rotating cylinders (Fig. 5.11). Assume the upper layer is slightly faster (because beads are neutrally buoyant there is no gravity and the "vertical direction" is just introduced for pictorial purposes; in reality the shear rate gradient is directed horizontally in Bagnold's original experiment). The normal pressure can be written as

$$P = fQN \tag{5.55b}$$

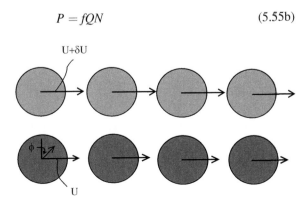

Fig. 5.11 Illustrating the calculation of the dispersive pressure

where f is the collision frequency between beads, N is the number of beads per unit surface, and Q is the average momentum exchanged at each collision. One bead shifting position from a lower to an upper layer acquires momentum, (see also the analysis in ←Sect. 3.2.4). During a collision, each bead carries an excess momentum along the direction perpendicular to flow of the order

$$Q = \delta U m \cos \phi \qquad (5.56)$$

where δU is the velocity difference between the two layers of Fig. 5.11 and ϕ is the angle of the velocity vector with respect to the vertical direction. The frequency of collisions f is

$$f = \frac{2R}{d^2} \bar{v} \qquad (5.57)$$

where R is the bead radius, d is the average distance between beads, and $\bar{v}$ is the velocity fluctuation. The quantities d and N are simply related by geometrical relations to the particle number density (number of particles per unit volume) $N = 1/d^3$. Substituting $\delta U = d(\partial U/\partial z)$ and the mass of each bead $m = (4/3)\pi \rho R^3$, the final result is

$$P_N = 2C\lambda \rho R \left(\frac{\partial U}{\partial y}\right) \bar{v} \cos \phi. \qquad (5.58)$$

Similarly, for the shear stress due to bead collisions it is found that

$$\tau = 2C\lambda \rho R \left(\frac{\partial U}{\partial y}\right) \sin \phi \qquad (5.59)$$

where $\lambda = 2R/(d - 2R)$ is the linear concentration of particles and C is a geometrical factor of the order 0.5. What value should be used for $\bar{v}$? Bagnold assumed that the fluctuation in the velocity $\bar{v}$ is proportional to the velocity difference between the two layers $2R\left(\frac{\partial U}{\partial y}\right)$, and so the normal and shear stresses become

$$P_N = 4C\lambda \rho R^2 \left(\frac{\partial U}{\partial y}\right)^2 \cos \phi$$

$$\tau = 4C\lambda \rho R^2 \left(\frac{\partial U}{\partial y}\right)^2 \sin \phi \qquad (5.60)$$

Note that the ratio between the two becomes $\tau/P_N = \tan \phi$. The normal pressure and shear stress of the form (5.60) are called dispersive. They both increase with the square of the particle diameter and with the square of the velocity gradient.

5.5.1.1 Velocity Profile of a Granular Flow Described by Bagnoldian Dispersive Pressure

If a granular flow can be modeled as a granular gas, the normal dispersive pressure must be in equilibrium with weight. Thus

$$P_N = 4C\lambda\rho R^2 \left(\frac{\partial U}{\partial y}\right)^2 \cos\phi = \bar{C}g(\rho - \rho_L)(D - y)\cos\beta \qquad (5.61)$$

where $\bar{C}$ is the mean grain concentration, related to the average particle distance d, D is the thickness of the granular flow, and y is the height from the base. From Eq. 5.61, the velocity profile is found by integration as

$$U = \frac{1}{3}\frac{1}{\lambda R}\sqrt{\frac{\bar{C}g(\rho - \rho_L)\cos\beta}{\rho C \cos\phi}}\left[D^{3/2} - (D-y)^{3/2}\right]. \qquad (5.62)$$

According to this model, the velocity thus changes fast at the bottom, while at the top the material flows more or less rigidly.

To account for the interstitial liquid, the Bagnold number is often introduced as

$$Ba = \frac{\text{Dispersive stress}}{\text{Viscous stress}} = \frac{4\rho(\partial U/\partial y)R^2 N^{1/6}}{\mu}. \qquad (5.63)$$

where μ is the viscosity of the interstitial fluid. If $Ba > 450$ the flow is dominated by the solid component and the presence of fluid can be neglected. If $Ba < 40$, corresponding to a flow regime called macroviscous, the energy is dissipated by viscous interaction between particles and the fluid.

Another used dimensionless number is the Savage number defined as the ratio between dispersive and hydrostatic pressures in the gravity field

$$Sa = \frac{\text{Dispersive pressure}}{\text{Hydrostatic pressure}} = \frac{4\rho(\partial U/\partial y)^2 R^2}{(\rho - \rho_F)g(D - y)} \qquad (5.64)$$

where ρ_F is the density of the interstitial fluid. If Sa is very large, pressure arises mostly from particle collisions while small Savage numbers indicate predominance of gravity-induced pressure. Bagnold used neutrally buoyant particles in his original experiments, which implies a very high Savage number. The reference value beyond which dispersive pressure due to particle collisions becomes important if $Sa = 0.1$.

5.5.1.2 Is the Concept of Dispersive Pressure Still Considered Valid?

The reasoning leading to Eq. 5.60 is to some extent akin to the demonstration leading to Eq. 3.38 for the viscosity of a gas (←Sect. 3.4.2). However, a fundamental difference between the two analyses relates to the expression for the velocity fluctuation. In a granular flow, the fluctuation velocity $\bar{v}$ and the average flow velocity are comparable in magnitude. This is reflected in Bagnold's view, where the fluctuation velocity $\bar{v}$ is proportional to the velocity difference between two layers of grains, $D\left(\frac{\partial U}{\partial y}\right)$. Conversely, in a gas the velocity fluctuation is due to thermal movement of the particles occurring at a much greater speed than the macroscopic velocity. This is the reason why the shear stress in a gas is proportional to the shear rate (←Sect. 3.4.2), while the Bagnoldian dispersive stresses (5.60) turns out to be proportional to the square of the shear rate. On purely theoretical grounds, the crudeness of the demonstration leading to Eq. 5.60 leaves the possibility that some other exponent of the velocity gradient could be more appropriate, perhaps something intermediate between the viscous ($n = 1$) and the Bagnold ($n = 2$).

This is what was suggested by Hunt et al. (2002) based on a reanalysis of the Bagnoldian experiments. It seems that the original experiments by Bagnold were plagued by too short an aspect ratio of the rheometer (the height to spacing between cylinders was only 4.6). Hunt et al. (2002) show that with this geometry, flow cells form between the top and the middle of the rheometer, a critical aspect that was unpredicted and not investigated by Bagnold. Thus the well-known dependence of the shear stress on the square of the shear rate could be an artifact of the geometry used by Bagnold.

5.5.2 Brazil Nuts and Inverse Grading

Certain clastic sedimentary deposits exhibit direct gradation, i.e., the average particle size decreases upward along a vertical section of the deposit. Typically graded deposits derive from turbidity currents[1] (→Sect. 10.5) consisting of a submarine turbulent flow of water mixed with a small percentage of sediment. Turbidity currents are driven by the small density difference with respect to ambient water. Particles keep flowing until the energy of the current becomes low, and at that point begin to settle. Because the settling velocity is greater for large particles, they are the first to reach the bottom, and a direct gradation will result. However, inverse grading characterized by the large particles at the top is also observed in sedimentary deposits. Typical deposits with inverse grading include sediments deposited by debris flow, in both submarine and subaerial settings (Fig. 5.12).

Let us consider a simple experiment where a glass sphere is placed in the bottom of a jar, filled with glass, and shaken. After some time, the glass sphere appears at the top. The experiment works with various kinds of materials, provided that the

---

[1]Other graded sediments are, for example, the varvae due to the deposition in glacial-lacustrine environment.

5.5 Dispersive Stresses and the Brazil Nuts Effect

Fig. 5.12 Inverse grading appearing in a debris flow deposit. Fjærland, western Norway

sphere is not too dense compared with sand. Another experiment can be performed by shaking a mixture of grains with different sizes. After little shaking, the large grains pop to the top (see the opening figure to this chapter). A similar behavior is observed in granular mixtures travelling along a flume (→Box 6.4). The effect of segregation has been termed "the Brazil nut effect" because of the tendency of Brazil nuts to pop up on a glass of mixed nuts. Because segregation is of practical importance also to the industry, much research has been dedicated to this problem.

5.5.2.1 Explanations of the Brazil Nuts Effect

At least three possible mechanisms have been suggested to explain this phenomenon.

1. The first explanation is based on dispersive pressure. Because dispersive pressure has a strong dependence on particle diameter, it has been speculated that it could affect differently particles of different size. However, this explanation is incorrect as it could also lead to the opposite conclusion that small particles raise to the top (Legros et al. 2000).
2. Another explanation suggested by both experiments and numerical simulation is based on the concept of *kinetic sieving*. When the granular mixture is shaken, voids form under the large grains. The voids are promptly filled by the small particles, causing the large grains to rise.

3. As stated earlier, in one type of experiment one single sphere is shaken in a granular medium. It has been observed that for this kind of experiment, convection cells are formed upon shaking. Sand in the centre of the container moves toward the top, and then descends from the walls. Pushed by sand, the sphere drifts to the top, where it remains trapped being too large in size to be affected by the descending convection flow. The use of particular geometries for the container and smooth walls confirm the importance of the confining walls in the rising of larger spheres. However, reverse grading appears also in granular avalanches where the effect of the confining walls must be insignificant.

General references Chapter 5: Duran (2003).

Chapter 6
Granular Flows and Rock Avalanches

"A huge mountain took apart from the others, and traveled through the entire valley to join the other peaks on the opposite side, so blanketing the villages with earth and stones." So a Polish monk describes the disaster of the Mount Grainer landslide in Savoy, France, 1248. The Alps have been several times the place of deadly large rock failures. Dante Alighieri in his Divine Comedy describes the scar of the Lavini di Marco landslides in Italy, speculating on the possible role of an earthquake or lack of support as a possible cause. Large deposits of much older landslides, historical and pre-historical, remind us that huge rock avalanches are an actual threat for mountain communities.

Rock avalanches result from the transformation of a rock slide (in itself characterized by limited disintegration) into a deeply disintegrated, rapid, and catastrophic flow of rock. Rock avalanches are among the most violent natural catastrophes. They may involve volumes of the order of several cubic kilometers and reach velocities higher than 100 km/h. The displaced rock can attain distances of tens of kilometers on land, and much more in the sea. Giant rock avalanches (also called Sturzstroms) are especially mobile and fast; they travel with much reduced friction. Numerous speculations have been put forward to explain this deviation from the Coulomb-frictional behavior. Another attribute of Sturzstroms observable in the deposits are in-place geological features demonstrating that a large part of a rock avalanche can move rigidly on top of a basal shear layer. Thus, the dynamics of rock avalanches is not akin to a chaotic granular flow as reproducible in the laboratory.

6 Granular Flows and Rock Avalanches

The opening figure to this chapter is an imaginative depiction of the Mount Granier landslide in France (from Hartman Schedel, Livre des croniques, Nurberg, 1493).

6.1 Rock Avalanches: An Introduction

6.1.1 Historical Note

Large rock avalanches in the mountains have plagued mankind throughout the centuries. Probably the first historical account dates back to the Greek historian Polybios (201–120 B.C.), who described the effects of a rock avalanche on Hannibal's army during the transit through the Alps. The Italian poet Dante Alighieri portrays the scar of the Lavini di Marco landslide, which probably occurred in the eighth century. He identified the reason for the mass movement in the lack of buttressing or in an earthquake. In a period where mankind could hardly influence the slope stability, these disasters were inevitable and natural, often associated to the wrath of God or the Evil.

Many ancient chronicles especially from the Alps describe the devastating effects of rock avalanches. In 1248, one of the most infamous catastrophes took place in Mount Granier, Savoy (southern France). A mass of limestone resting on a marl layer collapsed onto the valley underneath, leaving behind a scar still visible today. It partly transformed into a debris flow and killed between 1,000 and 10,000 people. Other tragic Alpine events occurred, for example, in 563, where a landslide caused overspill of the Lake Leman with consequent destruction, or in 1176 where the village of Donnas in Aosta Valley (Italy) was shattered by a rock avalanche. A tragic landslide that occurred in northern Italy eventually influenced the history of European architecture. Piuro (Plurs) was in 1618 a town of workers specialized in stylish building industry. When the tragedy took place, most workers were away for their summer work. The landslide had taken the lives of their families and destroyed their homes (Fig. 6.1). Many of them settled to what is now the Czech Republic, where they contributed to make of Prague one of the jewels of Baroque style (Kozak and Rybar 2003).

More complete accounts start from the eighteenth century, but it is only in the middle of the nineteenth century that a more scientific interest in landslides developed. Two circumstances prompted a renewed interest. Firstly, a number of disastrous landslides occurred in sequence, causing numerous fatalities, the last one being the famous Elm landslide of 1881. The second is the seminal contribution by the Swiss geologist Albert Heim, dated 1882. Heim's work stems from an initial interest for the Elm landslide, and then extended to other rock avalanches. The concepts developed in his papers were extensive and prepared the basis for further quantitative investigations on rock avalanches. His approach was not limited to the description of the landslide, but invested problems of the dynamics of the event, and of the causes that provoked it. Still today, the nomenclature of rock avalanches employs also German terms first introduced by Heim. In particular, Heim has introduced the German term *sturzstrom*, still used today as an equivalent term for large rock avalanche.

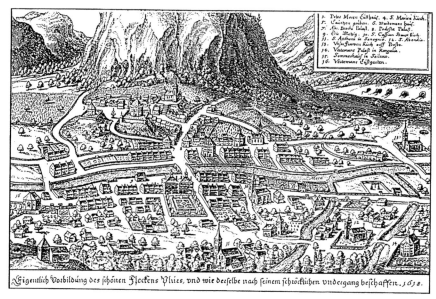

Fig. 6.1 Two famous engravings depict Piuro (Plurs) before and after the landslide of 1618. In contrast to the purely symbolic figure of the Mount Granier at the beginning of the chapter, this engraving is rich in details

6.1.2 Examples of Rock Avalanches: A Quick Glance

Here we examine very briefly some examples of rock avalanches and a selection of dynamical problems.

The *Blackhawk landslide* (volume $3 - 4 \times 10^8$ m^3) took place in the Californian desert in prehistoric time following the failure of a mountain peak of average altitude 2,000 m (the desert itself is at a level of about 1,000). In the first 2 km of run, the material flowed canalized along the valley flanks of the mountain, opening in a fan on the desert level surface (Fig. 6.2). The limestone in the deposit appears to be heavily fragmented. Although the bulk of the material stopped within 3 km from the

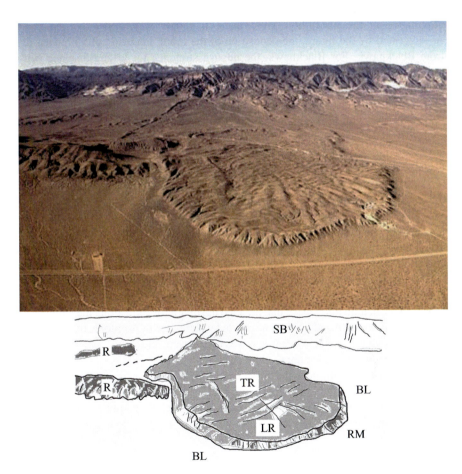

Fig. 6.2 *Upper*: the Blackhawk landslide in the Mojave desert. (Photograph by Kerry Sieh, image USGS of public domain reproduced from the USGS web site.) *Lower*: some basic features. *SB* San Bernardino mountains, *BL* main body of the Blackhawk landslide, *R* rock (not part of the landslide), *TR* transverse ridges, *LR* longitudinal ridges, *RM* raised margin

source, a relatively thick layer (5–10 m) continued to flow for 6 km along the desert, at an average slope angle of only 3°. To explain this extreme mobility, Shreve (1968) suggested the presence of a lubricating air layer underneath the landslide. However, Erismann and Abele (2001) propose water lubrication for the Blackhawk landslide, a possibility excluded by previous researchers based on the arid climate of the Californian desert. However, Erismann and Abele (2001) contend that when precipitation takes place in a desert, it often occurs in the form of intense rainstorm.

The *Frank landslide* (Fig. 6.3) occurred on April 29, 1903 in Alberta, Canada. A volume $3 - 4 \times 10^7$ m^3 of limestone collapsed from the peak called the Turtle Mountain against the entrance of a coal mine, and destroyed the southern part of the town of Frank before coming to rest on the valley plan. The trigger was most likely water infiltration, whereas the mine plant and tunnels are regarded as of minor importance in the instability. More than 76 people were killed by the slide. As suggested by the figure, the deposition of the Frank landslide differs markedly from Blackhawk. For Blackhawk, the deposit stretches for a long distance compared to the fall height. In contrast, for the Frank landslide the fall height and runout length are comparable. The rock avalanche ran up slightly on top of the opposite hills, which contributed to halt the front of the landslide.

The Val Pola landslide (Fig. 6.4) that occurred close to Bormio in northern Italy caused 29 deaths. The precipitation had been particularly intense during the days preceding the failure; combined with the annual effect of runoff from melting glaciers, this had resulted in an extremely unstable situation of the valley flanks in the Valtellina valley. Numerous debris flows had already hit the region and several thousands people had been evacuated from dangerous areas. In the worst scenario, geologists expected a gigantic slide at the flank of M. Zandila. This possibility was indicated by the presence of cracks and rockfall activity. The major mass movement occurred on July 28th, when a mass of 32–40 Mm^3 of gneiss, diorite, and gabbros collapsed. The mass movement hit seven workers engaged with the removal of a previous mass wasting deposit forming a river dam. The mass moving at high speed (74.3 m/s as estimated by Erismann and Abele 2001) displaced the water from a lake, causing a 95-m high flood wave that traveled upstream for 2 km along the valley. The other victims were caused by this unexpected water wave in Aquilone, the only town that had not been evacuated because it was considered sufficiently safe owing to its distance from the landslide area.

Falling to lower altitude, the mass spread mostly along the valley axis; thus the val Pola is classifiable as "low mobility" in the terminology of Nicoletti and Sorriso-Valvo (1991). There was a limited but well-discernible run-up at the opposite side of the valley, of about 290 m.

6.1.3 The Volumes of Rock Avalanches

Rock avalanches may differ dramatically regarding the volume (Fig. 6.4). While local failures along mountain roads can be harmful with a modest volume of some

6.1 Rock Avalanches: An Introduction

Fig. 6.3 *Top*: the Frank landslide. (Reproduced with permission of the Copyright act of Canada.) *Bottom*: The Val Pola landslide in Italy

cubic meters, most rock failures of some geological interest will involve at least one million cubic meters of material. The largest landslides on Earth, however, are in the range of some tens of cubic kilometers: the Baga Bogd (Mongolia, 5 Gm^3), the Green Lake landslide (New Zealand, 2.7 Gm^3), Saidmarreh (Iran, 2 Gm^3, Fig. 6.4), and Langtang in Nepal (3–15 Gm^3). Greater volumes are reached by subaqueous (→Chap. 9) and by extraterrestrial rock avalanches (→Chap. 7). Table 6.1 groups the rock avalanches in classes of increasing volumes. The volumes of rock avalanches depend on the material available for sliding, on the environmental conditions, or the magnitude of triggering earthquakes.

Fig. 6.4 Some examples of rock avalanches with increasing volumes. (**a**) A small rock avalanche in the Italian Alps (about 10,000 cubic meters, "extremely small" in the terminology of Table 6.1). Photograph courtesy of G.B. Crosta

Fig. 6.4 (continued) (**b**) The deposits of an old rock avalanche in Kirghizstan, of approximate volume in the range of 1 million cubic meters ("average" according to Table 6.1). Photograph courtesy of G.B. Crosta. (**c**) The hummocky deposits of the Saidmarreh landslide, Iran, perhaps the largest rock avalanche in the world. The scar of the landslide is visible on the crests of the mountain in the background (Unpublished photograph courtesy of Hassan Shaharivar)

Table 6.1 Volume classes of rock avalanches with some examples. The volumes are indicated in the three most used units of Mm^3, m^3, and km^3

Denomination based on volume	Volume (Mm^3)	Volume (m^3)	Volume (km^3 or Gm^3)	Typical values for the runout (km)	Example
Local		$V<100$		<0.01	Local failures along mountain roads; most rock falls
Extremely small	$0.0001<V<0.01$	$100<V<10^4$	$V<10^{-5}$	0.01–0.1	Example in Fig. 6.4a
Very small	$0.01<V<0.1$	$10^4<V<10^5$	$10^{-5}<V<10^{-4}$	0.1–0.3	Large rock falls like Yosemite
Small	$0.1<V<1$	$10^5<V<10^6$	$10^{-4}<V<10^{-3}$	0.2–0.5	
Average	$1<V<10$	$10^6<V<10^7$	$10^{-3}<V<10^{-2}$	0.5–5	Sherman (Alaska) Valpola
Large	$10<V<100$	$10^7<V<10^8$	$10^{-2}<V<10^{-1}$	1–7	Frank (Canada)
Very large	$100<V<1,000$	$10^8<V<10^9$	$10^{-1}<V<1$	1–10	Vaiont, Italy
Giant	$1,000<V<10,000$	$10^9<V<10^{10}$	$1<V<10$	2–10	Köfels (Austria)
Regional	$10^4<V<10^5$	$10^{10}<V<10^{11}$	$10<V<10^2$	5–30	Saidmarreh (Iran, Fig. 2.22)
Submarine or extraterrestrial	$V>10^5$	$V>10^{11}$	$V>10^2$	>30	

While monster landslides like Saidmarreh are luckily rare (all the known rock avalanche belonging to size class "regional" of Table 6.1 are prehistoric), everyone has observed smaller rock avalanches in the mountains, along riverbeds, or road cuts. Thus, like other geophysical phenomena, the frequency of rock avalanches increases with decreasing magnitude. Based on the analysis of databases or aerial photographs, there is good evidence that the volumes of rock avalanches are power-law distributed in the form

$$P(V) \propto V^{-\Gamma_V} \qquad (6.1)$$

where $P(V)$ is the relative number of rock avalanches of volume greater than V. The exponent gamma Γ_V has been determined for a number of data sets. For example, Evans (2006) finds the following exponents for a selected number of deposits: 0.770 (for the world, events of the twentieth century), 0.881 (for the European Alps, recent and historic) and 0.810 (in the Southern Alps of

New Zealand for the last 10 ka). Because areas are easier to determine than volumes on aerial and digital maps, the distribution of landslide areas are more often reported instead of the volume, which allows for the inclusion of more data and consequently better statistics

$$P(A) \propto A^{-\Gamma_A}. \tag{6.2}$$

A drawback of some of the reported analysis is that frequently landslides of different nature are included in the same counting. Exponents based on the areas usually vary between 0.7 and 1, but higher exponents (up to 2) are also reported. The interested reader may consult, e.g., Pelletier et al. (1997), Brunetti et al. (2009), Hergarten (2002), and the numerous references therein.

The epitome of power laws in geophysical sciences is the Gutenberg–Richter law of earthquake statistics that gives the number of earthquakes per unit time having magnitude greater than m as

$$\log_{10}(P(m)) = a - bm \tag{6.3}$$

where a and b are constants. Power laws like Eqs. 6.1, 6.2, and 6.3 are characterized by scale invariance, i.e., the ratio between the number of events of area greater than A to the number of events of area greater than $2A$ is independent of A, and is equal to $2^{-\Gamma_A}$ (the factor 2 has been chosen arbitrarily, but the reasoning is valid with any ratio). This signifies the absence of a characteristic scale of the phenomena, a statistics also called fractal. The power law of landslides and earthquakes is not the sole system exhibiting fractal statistics. The statistics of many processes also outside the geophysical sciences (even in biological system and human activities) are described by power laws. Different models have been put forward to understand the ubiquitous nature of power laws in natural systems. One of the problems is that some of the mathematical models reproducing power laws require a fine tuning of the model parameters, which is at variance with the apparent robustness of power laws. The most common model, called self-organized criticality (SOC), draws a comparison between the power laws observed in nature and the avalanches observed in a sand pile, which also exhibits fractal statistics. Because the avalanche statistics naturally tends to a critical state, the problem of fine tuning does not appear in SOC models (see Hergarten 2002).

6.2 Rock Avalanche Scars and Deposits

This part briefly considers some structures observed in the landslide scars and deposits. It is intended to convey basic facts rather than their explanation, and it is mostly descriptive. The interpretation of these structures in terms of the processes and dynamics of the landslide is spread in other sections of the chapter.

6.2.1 Rock Avalanche Deposits: Large-Scale Features

The route of a landslide is characterized by: (1) a scar, residue of the collapsed rock where the elevation has diminished after failure, (2) a path along which the landslide has traveled and that can partly be erosive and partly depositional, and (3) a distal zone of deposition.

The depth of the scar may depend on many variables: the size of the slide, the nature of rock, the depth of weak zones, and the possible layering of rocks. In some cases the rock pack is relatively thin but extended to a large area, like for the Lavini di Marco rockslide (Fig. 6.5top). In other cases the scar is thick but horizontally short, like for the Val Pola slide (Fig. 6.5bottom). The scarp, namely, the perimeter line of the scar (←Chap. 1) may be clearly discernible, while in other cases it is hardly noticeable.

Failure in the form of "Toreva blocks" has been described, for example, for the Toreva landslide (Reiche 1937) or for the Socompa debris avalanche. It consists in a large rock failure in the form of a slump, a geometry that is more common in homogeneous soil rather than rock.

Figure 6.6 shows schematically the planar view for a selection of rock avalanches deposits. Whereas some deposits appear tongue-shaped, others are more straight or follow closely the valley profile.

Nicoletti and Sorriso-Valvo (1991) have emphasized the role of the rock avalanche planar shape on the mobility, distinguishing three cases (Fig. 6.6): slides canalized along a valley (termed "high-mobility"), those falling without constraints ("intermediate mobility"), and finally the ones opening as an anvil due to impact against the opposite side of the valley ("low mobility"). As the names indicate, the mobility decreases from the first to the third kind. Also the longitudinal profiles of rock avalanches paths vary considerably. Some landslides have traveled along a long chute, while others are more akin to a collapse onto a flat surface.

Not all the deposits of ancient landslides are easily discernible. The prehistoric Langtang landslides in Nepal, for example, occupy a vast mountainous region of 14 km^2. Despite its huge volume (estimated to be of the order of some cubic kilometers for an average thickness of some 300 m) the deposits are not easy to study owing to postdepositional intense tectonic and glacial activity.

6.2.2 Rock Avalanche Deposits: Intermediate-Scale Features

More than other kinds of landslides, rock avalanche scars and deposits preserve their superficial features for a relatively long time. This is partly due to the depth of the scar and exposure of barren rock, which are harder to colonize by vegetation. *Transverse ridges* approximately parallel to the perimeter of the deposit are sometimes observed (see the figure for the case of the Elm landslide, →Box 6.1). Ridges may be several hundreds of meters long, some meters high, and some tens of

6.2 Rock Avalanche Scars and Deposits

Fig. 6.5 Two examples of superficial scars. *Top*: Even though the details of the preslide topography are unknown, based on the volume of the deposits it can be inferred that the Lavini di Marco landslide has probably affected a relatively thin layer of limestone (of the order 40 m) but extended to a considerable area of 3.6 km^2 (Genevois et al. 2002). The slide is the same described by Dante Alighieri in the Divine Comedy. The barren surface of early Jurassic limestone exhibits several tracks of dinosaur footprints (←Box 1.1). *Bottom*: the scar of the Val Pola landslide is deep but aerially limited

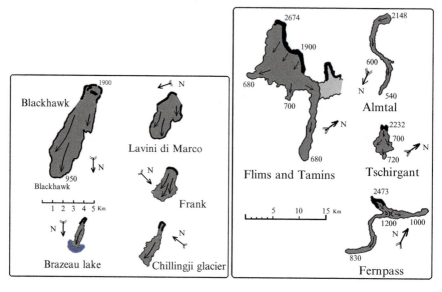

Fig. 6.6 Planar view of some rock avalanche scar, flow path, and deposits. *Left*: examples of noncanalized rock avalanches. *Right*: canalized avalanches

meters apart. Because ridges are perpendicular to the main flow direction, they are indicators of the velocity pattern, and show that the velocities at the sides are lower than in the centre of the rock avalanche. Similar features are formed when dirty water from a river enters into the open sea, and might also indicate that the material of the avalanche was flowing similar to a liquid.

Figure 6.7 shows the result of a simple experiment with fine sand. The sand is let to flow along a flume before coming to rest on a flat surface (→Box 6.4). Note the formation of longitudinal ridges similar to the ones of Elm. With much coarser sand ridges do not form, probably because the distance between ridges must be much greater than grain diameter for ridges to be discernable. Ridges may be interpreted as compression crests. The front of the rock avalanche stops first, thus causing the material behind to run against the front. The granular material is sheared along sliding planes cutting through the whole body of the landslide, that on the surface appear as linear ridges.

Longitudinal flow lines have been reported as well, for example, in the case of the Sherman landslide, but they are rarer (Dufresne and Davies 2009). They are probably linked to severe basal reduction of friction, like that occurring on icy surface.

In some cases (for example, Sherman →Chap. 7 or Blackhawk ←Sect. 6.1.2), the edge of the landslide forms a *raised margin* rising from the basal terrain. Lateral *levees*, which are very common in debris flows, form sporadically at the sides of rock avalanches.

6.2 Rock Avalanche Scars and Deposits

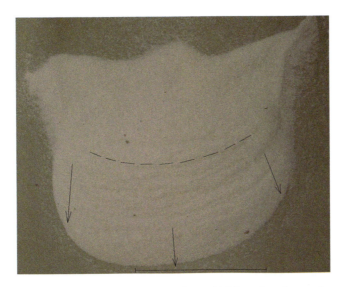

Fig. 6.7 Transverse ridges formed in the deposit of an artificial small-scale granular avalanche of fine sand. The sand travels along a short flume before reaching a flat area where it comes to stop (see Box 6.4). The arrows indicate the flow direction. The black segment is 10 cm long. Compare the compression ridges (parallel to the *dashed line*) with those of the Elm landslide (Fig. 6.9)

Artificial or natural cuts occasionally show in some detail the inner section of the landslide deposit. Figure 6.8 (top) shows the thick landslide deposits of the Saidmarreh landslide, in Iran. It is often observed that very small particles often coexist with huge boulders in the deposit of a very large rock avalanche. In may also be observed that geological features ranging from meters to some tens of meters have been perfectly preserved, despite the rock being intensely fragmented. Figure 6.8 (bottom) shows a diabase dike from a deposit in Köfels, an Austrian landslide ($\rightarrow$Box 6.8). In spite of a travel of about 2 km, the dike has been preserved in its original outline. Similar observations have been reported, for example, for the Flims landslide in Switzerland. This shows that despite the powerful fragmentation, the materials of the landslide have not been subject to significant relative displacement. Other features may form on the surface of a rock avalanche deposit after the main slide event: gullies, springs, distorted drainage and runoff patterns, tension cracks. Further collapse and creep within the rock avalanche deposit may occur as a secondary effect.

Material from the base may also be ploughed by the front of the moving landslide, like in the case of the Ananievo rock avalanche triggered by a 8.2 magnitude earthquake in Kyrgyzstan in 1911. An estimated volume of $\approx 10^7$ m^3, up to 100 m thick slid on a loamy substrate part of which was found at the front of the landslide deposit. It has been conjectured that the bulldozing effect delayed the landslide and reduced its mobility (Abrakhmatov and Strom 2006).

Fig. 6.8 *Top*: The river exposed thick landslide deposits of the Saidmarreh rock avalanche in Iran. (Original photograph courtesy of Hassan Sharharivar.) *Bottom*: A diabase dike on the distal part of the Kofels landslide appearing as the dark shadow against the whitish material. It demonstrates that despite the enormous energies associated with the landslide process and the long runouts, geological structures remained in place (Photograph summer 2004)

6.2 Rock Avalanche Scars and Deposits

Box 6.1 Brief Case Study: Elm and Rock-Falls Transforming into a Fast-Moving Avalanche

Salient data
Name: Elm
Location: Switzerland
Coordinates: 46 55' N; 9 10' E
Volume: (Mm^3): 10
Casualties: 115
Runout: 1.5 km
Year: 1881
Kind: rock avalanche
Material involved: slate
Cause: overcutting due to mining
Fahrböschung: 0.29 (16°).
Special characteristics: *it was historically the first rock avalanche to be studied in detail and for which the average velocity was measured; it exhibits sharp transverse ridges and fluid-like features.*

Poor mining technique in a slate quarry is at the origin of this famous rock avalanche that also marks the beginning of physically oriented studies of landslides (Hsü 1978). The 50-m deep scar in the quarry, excavated without the necessary precautions, was left without basal buttressing. A huge overhang above the scar began to creep. Small rockfalls signaled the instability. Miners kept away from the immediate reach of the boulders, without envisaging that the resulting failure would have traveled to unforeseen distances.

On 11 September 1881, the mass sank rapidly. Sliding in the initial phase, it then hit the floor of the quarry, disintegrated at once, sprouting debris horizontally toward north and north-west. The mass toward north ran up to the opposite side of the valley, coming to rest at a height of about 100 m. The rest of the mass pointed toward N-W and flowed along the Sernf valley, which is nearly horizontal, running for another 1.5 km (Fig. 6.9).

Witnesses were numerous. In correspondence of the quarry floor, they observed rock being cast in the air like water from a high waterfall sprouting against bedrock. Some people were partially buried but without major injury. Houses were pushed for some tens of meter by the front of the avalanche. People stated that the behavior of rock in the avalanche was akin to the flow of a fluid rather than to the sliding of a compact body. From the timing of distance run by some of the witnesses, it was possible to estimate the duration of the event: 45 s that implies an average velocity of about 50 m/s.

Today the deposit exhibits transverse ridges reminiscent of a lava flow, similar to the experimental features shown in Fig. 6.7. The stratigraphic sequence appears to be preserved.

(continued)

Box 6.1 (continued)

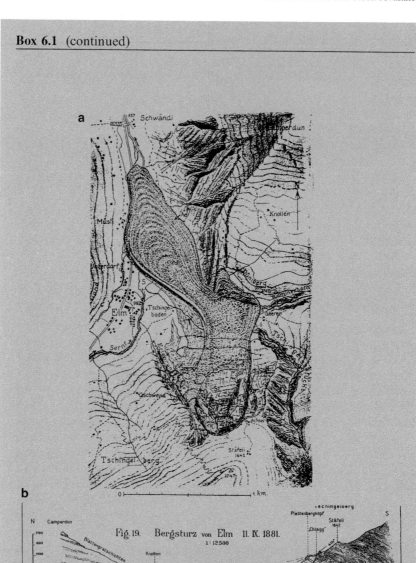

Fig. 6.9 The Elm landslide as from the original work by Heim (1932). (**a**) Planar view. The mass has traveled from the lower part of the figure to the top. (**b**) Cross section (From Heim 1932, slightly modified). The final deposit along the Sernf valley is the dotted area

6.2.3 Rock Avalanche Deposits: Some Small-Scale Features

The most evident feature of a rock avalanche deposit is the pronounced fragmentation of the rocky material. Figure 6.10 shows a handful of rock from the deposits of the Flims landslide in Switzerland. Clasts may range from submillimeter

Fig. 6.10 Fragmentation may reduce rock avalanche deposits to very small clasts. (Images from Flims landslide deposits, Switzerland.) *Top*: subcentimeter fragmented limestone. *Bottom:* Thoroughly fragmented deposits may acquire a conical shape due to erosion

sizes up to large boulders, with a tendency to decrease in size as a function of the depth (Crosta et al. 2007).

It is sometimes observed that blocks larger than some centimeters in diameter are broken in several pieces, but the different portions remain adjacent. This so-called *jigsaw-puzzle effect* indicates that although the shear and normal stresses have been sufficient to cause fragmentation, the collisions with other blocks have not been capable of separating the portions of broken rock. This is indicative of the low granular temperature present at the surface (←Sect. 5.4.5) and in the body of the rock avalanches, and of the limited space available.

Sometimes the deposits exhibit a regular change of their mean grain size along the vertical section. Inverse grading occurs when larger grains prevail in the upper part of the deposit, direct grading when larger grains appear at the bottom. Both tendencies may be observed in the deposits of rock avalanches and in debris flows. Inverse grading resulting from shaking of granular media is also called the Brazil nuts effect (←Sect. 5.5.2). In granular avalanches, the inverse grading is probably due to the increased fragmentation of the rocky mass in the lowermost layers of the landslide mass, where the normal and shear stresses are higher, and the velocity gradient is greater (→Sect. 6.10.3).

Other relevant observations on the deposit are the presence of levels where the shear rates have been very high. These include fused rock and a frictional gouge of minute particles (→Sect. 6.9). Cones of detritus with grain size smaller than the surrounding deposit have been reported, for example, for the Sherman (→Box 7.3) and the Blackhawk landslides (Shreve 1966). Small-scale erosion features can be present as well, such as grooves parallel to slide motion called slickensides.

6.3 Dynamical Properties of Rock Avalanches and Stages of Their Development

6.3.1 Velocity of a Rock Avalanche

The velocities of large rock avalanches are known only in a limited number of cases. The problem with measuring rock avalanche velocity is obviously the unpredictability. From time to time it has been possible to infer the average velocity from witness reports. During the Elm slide, a boy running for his life covered a known distance for the duration of the slide. This information gave an average velocity of 50 m/s, or 180 km/h. In the highly destructive case of Huascaran, velocities of the order 300 km/h have been speculated. Some rock avalanches are surprisingly fast, with top velocities of the order of 100–350 km/h, even though moderate speeds are probably more common. Single rock blocks may be launched

6.3 Dynamical Properties of Rock Avalanches and Stages of Their Development

at enormous velocities, which at times reach the fantastic value of 800 km/h like in the case of the Nevados Huascaran.

In other cases, velocities are partly reconstructed by side effect (e.g., water displacement), seismic recordings, superelevation, estimates based on the knowledge of the apparent friction coefficient (→Sect. 6.4.1), or numerical simulations. Based on superelevation, the average velocity of the Pandemonium creek rock avalanche has been estimated of the order of 30 m/s, with top velocities of perhaps 100 m/s. When available, filming may provide more precise information. For the recent Thurwieser rock avalanche, Sosio et al. (2008) have deduced a punctual velocity of 60–65 m/s and a mean velocity of 38 m/s.

Rock avalanches have also been triggered artificially. Recently, some rock avalanches of volume between 0.1 and 1,000 Mm^3 have been studied in detail in the site on Novaya Zemlia, an island in the Barents sea used by the former Soviet Union as a testing site for nuclear explosions. The landslides triggered by underground nuclear explosions provide a very accurate measurement of the velocity except for the initial phase where the site is hidden from sight by the haze of pulverized rock (Adushkin 2006). For example, the "B-1" rock avalanche (volume 80 Mm^3, Fig. 6.11a) reached a peak velocity of about 60 m/s after a 400 m fall.

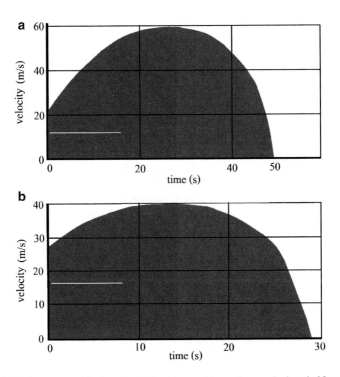

Fig. 6.11 Velocity measured for two landslides triggered by nuclear explosions in Novaya Zemlia test site. (**a**) The B-1 avalanche. (**b**) The Avalanche-1. The white line corresponds to the time interval when the landslide body was not visible because of the haze created by pulverized rock (From Adushkin 2006, redrawn)

Interrupting the Zhuravlevka river, it created the lake Nalivnoe. The total flowing time was about 50–55 s. The runout ratio (→Sect. 6.6.2), i.e., the ratio between vertical to horizontal displacement of the rock avalanche, was particularly low (0.21) possibly due to ice and water lubrication. Despite the similar fall height (350 m), a second slide of volume 1 Mm^3 attained lower velocities (less than 40 m/s) and a runout ratio of 0.37.

6.3.2 Stages in the Development of a Rock Avalanche

Rock avalanches usually result from the gravitational failure of a rock wedge (←Chap. 2). In general, the failure occurs in correspondence of joints and similar localized weak surfaces within the rock piles. Water circulating in the joints may facilitate failure increasing the pore pressure in the joints, and also by the action of freeze–thaw cycles. The erosion by glaciers or rivers and careless quarrying may also contribute to the instability by undercutting the base of a potential slide. Many rock avalanches have been produced by earthquakes. We can ideally recognize five stages of development of a rock avalanche:

0. A possible stage where the rocky surfaces move very slowly prior to catastrophic failure. If properly detected and interpreted, slow creep may indicate possible future catastrophic instability.
1. In the beginning, probably the rock slab slides and may partially travel in ballistic flight if the terrain is sufficiently steep.
2. The mass rapidly disintegrates. If the free fall releases a sufficient amount of energy, a shock wave may be created. Disintegration mechanisms probably comprise the bending of the rock slab beyond the tensile stress, the impact with the mountain surface or with topographic heights, particle–particle and particle–bedrock impacts, and crushing along force chains in the granular medium. Pulverized rock may remain suspended for hours.
3. The fragmented rock flows rapidly. Depending on its mass, it might slide on a thin shearing layer.
4. The flowing mass reaches a lower slope angle. If the rock avalanche has a small volume, it typically comes to a halt when the slope angle becomes as low as 28–35°. However, larger landslides (volume larger than 1–10 million cubic meters) may continue flowing to much lower slope angles, ending up with apparent friction coefficients of the order 0.2 or less.
5. Finally, the landslide comes to stop. Peculiar structures in the deposit may be revealing of the internal dynamics.

Box 6.2 Transition Between Slow Movement and Catastrophic Collapse of a Rock Avalanche

The transition between slow movement and catastrophic collapse of a rock avalanche has been the subject of much work. Failure to recognize the importance of the prefailure creep as the signature of a massive, fast failure resulted in a high number of casualties during the collapse of the Toc Mountain onto the Vaiont artificial water basin in Italy (→Box 7.1). The collapse of the Vaiont landslide has provided a very detailed picture of the creep phase. However, there are many difficulties in predicting the catastrophic evolution of a slow landslide and the time to failure. Not only was the final detachment of the Vaiont landslide not foreseen, the acceleration phases at the end of 1960 and 1962 did not produce any slope failure. In other cases, the prediction to slope failure has been more successful like for the Randa landslide in Switzerland. Seismic monitoring and measurement of crack openings showed an increase in the total displacement that reached 40 cm before the rock avalanche collapsed in May 1991. The timing permitted a prediction of one major failure.

Predicting the instant of collapse of a slow-moving landslide can be helpful in the evacuation of the affected areas. Extensometers (Fig. 6.12), geophones, radars, and laser systems are currently used to monitor suspect slopes. In Norway, some huge sectors of steep mountain walls dipping toward the fjords or valleys are constantly monitored (Fig. 6.13). Landslides of this kind have provoked large tsunamis in the past with numerous victims (→Sect. 7.2).

Fig. 6.12 An extensometer measuring the deformation of a wall in the La Verna landslide site in central Italy

(continued)

Box 6.2 (continued)

Fig. 6.13 This innocuous-looking mountain peak in western Norway is monitored because of its high creep rate. A failure will dam the river and give origin to a dangerous lake

Experiments and monitoring of slopes have shown a significant relationship between acceleration $\ddot{\varepsilon}$ and rate $\dot{\varepsilon}$ of the deformation ε

$$\ddot{\varepsilon} = A\dot{\varepsilon}^\alpha. \tag{6.4}$$

The integration of (6.4) gives

$$\begin{aligned}\dot{\varepsilon} &= [A(\alpha-1)]^{1/(1-\alpha)} \ (T-t)^{1/(1-\alpha)} &&\text{if } \alpha \neq 1 \\ \dot{\varepsilon} &= \dot{\varepsilon}_0 e^{At} &&\text{if } \alpha = 1\end{aligned} \tag{6.5}$$

where T is the time of failure. Data for creep collected from numerous sites indicate values $A \approx 0.047$; $\alpha \approx 1.49$ (Fig. 6.14). Knowing these parameters with precision for a particular landslide gives in principle T, the time to failure from Eq. 6.5.

(continued)

Box 6.2 (continued)

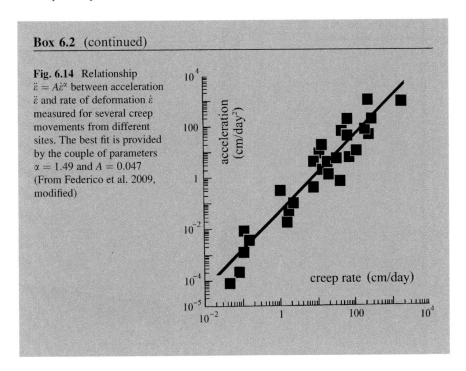

Fig. 6.14 Relationship $\ddot{\varepsilon} = A\dot{\varepsilon}^\alpha$ between acceleration $\ddot{\varepsilon}$ and rate of deformation $\dot{\varepsilon}$ measured for several creep movements from different sites. The best fit is provided by the couple of parameters $\alpha = 1.49$ and $A = 0.047$ (From Federico et al. 2009, modified)

6.4 Simple Lumped Mass and Slab Models for Rock Avalanches

As a simple toy model we can envisage a rock avalanche as a rigid, nondeformable object moving down slope and subject to the sole Coulomb friction. This avoids the difficult calculations of the internal deformations within the granular medium. In one type of model, called the *lumped mass model*, the whole mass is condensed to a single point. The equation of motion is calculated for this point, representative of the whole landslide. The centre of mass or the front of the landslide is a possible choice (Fig. 6.15a).

There are obvious drawbacks with a lumped mass model. A real landslide may cross a boundary between two media with different properties – for example, from a mountain with rocky flank to a glacier – so that the front of the landslide is subjected to a different resistance from the rest of the body. In addition, different parts of the landslide travel on terrain with different slope angle. This type of problems cannot be simulated with the lumped mass model where the whole landslide is collapsed to one point, but can be partially accounted for modeling the landslide as a slab and taking the average properties of the terrain. The slab model allows also introducing in a direct manner the resistance of the medium, called the drag force. For this reason, a slab model will be used especially for

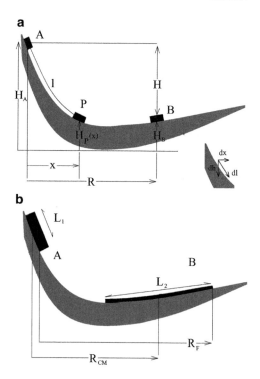

Fig. 6.15 (a) Elementary geometry of the lumped mass model. (b) Stretching slab model

subaqueous landslides, where the terrain is gentler and landslide properties do not vary much along the landslide body (→Chap. 9).

Note also that a rock avalanche usually changes its geometrical lengths during the flow. The reason is that the avalanche normally starts as a single slab unit, with very high cohesion. As it slides and strongly disintegrates, it is transformed into a deformable granular flow; the lateral pressure thus makes it to stretch, widening and flattening. While the centre of mass is likely unaffected by the process of fragmentation, the front may extend its reach. For this reason, a simple stretching slab model is introduced in →Sect. 6.5.2 (Fig. 6.15b).

6.4.1 A Simple Model of Landslide Movement

6.4.1.1 General

The equation of motion of a rigid mass M sliding down slope can be written as

$$M\frac{dU}{dt} = Mg\sin\beta - F_{RES} \tag{6.6}$$

where U is the velocity of the mass, β is the local slope angle, and F_{RES} is the resistance force. For the case of pure Coulomb friction (←Chap. 2),

6.4 Simple Lumped Mass and Slab Models for Rock Avalanches

$F_{RES} = Mg \cos \beta \tan \phi$. Particular physical situations might give a resistance term different from pure Coulomb friction. For example, if the landslide is moving on top of a fluid mud layer, the resistive force will be more likely of the form $F_{RES} = \mu S U/\delta$ where μ, δ are the viscosity and thickness of the mud, assumed Newtonian. The effect of the drag force or turbulence will result in a resistive term proportional to the square of the velocity. Moreover, the existence of a volume effect ($\rightarrow$Sect. 6.6) shows that the simple Coulomb model with constant friction coefficient must break down for large rock avalanches.

6.4.1.2 Basic Geometrical Definitions

Figure 6.15a shows the basic geometry for the lumped mass model. The horizontal coordinate is denoted as x. The distance along the slope path, also called the curvilinear distance, is denoted as l. The velocity calculated is the one parallel to the slope, and is called U.

If the generic force of resistance $F_{RES}(t)$ is given, the velocity can be calculated integrating the above equation to yield

$$U(t) = U(0) + \int_0^t \frac{dU}{dt} dt = U(0) + \int_0^t a(t) dt = U(0) + \frac{1}{M} \int_0^t F(t) dt \quad (6.7)$$

and a second integration gives the curvilinear length

$$L(t) = L(0) + \int_0^t U(t) dt. \quad (6.8)$$

However, for simple models of the resistance force the integration may not be necessary, as shown in the next sections.

6.4.1.3 Coulomb Frictional Behavior

We start considering the case in which the force is pure Coulomb friction, $F_{RES} = Mg \cos \beta \tan \phi$. The acceleration parallel to the slope is given by Eq. 6.5 as

$$\frac{dU}{dt} = g[\sin \beta - \cos \beta \tan \phi]. \quad (6.9)$$

As a starting point, we restrict ourselves to the simple slope path geometry of two straight lines. The slope angle is equal to a β_1 when the centre of mass of the slab has a horizontal distance x less than a distance R_1, and to an angle zero (i.e., a plateau) for distance greater than R_1. Note that this geometry would entail an impact and consequently energy dissipation at the slope break. Thus in situations like these we assume that the path is smoothed at the slope break, and also that the radius of curvature is sufficiently long to avoid centrifugal effects. There are also real cases in which the slope angle changes abruptly like, for example, in the Val Pola rock avalanche. Thus,

$$\begin{aligned} \frac{dU}{dt} &= g[\sin\beta_1 - \cos\beta_1 \tan\phi] \quad \text{if} \quad x < R1 \\ \frac{dU}{dt} &= -g\tan\phi \quad \text{if} \quad x \geqslant R_1. \end{aligned} \tag{6.10}$$

These equations can be easily integrated with Eqs. 6.7 and 6.8. The velocity as a function of time for the first part of the path is

$$U_1(t) = a_1 t = g[\sin\beta_1 - \cos\beta_1 \tan\phi]t. \tag{6.11}$$

The relationship with the horizontal distance is $x = L/\cos\beta_1$ (for $x \leqslant R_1$). Calling $g' = g[\sin\beta_1 - \cos\beta_1 \tan\phi]$ we also write $U = g't$ and $L = (1/2)g't^2$ from which eliminating the time from the equations we find the velocity as a function of the position x of the block

$$U = \sqrt{2g'L} = \sqrt{2g'x/\cos\beta} \tag{6.12}$$

and so the maximum velocity U_{max} reached at the point M of the slope break is

$$U_{max} = U(R_1) = \sqrt{2gL_1 \sin\beta_1 \left[1 - \frac{\tan\phi}{\tan\beta_1}\right]} = \sqrt{2gH_1 \left[1 - \frac{\tan\phi}{\tan\beta_1}\right]}.$$

After the landslide reaches the plateau, using as an initial condition for the velocity the calculated value for U_1, the total traveled distance along the path is calculated as

$$\begin{aligned} L_2 &= L_1 + U^2_1/(2g\tan\phi) \\ &= L_1 + (1 - \tan\phi/\tan\beta_1)R_1 \end{aligned} \tag{6.13}$$

while the total horizontal distance (runout) is

$$\begin{aligned} R &= R_1 + U^2_1/(2g\tan\phi) \\ &= R_1 \frac{\tan\beta_1}{\tan\phi} \end{aligned} \tag{6.14}$$

and so it is found that

$$\frac{H}{R} = \tan\phi. \tag{6.15}$$

This equation shows that the ratio between the fall height and the runout is equal to the internal friction coefficient. We will see that this remarkable result is general and independent of the specific slope path.

6.4.2 Use of Energy Conservation (1): Runout of a Coulomb Frictional Sliding Body

Instead of solving Newton's equations of motion, it is possible in the presence of the sole Coulomb friction to use energy arguments. Usually the conservation principle of mechanical energy is not applicable in the presence of energy dissipation, but the Coulomb frictional resistance is independent of the velocity.

Consider again a generic surface as slope path (Fig. 6.15a). If the landslide starts from rest (A), after sliding under the influence of gravity it will reach a final position downhill where the slope becomes gentler (B). Is it possible to predict the runout with any shape of the sliding surface? If we neglect the drag force with air, the answer is surprisingly simple.

Both the initial and final kinetic energy are zero, because the velocity of the slide is zero at start and at the stop. Thus, the change in the potential energy from the initial point A to the final point B must be equal to the work performed by the friction force along the path. Considering a length element dl as in the Fig. 6.15a, we have

$$MgH = \mu g M \int_A^B \cos\beta \, dl \tag{6.16}$$

where H is the total height fall in the gravity field and l is the length along the slide path. The integration is performed along the path between A and B as illustrated in the figure and at first it might appear that the exact knowledge of the slope path is necessary. However, the quantity $\cos\beta \, dl = dx$ is the infinitesimal horizontal distance, and so the integral becomes simply

$$\int_A^B \cos\beta \, dl = x_B - x_A \equiv R \tag{6.17}$$

where R is the total horizontal length traveled by the slide, also called the runout. Upon substitution in the first equation, we find the important result

$$R = \frac{H}{\mu}. \tag{6.18}$$

Thus, independently of the path form, the runout equals the fall height divided by the friction coefficient. This result stems from the double role played by the cosine of the slope angle: as part of the Coulomb friction law, and as geometrical element converting the curvilinear into horizontal displacement. Another useful form of the previous equation is

$$\frac{H}{R} = \mu = \tan\phi = \tan\alpha \tag{6.19}$$

showing that the tangent of the internal friction angle equals the ratio between the drop height and the runout, namely, the tangent of the average slope angle. Note that this result holds also if the terrain rises up as in the example of the figure. The above equation shows that the H/R ratio (also called the "runout ratio" or "apparent friction coefficient") is independent of the size of the slide, and depends only on the friction angle of the material. The reason for the adjective "apparent" is that, according to Eq. 6.19, the ratio H/R should be equal to the friction coefficient for rock. However, this is not verified for large rock avalanches ($\rightarrow$Sect. 6.8). With typical values for the friction angle of rock, one finds that the runout length should at most be about twice the height of fall in the gravity field.

6.4.3 Use of Energy Conservation (2): Calculation of the Velocity with Arbitrary Slope Path

The same principle can be used to calculate the velocity at any point P of an arbitrary slope path.

Let $H_P(x)$ be the height of the terrain at the point x (Fig. 6.15a). By energy conservation we can write

$$MgH_A = MgH_P(x) + \int_A^P \mu g M \cos \beta(x)\, dl + \frac{1}{2}MU^2 \qquad (6.20)$$

where now $\beta(x)$ is the local slope angle, a function of x, and $\frac{1}{2}MU^2$ is the kinetic energy. The integral simply becomes

$$\mu g M \int_A^P \cos \beta(x)\, dl = \mu g M (x - x_A). \qquad (6.21)$$

and so we obtain a useful equation

$$U(x) = \sqrt{2g}\sqrt{[(H_A - H_P(x)) - \mu(x - x_A)]}. \qquad (6.22)$$

With this equation it is possible to calculate the velocity at any point based only on the topographic height $H_P(x)$. The cases with straight trajectories can be envisaged as special cases of this equation, for which $H_P(x)$ should be parameterized with simple trigonometry. The friction coefficient can also be found from the previous equation putting $U(R) = 0$, which gives $\mu = (H_0 - H(R))/R$.

6.4.3.1 A Simple Formula for the Maximum Velocity of a Rock Avalanche Along a Slope Path of Exponential Form

Let us consider for simplicity a slope path as an exponential of the form

$$H(x) = H_0 \exp(-x/L) \tag{6.23}$$

and assume $x_A = 0$ as the starting point. The formula 6.22 gives

$$U(x) = \sqrt{2g[H_0(1 - e^{-x/L}) - \mu x]}. \tag{6.24}$$

and substituting $\mu = (H_0 - H(R))/R$ the maximum velocity for $H_0 > L$ along the trajectory can be found by the condition $dU(x)/dx = 0$

$$U_{MAX} = \sqrt{2gH_0}\sqrt{1 - \frac{L}{R}\left(1 + \ln\frac{R}{L}\right)}. \tag{6.25}$$

This formula may be useful in quick estimates of the maximum velocity based on the usually known parameters of the trajectory (approximated as an exponential) and the runout.

6.4.3.2 Apparent Friction Angle

The apparent friction angle is the angle whose tangent gives the apparent friction coefficient. Thus, in the absence of forces other than Coulomb friction and gravity, we can immediately find graphically the relationship between final location of the landslide and the apparent friction angle. It is sufficient to draw a line like in the Fig. 6.16 making an angle with the horizontal equal to the internal friction angle. The intersection of this line with the slide path gives the final position of the landslide. This means that standing on the centre of mass of a landslide deposit we can estimate the apparent friction coefficient by measuring the angular distance of the far landslide scar from the line of the horizon.

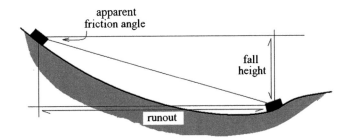

Fig. 6.16 Relationship between the apparent friction angle and the runout

6.4.3.3 Simple Example of Lumped Mass Model Calculation with Slope Path Made of Rectilinear Paths, with also the Possibility of Variable Friction Coefficient

If the slope path is built as the combination of straight lines, then instead of using Eqs. 6.7 and 6.8, it is simpler to impose energy conservation at each slope break. Consider the sliding along a path made of one first segment of horizontal length R_1 and slope angle β followed by an horizontal plane of length R_2, and finally by a run-up also with angle β (Fig. 6.17). Let the friction angle be ϕ_1 in the first and third part of the trajectory, and $\phi_2 < \phi_1$ in the middle part of the trajectory. The smaller value for the friction coefficient could be due, for example, to the presence of a glacier.

The frictional work dissipated after a distance x is $W = \mu g x M$. The kinetic energy in the first part of the trajectory is equal to the potential energy minus the energy dissipated by frictional forces

$$\frac{1}{2}U^2 = g(H_0 - H) - \tan\phi_1 g x \qquad (6.26)$$

from which we calculate the velocity

$$U = \sqrt{2g[(H_0 - H) - \tan\phi_1 x]}. \qquad (6.27)$$

Obviously, this is the same equation one would obtain by integration of the equations of motion. The maximum velocity is reached at the slope break and is obtained from $\frac{1}{2}U_1^2 = gH_1 - \tan\phi_1 g R_1$ which gives $U = \sqrt{2g[H_1 - \tan\phi_1 R_1]} = \sqrt{2gH_1\left[1 - \frac{\tan\phi_1}{\tan\beta}\right]}$. In the horizontal branch, a similar analysis leads to the velocity

$$\frac{1}{2}U^2 = \frac{1}{2}U_1^2 - \tan\phi_2 g(x - R_1). \qquad (6.28)$$

The landslide arrives at the second slope break with a velocity equal to $U_2 = \sqrt{U_1^2 - 2\tan\phi_2 g R_2}$. The velocity in the third part is calculated with a new condition of energy conservation

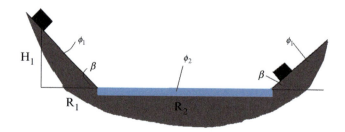

Fig. 6.17 Example for the case with variable friction coefficient

6.4 Simple Lumped Mass and Slab Models for Rock Avalanches

$$\frac{1}{2}U^2 = \frac{1}{2}U_2^2 - gH - \mu g(x - R_1 - R_2). \quad (6.29)$$

The final runout is obtained by imposing $U = 0$ in the above equation.

6.4.4 A Slab Model

A landslide traveling in ambient air or water is affected by an additional resistance, called the drag force. For this purpose, it is necessary to introduce the dimensions and shape of the landslide. Here the landslide is modeled as a slab of height D, length L, and width W.

6.4.4.1 Introduction of the Drag Force

The drag force is nonconservative and velocity dependent. This means that energy considerations such as those in the previous section cannot be used, and the solutions of the equation of motion require Newton's law, Eq. 6.6.

As the slide moves, resistance is generated at the front and at the lateral sides. The front and lateral resistances are also called front drag and skin friction, respectively, and are specified by two different friction coefficients C_D and C_{SF}. Each term of the resistance is thus of the form $\frac{1}{2}\rho_F C U^2 S$, where C is the drag coefficient (front or skin frictional), S is the area of the corresponding face and ρ_F is the density of the fluid.

In a simple approach, we sum independently the contributions from the front and the lateral faces in the following fashion

$$M\frac{dU}{dt} = M(g \sin \beta - \cos \beta \tan \phi) - \frac{1}{2}\rho_F C_D U^2 WD - \frac{1}{2}\rho_F C_{SF} U^2 WL$$
$$- \rho_F C_{SK} U^2 LD \quad (6.30)$$

Writing the mass as $M = \rho WLD$, it is found

$$\frac{dU}{dt} = g \sin \beta [1 - \tan \phi / \tan \beta] - \frac{1}{2}\frac{\rho_F}{\rho}\left[\frac{C_D}{L} + \frac{C_{SF}}{D} + 2\frac{C_{SF}}{W}\right]U^2. \quad (6.31)$$

The three separate terms in the second square bracket correspond to the drag force exerted by the front, the upper face, and the lateral faces.

Usually $C_D \gg C_{SK}$ ($\rightarrow$Sect. 9.4), but because a landslide is normally long and thin, the first two terms in second parenthesis of Eq. 6.31 could be equally important. If the effect of the front resistance is the largest, the equation simply becomes $\frac{dU}{dt} = g \sin \beta [1 - \tan \phi / \tan \beta] - (1/2)(C_D/L)(\rho_F/\rho)U^2$ and thus the resistance of the medium tends to decrease in importance as a function of the landslide length. If the resistance along the upper face is the most important, then the equation reduces to $\frac{dU}{dt} = g \sin \beta [1 - \tan \phi / \tan \beta] - \frac{1}{2} \frac{\rho_F}{\rho} \frac{C_{SF}}{D} U^2$ showing that the drag resistance is inversely proportional to the landslide thickness.

The expression above is essentially the Voellmy model, originally suggested for powder snow avalanches where the quadratic term represents the effect of air turbulence (Voellmy 1955).

Equation 6.31 can also be formally rewritten as

$$\frac{dU}{dt} = g[\sin \beta - \cos \beta \tan \phi] - \frac{1}{k} U^2 \qquad (6.32)$$

where k is a parameter of the dimension of a length,

$$\frac{1}{k} = \frac{1}{2} \left[\frac{C_D}{L} + C_{SF} \left(\frac{1}{D} + \frac{2}{W} \right) \right] \frac{\rho_F}{\rho}. \qquad (6.33)$$

In the general case this equation should be integrated numerically. However, an analytical solution can be worked out if the path is approximated with discrete straight segments. The solution along a discrete segment is

$$U = \frac{\lambda \left(\eta + e^{-2\lambda t/k} \right)}{\eta - e^{-2\lambda t/k}} \qquad (6.34)$$

where

$$\lambda = \sqrt{gk[\sin \beta - \cos \beta \tan \phi]}; \quad \eta = \frac{U_0 + \lambda}{U_0 - \lambda} \qquad (6.35)$$

and U_0 is the initial velocity of the landslide for a given segment. From second-order Taylor expansion of the solution and setting $U_0 = 0$ we also have $U \approx g't - \sqrt{\frac{g'^3}{k}} t^2$ where $g' = \sqrt{g[\sin \beta - \cos \beta \tan \phi]}$. The second term on the right hand side represents a correction term due to drag force.

We now estimate the importance of the drag term. From the definition of k (Eq. 6.35) and using the following values: $W = 100$ m; $H = 10$ m; $L = 300$ m; $C_F = 1.3$; $C_{SK} = 0.001$; $\rho_F/\rho = 0.0005$, it is found $k^{-1} = 1.1 \times 10^{-6}$ m^{-1}. With this numerical value, the effect of the drag force in air can be considered negligible in most cases. However, in water $1/k$ can be more than 1,000 times larger owing to the greater density of the medium.

A more thorough calculation including drag force is presented in the section devoted to submarine landslides ($\rightarrow$Chap. 9).

6.5 Application of the Models to Real Case Studies

6.5.1 Elm

As a simple application, we calculate the motion of the Elm landslide following Heim (1932) and Hsü (1978). The first segment has an inclination of about 45° (45°10′), and is followed by a straight terrain inclined of about 4° (3°54′), see Fig. 6.18a. The friction coefficient can be estimated from the runout as

$$\mu = H/R = \frac{613}{2,017} = 0.304. \quad (6.36)$$

In the first part of the trajectory, $H(x) \equiv H_A - H_P(x) = x \tan \beta$, $x_A = 0$ and hence

$$U(x) = \sqrt{2g}\sqrt{[(H_A - H(x)) - \mu(x - x_A)]} = \sqrt{2g}\sqrt{x[\tan \beta - \mu]}$$
$$\approx 3.703\sqrt{x}. \quad (6.37)$$

At the slope break at a distance $x = 507$ m the velocity is obtained as $U_{MAX} = 83.4$ m/s.

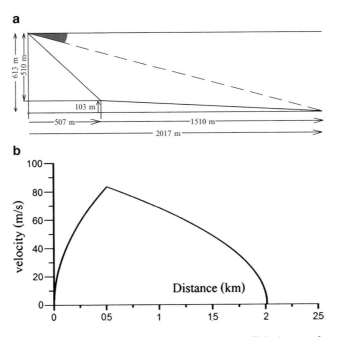

Fig. 6.18 Results for the Elm landslide. (**a**) The slope path. (**b**) Velocity as a function of the position

For the second part of the trajectory, it is readily found that $H(x) = H_A - H_P(x) = 103\frac{x-507}{1510} + 510 = 0.0682x + 475.4$, from which it is found

$$U(x) = \sqrt{2g}\sqrt{475.4 - 0.2358x} = \sqrt{9310 - 4.621x}. \tag{6.38}$$

The plot of the velocity as a function of the distance is shown in Fig. 6.18b. Note that the derivative of the velocity with respect to the position x is infinite at both the initial and final points, which means that the velocity changes very rapidly around these points. Thus according to this simple mathematical analysis the landslide stopped suddenly, as also reported by the surprised and relieved witnesses.

Note that Hsü (1978) considers the landslide front and not the center of mass as representative.

6.5.2 The Landslides of Novaya Zemlia Test Site

6.5.2.1 Motion of the Centre of Mass

In this section a simple model is introduced to account for the stretching of a rock avalanche. The model is tested against the avalanche "B-1" of the Novaya Zemlia test sites (Fig. 6.11).

Figure 6.19 shows the approximated slope and topography of the rock avalanche "B-1" of the Novaya Zemlia test sites. The slope trajectory has been approximated with two straight lines. As a first step we determine the apparent friction coefficient as the ratio between the fall height and the runout of the centre of mass

$$\mu = \frac{\text{Fall height centre of mass}}{\text{Runout centre of mass}} = \frac{250 \text{ m}}{950 \text{ m}} = 0.263. \tag{6.39}$$

To calculate the movement of the centre of mass, we refer to the energy equation, Eq. 6.26 and write

$$H_0 g = \frac{1}{2}U^2 + \mu g x + (R_0 - x)gH\tan(\beta). \tag{6.40}$$

and setting the geometrical values shown in Fig. 6.18, we find the velocity as a function of the position

$$U = 2.15\sqrt{x} \quad (x < R_0 = 500 \text{ m}). \tag{6.41}$$

The maximum velocity reached at a distance $x = 500$ m is so $U_0 = 48.2$ m/s, which is less than observed. The velocity U in the second part of the terrain is calculated solving the following equation for U

$$\frac{1}{2}U_0^2 = \frac{1}{2}U^2 + \mu g(x - R_0) \tag{6.42}$$

6.5 Application of the Models to Real Case Studies

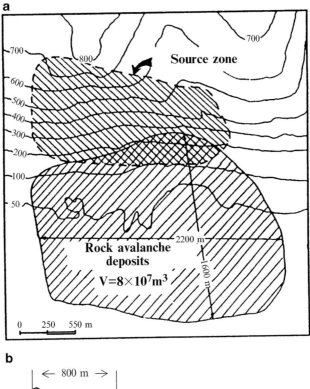

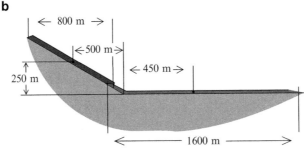

Fig. 6.19 (**a**) The rock avalanche "B-1" analyzed in the text (From Adushkin 2006. Permission to be obtained by Springer.) (**b**) The simplified geometry for the rock avalanche

from which

$$U = \sqrt{2323.24 - 2.57(x - 500)} \quad (x > R_0 = 500 \text{ m}) \quad (6.43)$$

6.5.2.2 A Simple Stretching Model Accounting for the Velocity of the Front

Landslides usually stretch during flow due to the effect of fragmentation that caused redistribution of the material on a larger area. The velocity of the front is thus higher than that of the center of mass. Because Fig. 6.11 reports the front velocity, in order to compare data with calculations it is necessary to estimate the front velocity.

From start to stop, the landslide has changed its horizontal length from $X_1 = 800$ m to $X_2 = 1,600$ m (Fig. 6.11). Assuming that the increase in the velocity is linear with landslide stretching, the velocity of the front is estimated as

$$U_{\text{FRONT}}(x) = U(x)\left[1 + \alpha \frac{(x - x_0)}{R_{\text{CM}}}\right]. \quad (6.44)$$

where $\alpha = (1/2)\Delta X/X_1$ is the total stretching (half the difference between the total landslide length variation $\Delta L = X_2 - X_1$ and the landslide length at start X_1). For this landslide, $\alpha = (1/2)\Delta X/X_1 \approx 0.50$ and so

$$\begin{aligned} U_{\text{FRONT}}(x) &= 2.15\sqrt{x}\left[1 + \tfrac{x}{1,900}\right] & (x<R_0 = 500 \text{ m}) \\ U_{\text{FRONT}}(x) &= \sqrt{2323.24 - 2.57(x - 500)}\left[1 + \tfrac{x}{1,900}\right] & (x>R_0 = 500 \text{ m}) \end{aligned} \quad (6.45)$$

The maximum velocity at $x = 500$ m is so $U_0 = 60.9$ m/s, which is closer to the maximum velocity observed (Fig. 6.11).

6.6 The Fahrböschung of a Rock Avalanche

6.6.1 The Importance of the Centre of Mass of the Landslide Distribution

It is clear from the equations of mechanics and from the above discussion that it is the centre of mass (or centre of gravity) which bears physical meaning in the calculation of the apparent friction coefficient. However, the centre of mass is usually difficult to determine because the distribution of the material prior and after the slide is not known with precision. For practical purposes, it is necessary to find a proxy for the apparent friction coefficient of a rock avalanche without having to determine the centre of mass.

6.6.1.1 A Simple Model of Layered Rock Avalanche Shows the Importance of the Centre of Mass and of the Basal Friction

Consider two slabs with the same thickness sliding on top of each other as a simple model for a slide (Fig. 6.20). The friction coefficient between the lower slab and the base and between the two slabs are denoted with μ_1 and $\mu_2<\mu_1$, respectively. It is a

6.6 The Fahrböschung of a Rock Avalanche

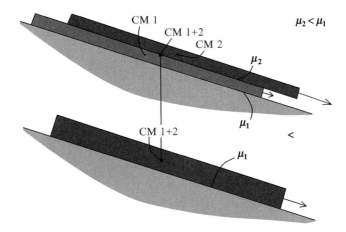

Fig. 6.20 Illustration of two slabs sliding along a slope (see text)

useful exercise to show mathematically that whereas the center of mass of the upper slab (CM2) travels farther than the center of mass CM1 of the lower slab (Fig. 6.20, upper), the center of mass of the composite system (CM 1 + 2) moves exactly as that of a single slab with friction coefficient μ_1 (Fig. 6.20, lower). Thus, the center of mass is influenced solely by the friction at the base and not by the relationship between the different superimposed units.

6.6.2 Fahrböschung of a Rock Avalanche

As a proxy to the apparent friction coefficient, field geologists frequently use the angle introduced by Heim and known as the *fahrböschung angle* (FB). This angle is built starting from two points: the higher part of the crown in the scar and the most advanced point of the deposit (Fig. 6.21). It is thus much easier to determine than the apparent friction coefficient based on the center of mass. Some researchers consider fahrböschung angle and fahrböschung as equivalent terms. Because the rock avalanche normally stretches during the flow, the *fahrböschung* often results to be lower than the apparent friction coefficient, but still a reasonable approximation of the rock avalanche mobility. Another term in the literature is that of *runout ratio*, used here as equivalent of the *fahrböschung*. The mobility of the landslide can also be defined as the inverse of the runout ratio. Following a definition by Heim, a large rock avalanche is also called with the German name of *sturzstrom*.

The Appendix GeoApp reports some data for rock avalanches including the runout ratio. It is interesting to plot the runout ratio as a function of the volume for the rock avalanches.[1] The resulting figure (Fig. 6.22) reveals a systematic decrease

---

[1] This approach was explicitly explored by A. Heim, A. Scheidegger, and others.

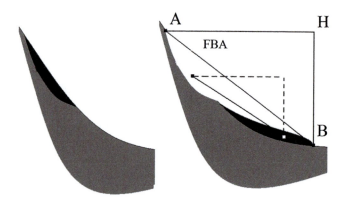

Fig. 6.21 Configuration before the landslide (*left*) and after the landslide (*right*). The fahrböschung is calculated from the upper part of the crown (point A) to the frontal part of the deposit (B). The FB is thus the ratio HB/AH. The FB angle (*FBA*) is the angle BAH. *Dashed lines* join the corresponding points for the center of mass

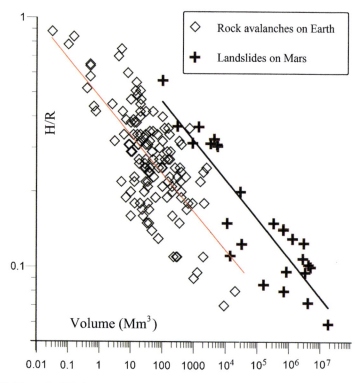

Fig. 6.22 The ratio H/R for a series of landslides on the Earth (*diamonds*) and Mars (*crosses*). Most data taken from the table in the Appendix

6.6 The Fahrböschung of a Rock Avalanche

of the runout ratio with increasing volume, showing that the landslide mobility increases with the volume. Consider, for example, the small Felsberg rock avalanche with a total volume of 100,000 m^3. The slide has a runout of 900 m and the difference in altitude between the niche of detachment and the final deposit is about 700 m. This gives an apparent friction coefficient of 0.78, corresponding to a *fahrböschung* angle of about 35°. A man sitting on the deposits would observe the source located a little lower than the midway between the zenith and the horizon. Now consider the huge Saidmarreh slide in the Zagros mountains in Iran, which has a runout of about 12 km. To produce such a long runout only 1,000 m of height were necessary, which is only a little more than the one for Felsberg. If the apparent friction angle were constant from one slide to another, Saidmarreh would have required a height of 10,000 to run for 12 km. The apparent friction angle as a function of the volume can be fitted by a power law[2]

$$\log\left(\frac{H}{R}\right) = A + \gamma \log V \qquad (6.46)$$

where A is a constant and γ is the slope of the curve in Fig. 6.22. Both A and γ need to be found empirically by fitting a straight line with data on the logarithmic plot. For example, if the volume is measured in units of 10^6 m^3, Nicoletti and Sorriso-Valvo (1991) report

$$A = -0.5277; \quad \gamma = -0.0847 \qquad (6.47)$$

while according to Scheidegger (1973)

$$A = 0.62419; \quad \gamma = -0.15666. \qquad (6.48)$$

Other empirical correlations have been suggested between runout and fall height, and between area of the deposits and runout. For example, Li Tianchi (1983) based on a fit with 76 events reports a relation of the form

$$\log R = 0.8117 + 0.3281 \log V_6 \qquad (6.49)$$

between the runout and the volume (V_6 is the volume in units of 10^6 m^3) and

$$\log S = 1.8807 + 0.5667 \log V_6 \qquad (6.50)$$

between the area and the volume.

---

[2] All the logarithms are to base 10.

Box 6.3 Brief Case Study: Flims and the Giant Postglacial Rock Avalanches

Salient data
Name: Flims landslide
Mount involved: Ringelspiz
Location: Switzerland
Coordinates: 46 50' N; 9 17' E
Volume: Giant (8–10 km³)
Age: Prehistoric (8,200–8,300 yBP)
Kind: rock avalanche
Cause: regarded as an example of failure caused by lack of buttressing during deglaciation, even though new dating locate the failure in a period well after the retreat of glaciers
Landslide material: Jurassic and Cretaceous limestone
Fahrböschung: 0.140 (8°)
Special characteristics: It is the largest rockslide in the Alps; it has a very small FB. Outcrops are well exposed by the dissection of the river Volderrhein.

The steep head scarp of the Flims landslide (Fig. 6.23a) still appears as a 10 km long overhang. The huge limestone block has traveled as a tabular unit

Fig. 6.23 (a) The edge of the Flims landslide scarp. Flimserstein is in the centre. The large landslide moved toward the observer. In 1939 a rock fall from the cliff killed 18 people in the village of Fidaz. (b) The landslide deposits cut by the river Volderrhein

(continued)

Box 6.3 (continued)

leaving no significant deposit for the first 5 km. The village of Flims is situated on the slope path of the landslide. Only after some km from the scarp, to deposits appear. The river Volderrhein has cut through the thick deposits, exposing the profound disintegration of the limestone (Fig. 6.23b).

The low FB of only 8° is compatible with a landslide of such enormous volume. However, the base of the landslide material is mostly unknown so that the determination of the centre of mass is not possible. The low inclination of the area is also well visible in Fig. 6.23a.

Near the town of Bonaduz, enigmatic coarse gravels (up to 0.6 km^3) appear to be somehow linked to the landslide (von Porchinger and Kippel 2008). Originally interpreted as glacial or fluvioglacial gravels, they have been recently reassessed as alluvial sediments squeezed by the body of the Flims landslide. The water in the gravels greatly increased in pressure during the flow of the landslide. A series of cylindrical vertical structures some decimeters in diameter and some meters in length, known as Pavoni tubes, have been interpreted as escape pipes of fluids from the Bonaduz gravel. It appeared that the failure occurred during glacial retreat as a consequence of the reduction of ice buttressing. This seemed to clarify certain features of moraines and glacial deposits. However, more recent radiocarbon dating suggests an age of about 8,000–9,000 yBP, much after the ice retreat from the area. This would entail a constant and concrete risk for large landslides in the Alps even today (von Porchinger 2002).

6.7 How Does a Rock Avalanche Travel?

At first, rock avalanches could be envisaged as the movement of particles in rapid motion, with a dynamics dominated by chaotic and energetic particle–particle collisions, similar to a granular medium tilted at much greater angle than the angle of repose (←Sect. 6.4.3). On the other hand, observations showing the absence of relative movement between grains are relatively common (←Sect. 6.2.2). Which scenario is closer to reality?

According to some researchers (Hsü 1978), the material does not appear to slide, but flows similar to a fluid. The first person to stress this point was Heim. Based on the report of a survivor of the Elm landslide, he wrote (reported from Hsü, 1978):

> The debris mass did not jump, did not skip, and did not fly in the air, but was pushed rapidly along the bottom like a torrential flood. The flow was a little higher at the front than in the rear, having a round bulgy head, and the mass moved in a wave motion. All the debris within the stream rolled confusedly as if it were boiling... I turned and ran, and a single jump saved me. When the Sturzstrom drove past me like a speeding train, its outer edge was only one meter away. During my last jump, small stones were whirling around my legs, being stirred up by the wind. Otherwise I was not hurt by any fallen stones, nor did I feel any particularly strong air pressure.

According to this observation, the behavior of the material resembles a fluid rather than a dense granular flow. One should be careful in inferring the behavior of the whole landslide based on what could be an observation of local validity. However, the remark is interesting in showing that the material was sufficiently disintegrated to flow.

A similar indication derives from laboratory experiments with granular flows (Box 6.4). Figure 6.24 shows the results of a simple flume experiment with grains of different size. Initially, the granular mixture is homogeneous. The examination of the artificial deposit after the experiments shows that the large grains tend to drift to the surface (Fig. 6.24). This shows that the "Brazil nuts effect" ($\leftarrow$Sect. 5.5.2) occurs also in an experimental rapid granular flow. Grains must be sufficiently mobile to allow for vertical particle segregation, indicating that particles enjoy some freedom.

This is in marked contrast with the observation reported in Fig. 6.8b, where the dike in the deposits of the large Köfels landslide has remained in relative position as it was before the landslip. Observations on the calcite veins in the calcareous landslide in Flims (Switzerland) show similar occurrence. The plug-like motion of a landslide is also confirmed by the common observation of jigsaw clasts abandoned on some landslide deposits ("jigsaw puzzle effect" according to Shreve 1968, $\leftarrow$Sect. 6.2.3). Thicker landslides like Flims may also have multiple shear zones (Pollet and Schneider 2004).

6.7.1 Shear Layer as an Ensemble of High-Speed Particles

A possible model for the shear layer is that shown in Fig. 6.26. The upper cap is made of broken rock (black circles) maintained aloft by a basal shear layer of particles impacting at high speed (white circles). This configuration is stable if a sufficient amount of momentum per unit time is communicated to the upper cap by collisions of particles from below. However, each collision in the shear layer also dissipates energy. The requirement of stability with limited dissipation implies that (De Blasio and Elverhøi 2008)

$$\tan \beta \gg \frac{2(1-\varepsilon)}{\eta} \quad (6.51)$$

where η is the ratio between the mass of the bouncing particles in the shear layer and the mass of the upper cap, and ε is the coefficient of restitution. Even allowing for a large mass contained in the shear layer (for example $\eta \approx 1$ corresponding to the mass of the bouncing grains comparable to the mass of the slab) and with $\varepsilon \approx 0.5$ it is found that the terrain should slope at least 45° for the upper cap to remain suspended. To summarize, the model of a slab suspended on bouncing particles runs into big difficulties.

6.7 How Does a Rock Avalanche Travel?

Box 6.4 Simple Views: Small-Scale Didactic Experiments with Granular Material

There are numerous instructive experiments that can be carried out with a small flume for teaching, such as the demonstration of the independence of the *H/R* ratio on the volume. For this purpose, one can execute a series of runs with increasing amount of granular material (decorative sand will serve perfectly). By opening a gate, the material is let to flow along the flume and comes to halt on a flat base. To establish the position of the centre of mass of the final distribution, a laser gauge may be used to read the height and so reconstruct the shape of the final deposit. As a cheaper alternative, one can use needles to dip into the granular material (fine sand will adhere on wet needles). At this point, one should determine the fall height and the runout of the centre of mass. The results show independence of the *H/R* ratio with the volume. For example, experiments with a 1-m long Plexiglas flume and decorative sand show a constant friction angle of about 39° even if the mass of the granular avalanche is varied by a factor 25. Similar results are obtained considering the fahrböschung instead of the centre of mass runout ratio.

Figure 6.24 shows the deposit resulting from another experiment in which the granular medium is initially composed of homogeneous mixture of sand

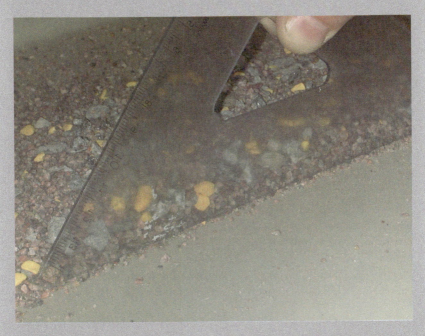

Fig. 6.24 The final deposit of a small experimental granular flow initially starting as a homogeneous mixture of grains with different sizes

(continued)

Box 6.4 (continued)

of different size. The artificial deposit clearly exhibits an inverse grading, which implies that particles have sufficient freedom to drift within the granular medium.

Box 6.5 Computer Simulation of Rock Avalanche Based on Molecular Dynamics

A direct way to simulate the flow of granular materials and of sturzstroms is to use advanced computer models such as molecular dynamics codes (Campbell 1990; Campbell et al. 1995). In this kind of algorithm, blocks are approximated as interacting disks. The equation of motion for each disk is calculated in time based on the Newton's equations of motion. Particles mutually interact with short-range repulsive forces of the Hertzian type (←Sect. 5.1.3). The Coulomb friction is also introduced, and blocks can spin around their axis. The accelerations of one disk in a position $\vec{P_i}$ is

$$\frac{d^2 \vec{P_i}}{dt^2} = \vec{g} + \frac{1}{M_i} \sum_j \vec{F}(\vec{x_i} - \vec{x_j}) \qquad (6.52)$$

where the right hand side includes the gravity acceleration $\vec{g}$, and the impulsive force due to collisions with other disks, $\vec{F}(\vec{x_i} - \vec{x_j})$. These equations are integrated to obtain the velocities and position as a function of time. In this kind of simulations, the dissipative force is usually defined to return appropriate restitution coefficients upon collision. This is accomplished by summing an appropriate velocity-dependent force of the kind

$$D_n(\vec{x_i} - \vec{x_j}) = -\delta \sqrt{\frac{1}{2} M k U_n} \mathrm{sgn}(U_n) \qquad (6.53)$$

where U_n is the normal velocity and δ is a coefficient accounting for the dissipation. The bedrock is usually simulated as glued particles along the terrain.

Figure 6.25 shows the results obtained with the computer code of Campbell et al. (1995). Note how the simulations preserve the original layering.

(continued)

6.7 How Does a Rock Avalanche Travel?

Box 6.5 (continued)

Fig. 6.25 Simulation of a rock avalanche with molecular dynamics. (**a–d**) 30,000 disks at different times: (**a**) $t=0$; (**b**) $t=15$, (**c**) $t=31$, (**d**) $t=46$. Time is measured in dimensionless units of $\sqrt{D/g}$, where D is the particle diameter. (**e**) Deposit obtained with one million disks (From Campbell et al. 1995, reproduced with permission according to the AGU rules, http://www.agu.org/pubs/terms-of-use.shtml)

Box 6.6 Simple Views: Transmission of Momentum Along a Pile of Spheres

A simple argument illustrates one possible reason for the compactness of the upper part of a rock avalanche and consequently for the lack of segregation and relative displacement. Consider a one-dimensional column of N spheres of radius R. Let us assume that the spheres are not directly in contact. By giving the lowermost sphere a vertical upward velocity v_1, a certain amount of momentum is transmitted through the chain. Each sphere acquires a velocity from the neighbor below and transmits a fraction of the momentum received to the particle above. A simple analysis based on Eq. 5.33 shows that the final velocity of each particle after the momentum has been transmitted throughout the chain is

$$v_n = v_1 \left(\frac{1+\varepsilon}{2}\right)^{n-1}; \quad v'_n = v_1 \frac{1-\varepsilon}{2}\left(\frac{1+\varepsilon}{2}\right)^{n-1};$$
$$v'_N = v_1 \left(\frac{1+\varepsilon}{2}\right)^{N-1}. \tag{6.54}$$

Notice that the last grain does not transmit momentum to any other particle. The velocity difference between two neighbours is then

$$v'_{n+1} - v'_n = -v_1 \left(\frac{1-\varepsilon}{2}\right)^2 \left(\frac{1+\varepsilon}{2}\right)^{n-1} \tag{6.55}$$

Even choosing a high value of 0.8 for the coefficient of restitution, it is found $(v'_{n+1} - v'_n)/v_1 \approx 0.0039$ for $n = 10$ and $(v'_{n+1} - v'_n)/v_1 \approx 3 \times 10^{-7}$ for $n = 100$. Hence, a perturbation from the bottom of the chain damps out very rapidly with height. Although this argument is much simplified, it shows four interesting features:

1. The derivative of the velocity that can be approximated as

$$\frac{dv}{dy} \approx -\frac{v_1}{d}\left(\frac{1-\varepsilon}{2}\right)^2 \left(\frac{1+\varepsilon}{2}\right)^{(y/2R)-1} \tag{6.56}$$

goes rapidly to zero as a function of the height y. The velocity v_1 can be interpreted as the velocity at which the lowermost part of the avalanche hits the roughness on the ground.

(continued)

Box 6.6 (continued)

2. Equation 6.55 shows that smaller grains are more effective in damping out the velocity differences for a given height. The damping of the velocity is very sensitive to the ratio $y/(2R)$ between flow height and particle diameter.
3. Because a rock avalanche disintegrates to very small grains, as a corollary to point (2) it follows that the avalanche becomes more efficient in damping out velocity differences during the flow.

Let us now assume that the pile is moving along slope, with the lowermost particle being agitated by periodic collisions with the ground. In a simplified treatment, we assume that the pile is moving at a certain velocity $v_=$, and that the velocity along the pile is proportional to it, $v_= = kv_1$ where k is a constant. The roughness on the ground is distributed with a certain length λ; the number of collisions per unit time is so $v_=/\lambda$. For each collision a certain amount of momentum mv_1 is given vertically to the column. The momentum transferred per unit time by the generic particle n to the upper cap is given as

$$m(v_n - v'_n)\frac{v_=}{\lambda} = mv_{n+1}\frac{v_=}{\lambda} = mv_1\left(\frac{1+\varepsilon}{2}\right)^n \frac{v_=}{\lambda} \qquad (6.57)$$

The part of the landslide above the level n collapses if this momentum is lower than the gravity force of all the spheres above, given by $m(N-n)g\cos\beta$. Using the fact that $v_n - v'_n = v'_{n+1}$, the condition so becomes

$$\left(\frac{1+\varepsilon}{2}\right)^n < \frac{(N-n)g\lambda}{kv_=^2}\cos\beta. \qquad (6.58)$$

With $\varepsilon = 0.8$, it is obtained that if $n \ll N$, then $n \approx 21.7(3.4 - \log N)$ and for $N = 200$ it follows $n \approx 24$. For $N \geqslant 2,500$ it results that $n \leqslant 0$, which means that there is no level at which the bouncing of the spheres can withstand the weight. This result, obtained with much simplified dynamics, nevertheless elucidates a possible reason why in the laboratory flows (where the number of particles n in a pile is small) the particles maintain a vertical component of the velocity that is absent in large rock avalanches.

6.7.1.1 Flow of a Rock Avalanche

The above examples show that in a small granular flow, grains enjoy greater freedom than in a large sturzstrom. In a small flow, the energy generated at the base is shared among relatively few particles. Particles maintain a sufficient energy (the "granular temperature" in the terminology of ←Chap. 5) to create voids around them.

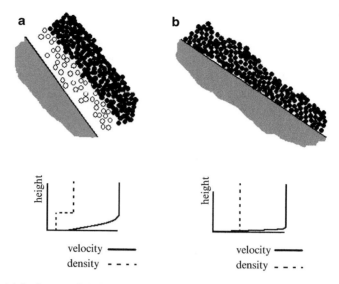

Fig. 6.26 (a) In these models for a rock avalanche, the upper cap travels without deforming. In (a), a hypothetical shear layer of free particles keeps the upper cap in the air with high-speed collisions. The resulting particle velocity and density are also shown schematically. In (b), the upper cap directly slides on the terrain, which becomes a region of concentrated shear rate

However, for thicker rock avalanches it becomes impossible for grains to stay separated. As a consequence, the base of a large rock avalanche travels as shown in Fig. 6.26b, in which rock slips directly on the terrain where high shear rate develops.

6.8 The Problem of the Anomalous Mobility of Large Rock Avalanches

6.8.1 Statement of the Problem

It has been shown that that the decrease of the apparent friction angle with the volume (Fig. 6.22) is inconsistent with the Coulomb friction model, according to which the friction angle should be independent of the mass and thus also of the volume. This means that in some way the material resistance is weakened during the flow of large rock avalanches. At a first glance, one could think of a kind of internal failing. An energetic rock avalanche is likely to produce bouncing rocks whose average distance becomes much smaller as the material acquires more kinetic energy. In fact, it is known that granular materials expand during the flow.

6.8 The Problem of the Anomalous Mobility of Large Rock Avalanches

It might seem that a reduction of internal friction due to diminished density might do the work. However, basic physics tells us that in order to decrease the resistance, something must be happening at the interface between the slide and the valley bottom, rather than internally to the slide. This is clearly seen with the model of Fig. 6.20. The centre of mass acceleration is independent of the internal forces, and is affected only by external forces.

6.8.1.1 Centre of Mass Versus Fahrböschung

Recent compilations show that the volume effect persists also when the change of the centre of mass position is properly considered in the calculations instead of the more common fahböschung (e.g., Straub 2001). Thus, the stretching of a landslide alone cannot account for the volume effect.

6.8.2 A List of Possible Explanations

Several mechanisms have been suggested to explain the apparent decrease of the friction angle with increasing volume displayed by rock avalanches. These explanations can be listed as follows:

- A highly energetic granular avalanche "yields" for very high shear stresses.
- The decrease in the friction coefficient is a process inherent to the dynamics of a granular material. Thus, a rapid granular flow should naturally exhibit a decrease in the apparent friction angle with increasing volume. Some researchers refer to *mechanical fluidization* for the hypothetical process behind this explanation.
- The functional form of the Coulomb friction changes somehow during flow, for example, acquiring a velocity dependence.
- Air is trapped underneath the rock avalanche and acts as a lubricant.
- As a consequence of intense friction, high temperatures are reached at the interface between the bedrock and the landslide, which may result in the melting of rock or the formation of gas. Also water could in principle become vaporized by the frictional increase in temperature. The ensuing layer of liquid rock or volatiles might provide a kind of natural lubrication.
- Acoustic fluidization, whereby high-frequency acoustic waves provide a temporary release of the stress in some points of contact between the slide and the bed, might boost the flow.
- Water present could act as a lubricant (without necessarily increasing its state from liquid to gaseous).
- Heat-generated pore pressure.
- Dust created during the disintegration process.
- Mass exchange between the slide and the valley bottom may boost the slide.
- If the landslide starts with a very high aspect ratio and ends up with a small aspect ratio, then much of the energy does not come from the fall of the centre of

mass along the path, but rather by the change of shape of the deposit. In this case, a volume effect would be possible.
- The volume effect might be related to the disintegration of the rocky mass.

Note that more than one mechanism could be in principle involved in high mobility and the volume effect. We now consider in more depth some of the most plausible explanations.

6.8.3 Explanations That Do Not Require Liquid or Gaseous Phases

6.8.3.1 Granular Material Yields at High Shear Stress[3]

In essence, the reason why a landslide should have a fahrböschung independent of the mass is that both the pulling force (gravity) and the resistance force (friction) are proportional to the mass and to the gravity field, and are independent of the velocity. In a purely frictional material, the two effects compensate perfectly. However, let us assume that for some reason the resistive stress cannot exceed a certain limit τ_y. In a simple scenario, a slide travels with Coulomb frictional resistance only for shear stress $\tau < \tau_y$, where $\tau = \rho g H \sin\beta$ is the basal stress. When the shear stress is larger than τ_y, the resistance is set equal to τ_y and so

$$\frac{dU}{dt} = g[\sin\beta - \mu\cos\beta] \qquad \text{for } \tau \leq \tau_y$$
$$\frac{dU}{dt} = g\sin\beta - \frac{\tau_y}{D\rho\cos\beta} = g\left[\sin\beta - \frac{\tau_y}{\tau}\right] \qquad \text{for } \tau > \tau_y \qquad (6.59)$$

Note that the resistance diminishes with the landslide thickness. Thus, larger slides (which will normally have a greater thickness D) flow with diminished resistance. Dade and Huppert (1998) find a good fit with data for rock avalanches with a value $\tau_y = 48 \pm 15$ kPa. The resistive stress τ_y is reminiscent of the Bingham fluid model with τ_y playing the role of the shear strength. Trunk et al. (1985) simulated the north-American Madison Canyon slide and found satisfactory results with a yield stress of the order 500–600 kPa.

Why should a granular avalanche exhibit a shear strength-like behavior? The absence of clear-cut arguments is a weak point of this explanation. Shear strength behavior is characteristic of fluids that can form bonds, which does not seem to be the case for dry granular material. Additionally, the model is not based on a microscopic understanding of granular materials at high energy and shear stress.

---

[3] Some references: Dade and Huppert (1998), Hsü (1978), Davies (1982).

6.8.3.2 Rock Disintegration

A missing ingredient in the models of rock avalanches is the disintegration of the rocky mass. It is thus natural to speculate that the disintegration might play a role in increasing the mobility of rock avalanches. Experiments seem to indicate that the friction coefficient decreases as a consequence of fragmentation (Deganutti 2008), albeit at a low level to be able to explain the volume effect. The increase in mobility was attributed by Davies and McSaveney (1999) to an increase of the earth pressure force, and in a second examination to the energy released during rock disintegration processes (McSaveney and Davies 2006). In their explanation, when a rock breaks apart into a myriad of fragments, the high kinetic energy acquired by fragments may assist in decreasing the basal resistance. However, the process of disintegration absorbs energy, so that the relationship between fragmentation and mobility gain of the rock avalanche needs some additional study.

6.8.3.3 Acoustic Fluidization[4]

It is likely that the stress inside a granular avalanche fluctuates highly in space and time (Davies 1982). As a consequence, the Coulomb law does not hold exactly everywhere, but only on the average. According to Melosh (1979), Collins and Melosh (2003) and others, the spaces where the overburden pressure is below the average value may slip more easily. Fluctuations are due to acoustic waves, and for this reason the hypothetical process is called acoustic fluidization. The effective viscosity at high frequency of an acoustically fluidized granular medium is, according to Melosh,

$$\mu_{\text{eff}} \approx \frac{\lambda G}{2c_P} \qquad (6.60)$$

where c_P is the speed of compressive wave, λ is the wavelength, and G is the bulk modulus of the material. In a first approximation, the material behaves like a Newtonian fluid with an effective viscosity. At low frequencies, the above equation does not hold and acoustic fluidization becomes ineffective.

Acoustic fluidization may also be described in analogy with the flow of a granular bed when the ground is shaking, like during the operation of heavy vibrating machines or in vibrated tables. Erismann and Abele (2001) provide a pictorial view of the effect that we follow here with slight modifications. Imagine shedding uniformly some amount of coarse sand on a tilted table. Then place a book on the table on top of the sand layer. Sand grains maintain the book at a distance from the table. High frequency acoustic waves can be obtained by hammering the table. During high-frequency shaking, the book loses contact with some of the grains and is so capable to advance a bit. A problem with the concept of acoustic fluidization is that the vibration of both the ground and the medium must be

---

[4] Melosh (1979).

produced by the landslide itself, and it is not clear whether the energy dissipation will damp out vibration energy. As also observed by Erismann and Abele, a fluidized landslide should flow as a Newtonian fluid. This implies a strong shearing in the material at all levels, whereas the good preservation of superficial landslide deposits demonstrates that the upper cap of sturzstroms move like plugs. Mechanical self-fluidization lacks experimental support, as the effect is obtained only through an external generation of acoustic waves. Perhaps acoustic fluidization might become effective during ground shaking produced by an earthquake. It has been suggested that earthquake-generated landslide might become particularly mobile because of the shaking effect of the earthquake (Genevois et al. 2002).

6.8.4 Explanation of the Anomalous Mobility of Rock Avalanches Invoking Exotic Mechanisms and New Phases

6.8.4.1 Air Lubrication[5]

Shreve (1968) has suggested that rock avalanches could slide on top of a lubricating air layer. This would reduce the chance of dislocation between neighboring blocks (so produce the jigsaw-puzzle effect ←Sect. 6.2.3) and at the same time provide a lubrication mechanism. The original scheme, applied especially to the Sherman landslide in Alaska (→Box 7.3) and to the Blackhawk landslide (←Sect. 6.1.2), is that the frontal portion of a rock avalanche could jump against an overhang on the valley floor, ingesting air. The air layer beneath the avalanche would provide, according to Shreve, a lubrication effect similar to the one experienced by playing cards in a simple experiment (→Box 6.7). Witness observations of very strong winds at the front of some landslides could be interpreted as the final ejection of the air carpet when the landslide comes to rest. Is air lubrication appropriate to explain the mobility of rock avalanches?

The model can be challenged on several grounds. First, the material should be impermeable to air passage, which is not obvious for a granular flow. In addition, the process of air lubrication does not depend in an obvious way on the mass of the slide. Hence, the volume effect is not explained by this model. The Sherman landslide, for which the air lubrication model had been originally suggested, has slid on a glacier. It seems more reasonable to invoke a lubrication effect supported by molten water from the scraped glacier surface, without need for more exotic mechanisms (Erismann and Abele). Finally, and perhaps above all, the long runout of giant avalanches has been observed also for landslides on Mars and other celestial bodies where the amount of volatiles is small or negligible (→Sect. 7.4). For this reason, the idea of air cushion has been abandoned, at least as a general process.

---

[5] References: Kent (1996), Shreve (1968).

6.8 The Problem of the Anomalous Mobility of Large Rock Avalanches

Box 6.7 Simple Views: An Instructive Experiment with Playing Cards Illustrates the Physics of Air- and Water-Lubricated Landslides

The mechanism of air lubrication may be illustrated with a very simple classroom experiment making use of a deck of cards. Cards are gently placed one at a time on the upper end of an inclined flexible plane (Fig. 6.27). After the initial acceleration on the plane (A–B), one card meets a flat area (C) where it unexpectedly travels suspended on a thin air layer. In this way, cards may reach surprisingly long runouts (more than 5 m for a fall height of some 50–100 cm).

The condition for the card to remain suspended in air involves a balance between the gravity force acting on the card and the air dynamic pressure (see more in →Sect. 9.6.2). The condition for the onset of air lubrication then reads $P_{dyn} = (1/2)\rho_F U^2 > \rho g D$, where D and ρ are the card thickness and density. Solving for the velocity it is found $U > \sqrt{2\Delta\rho g D/\rho_F}$ which shows that a minimum velocity is necessary for air lubrication. Too thick cards will not produce the effect. This can be directly tested gluing more cards together. This result indicates that air lubrication would be more effective for small, thin landslides than for thick ones, in contradiction with the volume effect. A similar effect with water instead of air is possible for subaqueous slides owing to the much greater density of water (→Sect. 9.6.2).

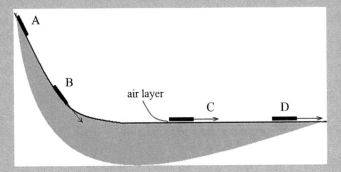

Fig. 6.27 A simple experiment with cards. Indicative height and length are 1 and 5 m, respectively

6.8.4.2 Water and Mud Lubrication

Water is almost invariably present in the superficial layers of the Earth. Prior to the detachment of a landslide, water often impregnates soils and seeps through rocks openings. Water might have contributed to the mobility of prehistoric landslides even in desert areas, like in the case of the Blackhawk landslide in the Mojave desert (←Sect. 6.1.3). There are examples of rock avalanches with long runout that were mobilized by a soft wet soil (Fig. 6.28).

Fig. 6.28 Landslides may acquire mobility due to a soft, cohesive substrate like for this old rock avalanche deposit in western Norway

It is thus possible that dry rock avalanches are in reality never really dry (Legros 2002). Water will likely reside in the pores between rock grains during fragmentation, saturating the granular material. The pore water pressure will increase, with the result of diminishing the effective friction. Furthermore, the intense stresses on a rock avalanche produce a large amount of fines embedding coarser material. Mixed with water, the fines will form a slurry with some unspecified non-Newtonian rheology. The viscosity μ and the shear strength of a suspension of noncolloidal particles with a solid fraction $1 - \phi$ is given in terms of the equation by Krieger and Dougherty (Eq. 4.10) This transition might result in a load-independent shear resistance and hence in a volume effect. Rheological flows do exhibit a volume effect (Corominas 1996).

Adopting the equation of motion of a Bingham fluid ($\leftarrow$Sect. 4.7.2)

$$\frac{dU}{dt} \approx g \sin \beta - \frac{\tau_y}{\rho D} - 2 \frac{\mu U}{\rho D^2} \qquad (6.61)$$

where ρ is the density of the material and earth-pressure force is neglected. When the landslide comes to a halt, the last term on the right hand side of Eq. 6.61 becomes negligible compared to the second one, and the final thickness of the landslide is obtained putting the right hand side of Eq. 6.61 to zero, which yields Johnson's criterion for the thickness of a Bingham fluid $D \approx \frac{\tau_y}{\rho g \sin \beta}$. Thus, the Bingham fluid leaves a deposition layer upon passage of thickness directly proportional to the shear strength and inversely proportional to the sinus of the slope angle. Along a slope path which is concave upward, the thickness increases toward the deposit front.

This results in a *H/R* ratio decreasing with the volume, in accordance with data (De Blasio 2009). The model, however, fails to explain the typical absence of deposition in the proximal region, as a Bingham fluid leaves a deposit along its path. To conclude, the volume effect can still be considered an open question.

6.9 Frictionites, Frictional Gouge, Thermal Effects, and Behavior of Rocks at High Shear Rates

6.9.1 Frictionite, Melt Lubrication, and the Kofels Landslide

6.9.1.1 The Enigmatic Frictionite and Melt Lubrication

In the second half of the last century, a rock similar to pumice was brought to the attention of the scientists. The mysterious material, known for generations by local craftsmen, was initially recovered from an outcrop along the section of a valley exposed by river erosion. The location was near the village of Köfels in the Austrian Otz valley. The porous, bubble-rich rock was only one of at least two rock types. A second rarer glassy rock was also found. The general opinion was that the pumice and glass derived from an inactive volcano located somewhere in the valley. Others suggested the hypothesis of a meteoritic impact, an idea that still persists. A geological survey showed unambiguously that a mass movement had occurred in the area. Thus, it appears that this rock (also called frictionite) is associated with the landslide Masch et al. (1985). Except for other few cases (notably in the Himalayas, Peru, and Korea) it is very rare to find this kind of rock in a landslide site. How could the landslide produce molten rock?

> **Box 6.8** Brief Case Study: Köfels and Melt-Forming Landslides
>
> *Salient data*
> *Name: Köfels landslide*
> *Location: Austria, near the boundary to Italy*
> *Coordinates: 47 07 N; 10 55 E.*
> *Volume: Giant (2,100–2,500 Mm^3)*
> *Age: Prehistoric (dated to 8,700 years BP with radiocarbon)*
> *Kind: rock avalanche*
> *Casualties: unknown*
> *Cause: unknown*
> *Landslide material: alkali-feldspar gneiss*
> *Fahrböschung: 0.2*
> *Special characteristics: it is one of the few examples of rock avalanches from which frictionite has been recovered.*
>
> Traveling along the Otzel valley in Austria, near Köfels one is struck by a huge obstruction on the middle of the valley about 3 km long and up to half a kilometer thick (Fig. 6.29). It is the deposit of one of the largest landslides in
>
> (continued)

Box 6.8 (continued)

Fig. 6.29 The main body of the Köfels landslide seen from the scarp area. The landslide traveled away from the observer, toward the opposite valley

the Alps. The Kofels landslide has a volume of about 2,100–2,500 Mm^3 and stretches along the valley from east to west. The mass movement occurred at about the time of deglaciation (8,700 years b.p.), a period of great instability to landsliding in the Alps.

The Otz river has cut through the landslide material for about 500 m. The glassy like frictionite has been extracted from some of the river incisions.

Figure 6.30 shows a section of the valley, perpendicular to the river and parallel to the main landslide movement. The pumice-like frictionite has been found in the region shown with an arrow. The figure (after Erismann and Abele 2001) shows a sketch of the possible evolution of the landslide. The shoulder protruding at a distance of 4 km in the valley morphology had, according to Erismann and Abele (2001), a key role in the evolution of the sturzstrom. During the initial phase of landslide movement, the lower part of the landslide hit the shoulder and stopped suddenly in a violent impact. The upper part continued its movement uninterrupted, detaching from the underlying material. The sturzstrom was so cut into two pieces during flow; intense frictional heat was developed along the plane between the two portions, giving origin to rock melting solidifying into the pumice-like frictionite.

(continued)

Box 6.8 (continued)

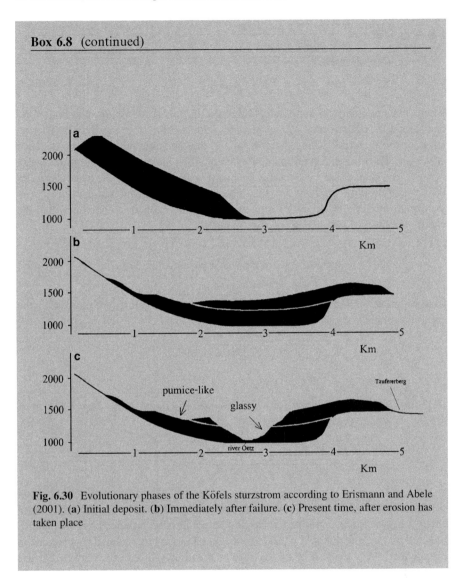

Fig. 6.30 Evolutionary phases of the Köfels sturzstrom according to Erismann and Abele (2001). (**a**) Initial deposit. (**b**) Immediately after failure. (**c**) Present time, after erosion has taken place

6.9.1.2 Thermodynamical Estimates and Heat Involved in Deep-Seated Friction

In the following, we examine some simple estimates of melt generation at the boundary between two sliding surfaces beneath sturzstroms. Let us consider a square meter of material at the interface between a rock slab and the underlying surface. The energy dissipated by friction per unit time and unit surface that is converted into heat is about

$$j_0 = \frac{\text{Work performed by dissipation force}}{\text{Area} \times \text{time}} = \left(\frac{\text{Force}}{\text{Area}}\right) \times \left(\frac{\text{Horizontal displacement}}{\text{Time}}\right)$$
$$= e(\rho g D \cos \beta \tan \phi)(U) \tag{6.62}$$

where e is a geometrical factor accounting for the fact that only one fraction of the heat produces molten rock, U is the slide velocity, and ϕ is the internal friction angle of the material. After a certain time τ, the landslide has traveled a distance $\Delta L = U\tau$. To bring a certain mass m of material to melting, an amount of energy $\lambda m + cm\Delta T$ is necessary, where λ is the latent heat, c is the specific heat, and ΔT is the temperature range between the initial and the melting temperatures. The maximum possible thickness of molten rock can be estimated equating $j_0\tau$ to the energy necessary to melt 1 m² of rock to a depth δ, or

$$j_0\tau = e\rho g D \cos \beta \tan \phi \Delta L = \rho \delta (\lambda + c\Delta T) \tag{6.63}$$

from which we obtain the thickness of melt produced per unit length traveled by the rock avalanche

$$\frac{\delta}{\Delta L} = \frac{eg D \cos \beta \tan \phi}{\lambda + c\Delta T}. \tag{6.64}$$

Using values $\lambda = 3 \times 10^5$ J kg^{-1}, $\Delta T = 900$ K, and $c = 10^3$ J kg^{-1} K^{-1} it is found

$$\frac{\delta(\text{mm})}{\Delta L(\text{m})} \approx 4 \times 10^{-3} D(\text{m}). \tag{6.65}$$

Thus for a landslide of the thickness comparable to Köfels, $D = 400$ m, a layer about 1.6 mm thick or molten rock can be potentially produced per one meter traveled by the landslide. After 100 m the layer amounts to 16 cm. This estimate is exaggerated, because part of the energy, as previously discussed, may be carried away from the zone of melting ($e < 1$). The presence of a melt likely alters the shear resistance between sliding surfaces; in addition, part of the heat is used to heat up an already molten layer rather than in melting more rock. The thickness of the molten layer will be lower if heat travels as a diffusive process. Thus, another useful estimate is the distance reached by the heat wave perpendicular to the sliding plane for a constant landslide velocity U. The heat travels perpendicular to the slope as a diffusion process and after a time τ has reached a distance δ given as

$$\delta^2 \approx \frac{\chi}{C\rho}\tau \tag{6.66}$$

where χ, C are, respectively, the thermal conductivity and the specific heat. Using values ($\chi = 2.5$ W/mK; $C = 1,000$ J/kg K) it is found $\delta(\text{mm}) \approx \sqrt{\tau(s)}$. The increase in temperature in this layer is thus

$$\Delta T \approx \frac{1}{2}\frac{j_0\tau}{\delta\rho C} \approx \frac{gD\cos\beta\tan\phi\, U}{2}\frac{\tau}{\delta\rho C} \tag{6.67}$$

and numerically $\Delta T \approx 4eD\,(\text{m})\sqrt{\tau}$ K. Setting now $\Delta T \approx 500 - 600$ K as the temperature increase necessary for melting, it follows that a few seconds are necessary to reach melting for a sturzstrom of the thickness of Köfels. If the landslide starts from zero velocity, after a time τ it will have reached a velocity $U = (\sin\beta - \cos\beta\tan\phi_0)\tau$ and a temperature

$$\Delta T \approx \frac{1}{2}eg^2 H\cos\beta\tan\phi_0(\sin\beta - \cos\beta\tan\phi_0)\left(\frac{\rho}{\chi C}\right)^{1/2}\tau^{3/2} \tag{6.68}$$

6.9.1.3 Self-Lubrication

Erismann (1979) and Erismann and Abele (2001) have suggested that the molten layer may lubricate the flow by replacing the Coulomb-frictional resistance $\approx \rho gH\cos\beta\tan\phi$ with a viscous-like resistance $\approx \mu U/\delta$ where μ is the viscosity of the newly produced melt. One problem is that the viscosity of a melt of composition similar to the Köfels gneiss becomes very large in proximity of the melting point. The viscosity has the form ($\leftarrow$Eq. 4.10)

$$\mu = \mu_0 \exp[a/T]\left(1 - \frac{C}{C_*}\right)^{-2.5\,C_*} \tag{6.69}$$

where μ_0, a are constants of the melt. The last part of Eq. 6.69 depends on the amount of solid particles of the gouge (see also $\rightarrow$Sect. 10.1 for melt viscosity). Even assuming the gouge rock to be completely molten and so setting the last term of Eq. 6.69 to 1, the viscosity of a felsic magma in proximity of the melting point is of the order $\mu \approx 10^{10} - 10^{11}$ Pa s, which implies that the shear stress $\mu U/\delta$ is of the order $\approx 10^{12} - 10^{13}$ Pa with a thickness of the molten layer of some mm, much greater than the shear stress due to Coulomb friction $\rho gH\cos\beta\tan\phi \approx 10^7$ Pa. Thus, for felsic composition the melt behaves more like glue than a lubricant. Melts of intermediate compositions $\mu \approx 10^5 - 10^6$ Pa s will result in a much lower resistance, $\mu U/\delta \approx 10^7 - 10^8$ Pa and a lubrication effect may begin to take place. For mafic compositions ($\mu \approx 10 - 1,000$ Pa s) a strong lubrication effect may occur. Ironically, the sturzstrom for which melt lubrication has been originally suggested has a felsic composition.

One appealing feature of melt lubrication is that it can potentially explain the volume effect. This is because, as Eq. 6.65 shows, the amount of melt increases with the thickness of the landslide. Erismann (1985) has built a rotary apparatus to study the frictional generation of rock melt, and found that the effective friction coefficient decreases with simulated overburden pressure, thus demonstrating both lubrication and the experimental equivalent of a volume effect also for felsic rocks. For example, at a simulated depth of 20, 200, and 750 m, the measured friction coefficient is respectively >1, 0.4 and 0.18. Figure 6.31 shows a specimen of Köfels frictionite.

Fig. 6.31 Köfels friction rock (frictionite) compared with original rock. (**a**) The original gneiss at the left, frictionite on the right. (**b**) Thin sections with cross nicols of the original rock and of the frictionite (**c**). Note how crystals are much smaller in the frictionite (Thin sections photographs (**b**) and (**c**) courtesy of Luca Medici)

The microscopic examination shows the presence of intensely comminuted rock, called gouge, which melts only partially. The discrepancy with the high values of the viscosity demonstrates a need for better theorical understanding of the transitions rock → gouge → gouge + melt. Frictionite is known also from fault zones (also called pseudotachylyte in the petrological jargon).

6.9.2 Vapor or Gas at High Pressure

6.9.2.1 Possible Transformation of Pore Water into Vapor

If frictional heat generated at the slippage plane of a large rock avalanche is sufficient to melt the rock, it is even more likely that water, if present in the pores, will be transformed into steam. From Eq. 6.68 it is found that the time needed to increase the temperature by $\langle \Delta T \rangle = 100$ K around the area of the slippage plane is less than 1 s for a 100 m thick rock avalanche, and some $\tau \approx 1-5$ s for a thinner landslide. The energy flux after this elapsed time is of the order of some tens to hundreds of kW m^{-2}.

After the transition, both temperature and pressure of the vapor increases as heat continues to be pumped in the system. Pressure P_{gas} and temperature are mutually proportional according to the equation of state of perfect gases

$$P_{gas} = k \rho_{gas} T. \tag{6.70}$$

where $k = 460$ J kg^{-1} K^{-1} and ρ_{gas} is the density of the vapor. The effective friction coefficient in the granular medium so becomes

$$\tan \phi_0 \rightarrow \tan \phi_0 \left[1 - \frac{P_{gas}}{P_N} \right]. \tag{6.71}$$

where P_N is the normal pressure.

The gas at high pressure may percolate through the permeable granular medium, obliterating the effect. Thus, only below a certain permeability will vapor build up sufficient pressure to become an effective lubricant (Goguel 1978; Habib 1975). More recent calculations (De Blasio 2007, see also Fig. 6.32) show that vapor lubrication may occur also for thin landslides (some tens of meter). The permeability should be lower than 10^{-16} m^2 in the presence of little water (<2 kg/m^3 of water) while for greater water amounts (>3 kg/m^3) the effect may take place also with permeability of the order 10^{-12} m^2. An increase in the volume will imply more heat, but also more weight to be lifted, so in principle it is not known if this kind of lubrication gives a volume effect.

6.9.2.2 Gas Generated by Frictional Heating on Carbonate Rocks

Carbonate rocks subject to intense heating will generate gas also in the absence of water. The process is due to the following reaction

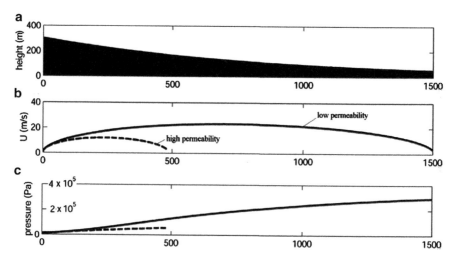

Fig. 6.32 Slope path (*A*), velocity, temperature, pressure, and gain in mobility simulated for a thin (20 m) model landslide lubricated by vapor. "Low permeability" and "high permeability" indicate permeabilities of 10^{-17} m^2 and 10^{-12} m^2 respectively (From De Blasio 2007)

$$CaCO_3 \rightarrow CaO(solid) + CO_2(gas). \tag{6.72}$$

Erismann (1979) has suggested that for landslides composed of carbonate rocks, the carbon dioxide may heat up and provide a gas lubrication effect similar to that of vapor.

General references Chapter 6: Erismann and Abele (2001); Evans and DeGraff (2002).

Chapter 7
Landslides in Peculiar Environments

Several disasters have been caused by landslides and rock avalanches falling onto natural or artificial water basins. The most studied case occurred in northern Italy on October 9th, 1963, when a volume of 270×10^6 m^3 *of limestone collapsed into the artificially dammed Vaiont lake. The dam survived the impact, but water overtopped the dam by about 200 m; the ensuing water wave took 2,000 lives. Many investigations have been devoted to the Vaiont failure, mostly related to the hydraulic and mechanical history prior to the landslide. Probably the disaster could have been avoided if the danger of landslides falling at high speed in water reservoirs and their capability of displacing the water had been recognized.*

When a failure occurs along a steep coast, the resulting landslide may start subaerial and continue traveling underwater, creating hazardous water waves upon impact.

Glaciers are another peculiar environment affected by landslides. Although landslides traveling on glaciers are relatively rare due to the limited extension of ice-covered mountain areas, it is likely that their frequency will increase in the future due to global warming, a trend that destabilizes the mountain environment. Landslides traveling on glaciers exhibit peculiar characteristics, the most significant being the enhanced mobility.

The chapter closes with a short introduction to landslides on other planets and moons of the Solar System. This is a fascinating subject with important implications for planetary sciences. The state of exploration of the Solar System is now of such a high quality, especially concerning Mars, that the subject of extraterrestrial landslides can be tackled from a much deeper level than it was possible only a couple of decades ago. Because landslide dynamics is altered by ice or water, the structure of ancient landslide deposits on Mars may indicate the presence of these media in the past of the planets, which is a very relevant question for the reconstruction of the past climate of the planet and eventually for the search of life. A physical approach to landslides is even more valuable for extraterrestrial landslides, where sites are inaccessible and landslide deposits are often conserved with astonishing detail after long time.

The photograph below shows the Vaiont dam as of Summer 2008. The front of the Monte Toc landslide fallen in the Vaiont reservoir in Italy in October 1963 is visible to the left.

7.1 Landslides Falling into Water Reservoirs

7.1.1 General Classification

The fall of a landslide into a water reservoir may displace a large amount of water. If the water reservoir is large (a vast lake, a fjord, or the open sea) waves created can travel fast as a tsunami and discharge their power in distant locations. However, also landslides falling in small water reservoirs may cause local damage. Several physical-dynamical problems arise during the process of landslide impact with water. A first plain classification of landslides falling on water reservoirs may be the following (Fig. 7.1).

1. First kind. If the mass of displaced water is much less than the landslide mass, the latter will be only weakly affected by the water. Water is totally displaced. Example: Val Pola (←Sect. 6.1.2) and Vaiont.
2. Second kind. If the amount of water is comparable to the landslide mass, the movement of the slide and the water displacement are interconnected. The landslide will be more delayed by water than in the previous case.
3. Third kind. If the mass of the landslide is much less than that of water, the movement of the landslide is strongly affected by water; water is displaced

7.1 Landslides Falling into Water Reservoirs

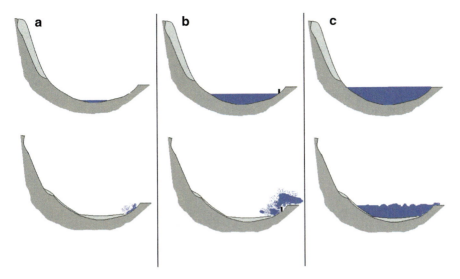

Fig. 7.1 Possible outcomes of the collapse of a landslide into a water reservoir. (**a**) Landslide mass much greater than the mass of water in the reservoir. (**b**) Landslide mass comparable to the water mass. (**c**) Landslide mass much smaller than the water mass (in (**c**) it is assumed that water extends in the direction perpendicular to the page)

locally but not permanently, and will promptly return to the equilibrium level. So doing, depending on the speed of the landslide and on the local geometry of the basin, it may generate a tsunami (→Sect. 7.2).

The first two cases are briefly addressed in the remainder of the present section. The case of large water volume is treated in the next section →Sect. 7.2, together with coastal landslides.

Following Erismann and Abele (2001) the ratio between the water mass in the reservoir and the landslide mass is considered as a critical parameter for assessing the behavior of a landslide falling onto a water reservoir

$$C = \frac{\text{Mass of water in the reservoir}}{\text{Mass of the landslide}}. \tag{7.1}$$

7.1.2 Limit $C \ll 1$ (Mass of the Landslide Much Greater Than the Water Mass)

In this limit the landslide is little affected by the comparatively small water mass (Fig. 7.1a). The water mass will be totally displaced by the landslide and the critical question becomes the velocity acquired by the water splash U_W. Following Erismann and Abele (2001) we can write

$$\frac{U_W}{U} = 1 + J \tag{7.2}$$

where U is the velocity of the landslide at the moment of impact, and J is the equivalent of a coefficient of restitution, introduced to account for energy dissipation during impact. In the absence of any dissipation, the velocity acquired by water would be twice the landslide velocity. Following Erismann and Abele (2001), this simple result can be grasped thinking at the description of the impact seen from an observer sitting on top of the landslide. She would perceive water impacting against the landslide with a velocity $-U$ and then bouncing back at the speed $+U$. From an observer at rest this would correspond to a velocity acquired equal to twice the initial velocity of the landslide. The Vaiont landslide, for which $C \approx 0.05$, also belongs to this case. From Eq. 7.2 one can also obtain an estimate for the theoretical height ΔH reached by displaced water

$$\Delta H = \frac{1}{2g} U_W^2 = \frac{U^2(1+J)^2}{2g} \qquad (7.3)$$

and with a landslide velocity 25 m/s, a height of 32 m for $J = 0$ is obtained, and nearly 128 m for $J = 1$.

Box 7.1 Brief Case Study: Monte Toc (Vaiont): A Landslide Fallen onto a Water Reservoir

Salient data
Name: Vaiont landslide (from the name of the river)
Mount involved: Monte Toc
Location: Belluno (Northern Italy)
Coordinates:46 15′ 28″ N; 12 20′ 39″ E
Volume: very Large (230 Mm³)
Year:1963
Kind: rock avalanche
Casualties: 1996
Cause: the presence of a water reservoir greatly increased the flank instability in a geological setting already susceptible to failure
Landslide material: Jurassic and Cretaceous limestone
Special characteristics: the landslide falling into the dammed water reservoir generated a huge wave that overtopped the dam with a 200 m water front. It is one of the best studied landslides in the world. The time history is a source of unique information of creep anticipating the failure. The Vaiont disaster represents a major catastrophe in the twentieth century.

In early October 1963, technicians of the Vaiont dam noticed that the southern flank of the mount called Toc was sliding onto the artificial lake with a speed of 4 cm/h. Fearing a major landslide, they had been measuring the creep of the mountain flank for about 4 years. According to widespread views, the landslide should have been slow, somehow controllable. Still, a slow landslide would have had two major negative impacts. It would have created a dangerous lake fed by the waters of the Vaiont river, and made one of the largest hydroelectric plants unusable. However, few people had predicted that things would have turned much worse.

(continued)

Box 7.1 (continued)

Figure 7.2 shows the measurements of the reservoir level and of the measured landslide speed as a function of time. The level of the reservoir was raised during the whole year 1960. In November a landslide of 700,000 m^3 plunged into the artificial lake, questioning the stability of the area. People began to think more seriously of the possibility of a much larger landslide. However, the dominant idea was that the landslide would have been slow, much like the flow of a glacier. The concern was that the body of a large landslide would have dammed the river. To circumvent this risk, a by-pass tunnel was built to drain the lake that would have resulted from the damming. In the meantime, a series of corings demonstrated that the zone was intensely fractured and that clay beds were present between the Upper Jurassic limestone and the Cretaceous, destined to become the weak layer along which the landslide took place. An old landslide was also found (Semenza 2001). The observations gathered during the period of tunnel construction are interesting in demonstrating that the creep rate of the slide

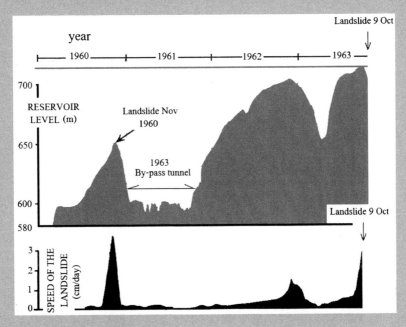

Fig. 7.2 The water level in the artificial Vaiont reservoir and the corresponding measured speeds between 1960 and 9 October 1963 (From Müller 1964, modified and greatly simplified)

(continued)

Box 7.1 (continued)

diminished dramatically (Fig. 7.2). The creep rate started to increase enormously (up to 1 cm/day) when, after completion of the bypass tunnel, the water level was increased to a level of 700 m toward the end of 1962. When, in the first months of 1963, the water level was lowered to 650 m, the creep movement decreased by one order of magnitude, which strengthened the impression that the creeping was controllable. The level was then raised again to the operative value of 710 m. The creep increased enormously reaching values of some cm/day. The huge mass involved also became evident. The decision to lower the level was taken late and was also delayed by the intense rains of those critical days.

Despite the immediate order to empty the basin, the creeping motion suddenly accelerated irreversibly, and turned into a sliding regime more similar to a rock avalanche than to creep. Velocities reaching up to 90 km/h displaced a large amount of water that overtopped the dam, and plunging onto the valley below wiped out the town of Longarone and other smaller settlements claiming about 2,000 lives. The water wave, rising along the basin, also produced some damage. The plunge of the landslide was

Fig. 7.3 The eastern landslide scar

(continued)

Box 7.1 (continued)

Fig. 7.4 a) The landslide scar seen from Erto. b) Casso was spared by the water wave

(continued)

Box 7.1 (continued)

Fig. 7.5 Mosaic of the Vaiont landslide scar

(continued)

Box 7.1 (continued)

dramatically recorded by the seismic recordings at the Pieve di Cadore seismic station. The duration of the event was so established between 20 and 25 s, corresponding to an average speed of 25 m/s.

Both the geological history of the valley, by that time studied in detail, and the 4-years experience matured during the construction of the dam, including the precise measurement of creeping flow, left little doubt as to the unstable nature of the terrain. The high speed and consequent water wave were not predicted, even though the works by Heim (←Chap. 6) indicated the high speeds reachable by rock avalanches. Figures 7.3 and 7.4 show more details from the landslide site, and Fig. 7.5 a photomosaic of the scar.

The transition between creep and final detachment of the landslide has been the subject of several studies. The transition was perhaps the result of high pore pressure in the clay layers, whereas other specialists have regarded the thermal effects due to friction as an important component (see for example Voight and Faust 1982; Vardoulakis 2002).

Other references: Semenza (2001); Müller (1964); Hendron and Patton (1985); Broili (1967); Genevois and Ghirotti (2005).

Box 7.2 An Experimental Model of the Vaiont Landslide

In 1959, 4 years prior to the Vaiont disaster, a large landslide falling into the nearby reservoir of Pontesei had provoked 20-m high waves and killed a man. Another landslide, of volume around 30 Mm^3, had slid into the Vaiont reservoir, demonstrating the danger represented by fast landslides falling into reservoirs. Fearing a new and larger landslide, in 1961 the engineers at Padua University were entrusted with the task of building an accurate model of the Vaiont basin to simulate the landslide (or more landslides falling at the same time) and their effects on the water in the reservoir. The model was built to a scale 1:200.

To simulate the landslide with the right velocity, the Froude number (←3.7.1)

$$Fr = \frac{U}{\sqrt{gH}} \quad (7.4)$$

should be the same in the model and in the reservoir scale. Hence

$$\frac{U(\text{model})}{U(\text{Vaiont})} = \sqrt{\frac{H(\text{model})}{H(\text{Vaiont})}}. \quad (7.5)$$

(continued)

> **Box 7.2** (continued)
>
> resulting in a speed 14 times smaller in the model. A time fall of 60 s was considered as the most pessimistic limit, which implies a model landslide velocity of 2 m/s. To give the right scaled speed, the landslide was simulated using heaps of granular material contained in nets, and pulled by ropes. Tractors were used to obtain the desired speed. The height of the wave was relatively small. However, the speed reached by the real landslide turned out to be at least a factor two greater, which properly rescaled would have given a correct height of the wave (Datei 2005).

7.1.3 Mass of the Landslide Comparable to the Water Mass, $C \approx 1$

In this case (Fig. 7.1b) the momentum carried by the water is not negligible, and affects the dynamics of the landslide. An analysis identical to the inelastic impact of two spheres (Eq. 5.32, ←Sect. 5.3.2) leads to the following result by identifying with the index "1" the landslide and with "2" the water. Putting $U_{12} \equiv U$(velocity of the landslide prior to impact); $U_2 = 0$ (water initially at rest); $U_2^* = U_W$ (velocity acquired by water), and $\varepsilon_\perp \equiv J$, Eq. 5.32 becomes

$$\frac{U^*}{U} = 1 - \frac{C}{C+1}(1+J)$$
$$\frac{U_W}{U} = \frac{1}{C+1}(1+J). \tag{7.6}$$

Velocities are thus reduced compared to the previous case $C \ll 1$, but still of the order of the landslide velocity. The theoretical height reached by water upon impact with the landslide is so

$$\Delta H = \frac{1}{2g}U_W^2 = \frac{U_W}{U} = \frac{U^2(1+J)^2}{2g(C+1)^2}. \tag{7.7}$$

A relevant question concerns the impact dynamics if the landslide material is porous. Water will partially penetrate into a porous landslide, reducing the effect of the splash. A simple calculation shows that the velocity necessary to obtain a certain displacement when the material is porous is given as

$$U_{\text{POROUS}} \approx \frac{U_W}{\left(1 + \frac{v_F}{U}\right)} \tag{7.8}$$

where U is the velocity in the case of nonporous material and v_F is the velocity of the water in the porous material (Datei 2005). Assuming the front pressure of the order of the stagnation pressure ($\rightarrow$Sect. 9.6), it follows that

$$\frac{v_F}{U} \approx \frac{1}{2}\rho_F U \frac{K}{H\mu} \qquad (7.9)$$

where K is the permeability, μ is water viscosity, and H is a scale length for pressure difference. Using a value $K \approx 10^{-7}$ m^2 corresponding to a pervious material like gravel and a length H of 10 m, a ratio $v_F/U \approx 0.1$ is obtained with a velocity of 30 m/s which would indicate a limited role of the porosity.

7.2 Coastal Landslides and Landslides Falling onto Large Water Basins, $C \gg 1$

7.2.1 General Considerations

The last case to consider is that in which a landslide falls into a much greater water basin such as a large lake, a fjord, or the open sea. Based on the dynamics of the phenomenon, there may be different geometrical types.

1. Landslides may originate mostly subaerially and plunge into the water (Fig. 7.6). The impact with water and the dynamics at the air–water interface are significant in the subsequent movement of the slide and of the water. The impact with water may also produce a tsunami. We can recognize two subtypes.
   (a) The landslide acquires a limited velocity and remains in the deep basin (Fig. 7.6a).
   (b) A second subtype is typified by the event that occurred in Lituya bay in 1958, where a rock avalanche plummeting into an Alaskan fjord produced a tsunami several tens of meters high (Fig. 7.6b). In this geometry, the landslide front invaded the opposite side of the fjord, causing a run-up along the opposite side of the valley of over 500 m.
2. There are also examples of mixed subaerial–subaqueous landslides without frontal impact with water. This may occur either because the landslide starts half submerged and half subaerial (Fig. 7.6c) or because it starts underwater, and then propagates retrogressively over the sea level (Fig. 7.6d). These landslides may cause significant tsunamis, as demonstrated by the Stromboli landslide of 2002 ($\rightarrow$Sect. 7.2.3).
3. Some landslides develop from the cliff failure onto the tidal flat (Fig. 7.6e). The collapses of chalk cliffs along shallow seas are typical of southern England, Denmark, and northern France. Because the sea is very shallow, the landslide material frequently remains partly submerged at the base, with the upper cap exposed above sea level.

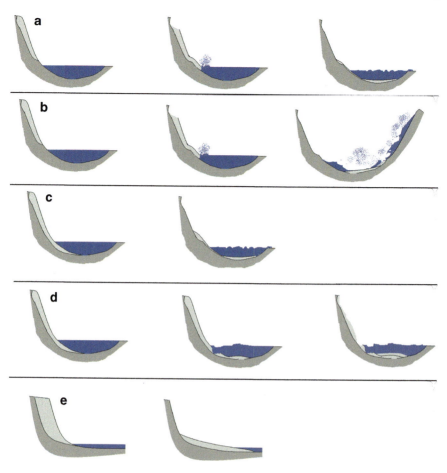

Fig. 7.6 Different possibilities for landslides falling in large water basins. See text for an explanation

The large rock avalanches of midocean islands such as the Canary islands and the Hawaii are also partly subaerial and partly subaqueous, but because the submarine fraction is dominant, these will be considered in the next chapter devoted to submarine landslides.

Table 7.1 shows some examples of landslides in large water reservoirs. Water waves formed at the surface of the water basin can travel as tsunamis to long distances. When reaching the shore, waves may markedly increase in height, and inundate inhabited areas. Due to its high and steep mountains enclosing long fjords, Norway has a remarkable history of tsunami waves generated in the fjords. For example, in 1756 a very large landslide (15 Mm^3) occurred in Tjelle, claiming the life of 32 people. In 1905 a smaller rockfall from the mountain called Ramnefjell (50,000 m^3) hit on loose sediment below, mobilizing a total of 300,000 m^3 into the

7.2 Coastal Landslides and Landslides Falling onto Large Water Basins

Table 7.1 Some examples of mixed subaerial–subaqueous landslides

Name	Year	Kind of water basin	Slide volume (Mm^3)	Max. wave height	Casualties	Run-up (m)	Type
Loen	1905	Lake	0.05–0.3		61	40	1A
Loen	1936	Lake	1		73	74	1A
Tafjord	1934	Fjord	1–1.5		41		1A
Stranda; Skafjell	1731		>0.1		17		1A
Årdalsvatnet, Stegane	1948	Lake	0.030	5	0		1A
Årdalsfjorden; Kleppura	1983	Fjord	0.150	7	0		1A
Sørefjorden; Katlenova	1998	Fjord	0.020–0.030	6	0		1A
Lituya Bay (Alaska)	1958	Fjord	30	170		520	1B
Pontesei	1959	Artificial lake	5		0	17–20	1A
Vaiont (small)	1960	Artificial lake	0.7	2	0	10	1A
Paatuut (Greenland)	2000		30			50	
Scilla	1783	Open sea			1,500	16	1A
Stromboli	2002	Open sea				10	2
Finneidfjord	1996	Fjord	0	4	0		2

Note: Type 1 and 2 listed at the beginning of Sect. 7.2.1

lake Loen (Fig. 7.7). The consequent tsunami reached 40 m high waves, killing 61 people. Also the coastlines and cliffs along the open sea are often subject to failures, sometimes catastrophic.

7.2.2 Lituya Bay

On July 9, 1958, a strong 7.5–8 quake along the Fairweather Fault shook the region of the Glacial Bay National Park. The deep fjord in Alaska (USA) terminating in a bay known as Lituya Bay was only 20 km away from the epicenter. The ensuing landslide created one of the highest water splashes ever recorded. The rockslide of 60 Mm^3, which originated entirely above the sea level, plunged into the fjord, displacing water to the height of 150 m. The area of origination of the slide has been highlighted in Fig. 7.8, with a white circle. The tsunami wave, running along the fjord, swept away the soil and forest from the zones adjacent to the sea (arrow in Fig 7.8). However, this water height was even modest compared to the inordinate run-up of 524 m over the ridge in front of the landslide path, on the opposite side of the fjord. Some boats were swept against the cliffs; the crew of one was killed. Along the bay, scientists noticed a series of parallel trimlines indicative of levels where the forest had been wiped out in earlier times and grown again. Trimlines demonstrated that these occurrences are not rare in the Lituya Bay. With tree rings

Fig. 7.7 The mountain Ramnefjell in Norway. The 1905 rockfall detached from the middle part of the scar visible in the photo: that of 1936 was more extensive and involved part of the upper scar

counting it was possible to date a couple of events: the answers were that major catastrophes had happened in 1936 and 1874. The trimlines corresponding to the older run-ups are higher than that of the 1958 event, showing that the latter was not the largest in recent times.

Because the landslide entered the basin at very high speed, only the water in the front of the landslide was pushed by the front of the landslide into land.

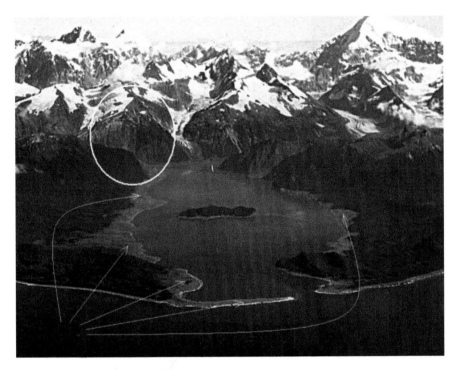

Fig. 7.8 The Lituya Bay landslide. The landslide originated from the *encircled area*. The maximum run-up of over 500 m was observed on the opposite side. The *arrows* show the area inundated by the tsunami (Photograph from the USGS site of public domain)

As a consequence, only some of the water of the basin was mobilized by the landslide; a total amount of something like 4 Mm^3 was in fact involved in the splash. Thus, although the water basin had virtually infinite volume, Lituya Bay represents a case with $C \ll 1$.

It is possible to estimate the run-up of the Lituya Bay up splash by a little dynamical reasoning on the geometry (Fig. 7.9). In an approximated calculation, we can estimate the velocity with the simple formula ($\leftarrow$Chap. 6): $U = \sqrt{2g}\sqrt{H_0 - \mu x_0}$ where the maximum vertical displacement of the center of mass $H_0 = 686$ m, accounting for the fact that the landslide center of mass in the final position is partly submerged, and x_0 is the horizontal displacement of the center of mass. With a friction coefficient μ of 0.2, the maximum velocity is found in the order 105 m/s, in the hypothesis that the friction with the bottom of the fjord is negligible. The fact that the bottom friction of landslides in water is reduced derives from numerous observations of submarine landslides ($\rightarrow$Chap. 9). Assuming that only the water in front of the landslide is displaced, from momentum conservation the final velocity of water is $U_W = (\rho/\rho_W)(L/L_{BAS})U$ where $L = 970$ m; $L_{BAS} = 1,342$ m are the lengths of the landslide and of the basin, respectively. This gives a velocity of water of about 152 m/s. So the height reached by the splash is

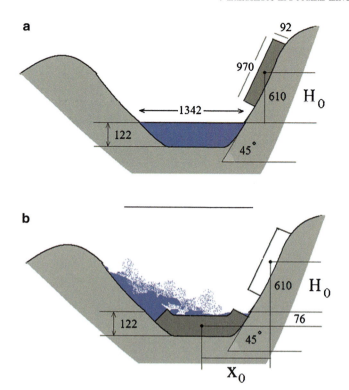

Fig. 7.9 Simple model for the Lituya Bay run-up

$$\Delta H = \frac{U_W^2}{2g} = \left(\frac{\rho}{\rho_W}\right)^2 \left(\frac{L}{L_{BAS}}\right)^2 (H_0 - \mu x_0) \qquad (7.10)$$

that with the parameters used gives more than 1,000 m. If the friction in the fjord is assumed equal to that on land (0.2), a more moderate height for the splash is obtained: 400 m. It is clear that this calculation is approximated and answers are uncertain, but it has a value in showing that it is possible to explain the run-up heights observed with the Lituya Bay geometry. The key role is played by the high velocity prior to impact into the fjord. Because the landslide traveled very fast, the water mass thrust by the landslide front was incapable of escaping from the sides.

7.2.3 Landslides Propagating Retrogressively from the Sea to Land

The landslide of 30th of December 2002 of the Sciara del Fuoco (Stromboli, Italy) started as a 20 Mm³ subaqueous mass movement. A second landslide, between two and five times smaller, followed a few minutes later from a niche about 500 m above sea level, apparently in retrogressive continuation to the first one. Both landslides produced a tsunami that caused severe damage in Stromboli, and traveled to the nearby isle of Panarea (Tinti et al. 2003, 2005; Tommasi et al. 2006).

Another interesting example of this kind of slide is provided by the landslide that occurred in the southern tip of Calabria (southern Italy) in 1783 in the neighbourhood of Scilla, a small town in Calabria facing Sicily (Fig. 7.10) (Mazzanti 2008). The landslide was set in motion by a strong earthquake belonging to a seismic swarm. Because of the numerous earthquakes, people had temporarily abandoned their home to overnight along the beach of Marina Grande. A little after midnight between 5th and 6th February 1783, another quake with estimated magnitude of 5.8 caused the collapse of a part of the Mount Paci into the sea. The landslide generated a 16-m high tsunami that devastated the near coasts, and killed more than 1,500 people.

Simulations with the slab model for submarines landslides ($\rightarrow$Sect. 9.5.1) gives the results shown in Fig. 7.10c–d. The friction coefficients that reproduce the observed runout are of the order 0.1, which points to a problem addressed in detail in Chap. 9, namely the low apparent friction coefficient for subaqueous slides. Top velocities calculated with this simple model are over 40 m/s, which is consistent with the devastating character of the tsunami developed by the Scilla landslide.

7.2.4 Landslides Falling on a Tidal Flat

Nearly vertical cliffs of clastic limestone called "chalk" of late Cretaceous–early Tertiary age are relatively common along the shores of Northern Europe (Fig. 7.11). They often collapse producing small tongue-shaped landslides. In addition to the usual causes of instability, cliffs are exposed to the additional battering of the sea that is effective in weakening at the base. Table 7.2 reports some data from Hutchinson (2002). Note that the FB is relatively low for landslides of this small volume. Because these landslides are geometrically of the "Frank" type (i.e., they just collapse onto a flat area), a role in their mobility may be played by the lateral earth pressure force (see also Lucas and Mangeney 2007). Probably this kind of material behaves partly as frictional, and partly as cohesive when wetted. The FB is certainly lower than that found in experimental collapse of granular media of similar geometry (Lejeunesse et al. 2006).

Perhaps the greater mobility of cliff failures compared to granular experiments can be explained by the lubricating effect of water along the shore. Water can dramatically affect mobility in at least two ways. First, the soft rock comminuted by the rapid collapse may create a rocky flour composed of very small grains, probably in the range size of silt and clay. Underneath the landslide, this fine material will form a cushion of impermeable and cohesive water-rich material. This slurry may be very mobile and explain the long runouts and the low apparent friction. In addition, if the front of the landslide moves at sufficient speed, it may advantage from hydroplaning ($\rightarrow$Chap. 9). Note that the terrestrial Frank landslide, also falling onto a flat area, had a much higher FB, probably owing to the lack of

Fig. 7.10 (**a**) The scar and slope path of the Scilla landslide seen from the top. The *arrow* indicates the direction of flow. (Original photograph courtesy of Paolo Mazzanti.) (**b**) An engraving of the Dominican monk Minasi reproducing the event. (Photo courtesy of Paolo Mazzanti.)

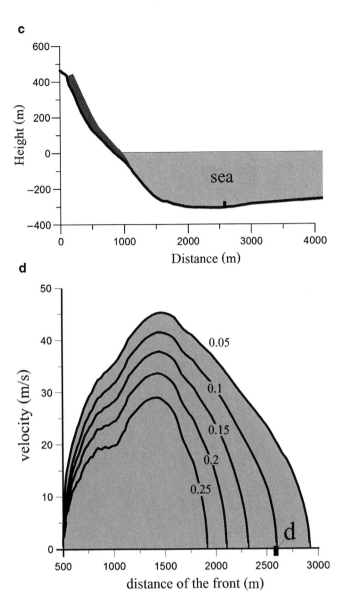

Fig. 7.10 (continued) (**c**) The slope profile of the Scilla landslide. The hypothetical profile of the landslide before the event is shown in *dark grey*. The *black square* denotes the (partly uncertain) position of the front deposit. (**d**) Calculation of the front velocity with the slab model (→Sect. 9.5)

water as lubricating medium. Hutchinson (2002) in his comprehensive article does not consider water as potential lubricant, but notices that the more porous chalk is correlated to higher mobility.

A clue as to the importance of water in the mobility of coastal landslides comes from the Bulganuc landslide, a small failure of a steep cliff that occurred

Fig. 7.11 Møns Klint (Denmark) has a long record of vertical cliff collapses

in March 1997 along the coastline of Maine (USA). Several outrunner block traveled a long distance offshore, with a minimum FB of 0.22 (Fig. 7.12). The evidence for water lubrication here is compelling. In contrast to the classical hydroplaning model (→Chap. 9), requiring a high pressure at the front of a

Table 7.2 Example of cliff failures on a flat tidal flat ("kind 3" in the present denomination)

Localities	Event (Hutchinson 2002)	L (m)	H (m)	H/L	Material	Year
Folkestone Warren (Kent)	E2	628	150	0.238	Chalk	1915
Kent	E7	442	145	0.328	Chalk	1988
Saint Margret's Bay (Kent)	E17	405	68	0.168	Chalk	1970
Northwest France	F2	320	66	0.206	Chalk	
Bunganuc (Maine, USA)	–	13	50	0.22	Glacial clay, sand and rubble	1997

Note: See Hutchinson (2002) for the first four slides

landslide, here the blocks probably traveled on very shallow water. Thus Froude criteria invoked to explain the onset of hydroplaning in fully submerged landslides are probably invalid in this case.

7.2.5 Generation and Propagation of the Tsunami in Lakes and Fjords

The subject of landslide-generated tsunamis and the traveling of tsunami waves in the sea is shortly treated in (→Chap. 9). Here, we limit ourselves in anticipating a few basic facts concerning lakes and fjords.

Tsunamis generated by landslides in fjords are characterized by early generation of very high waves in the zone of landslide impact. Wave amplitude promptly damps out, as they propagate away from the impact region with a velocity given by the formula $c = \sqrt{gH(1 + h/H)}$ where h and H are the wave height above unperturbed level and water depth, respectively. Because usually $h \ll H$, we can also write $c \approx \sqrt{gH}$.

This approximation is valid for wavelengths much greater than the water depth, typically with large landslides characteristic of the deep ocean. Smaller landslides in the sea and fjords (that are normally very deep owing to extreme glacial erosion) will generate waves with a more complex dispersion relation (→Box 9.5) and the associated phase velocity will be $C = \sqrt{\frac{g\lambda}{2\pi} \tan h \frac{2\pi H}{\lambda}}$. Thus, the velocity of tsunami waves increases with the wavelength, a phenomenon called dispersion.

Other possible effects in closed basins and fjords are reflections with the shore, interference, and standing wave oscillations that may lead to local strong variations of the run-ups and persistence of disturbed water level for several hours after the event (Harbitz et al. 1993).

To assess the hazard related to landslides in water basins, experiments simulating the generation of tsunami with a solid block falling on a small experimental water basin have been attempted. By numerous tests obtained varying several parameters, Vischer and Hager (1998) found that the height of the resulting waves G scales like

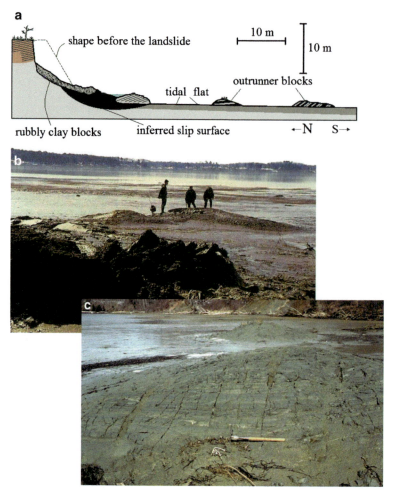

Fig. 7.12 An example of small-scale failure of a steep cliff with formation of outrunner blocks: the Bunganuc landslide of March 1997 along the coastline of Maine (USA). (**a**) general setting. (**b**) the distal outrunner blocks. (**c**) one of the blocks reveals cracks perpendicular to the flow direction, probably due to the tension and deformation during traveling. Cracks are filled with mud injected from the bottom of the sea (the infilling is clearly visible in the figure) (Scheme and photographs by H.N. Berry reproduced from the web page of the Maine Geological Survey, www.maine.gov; images of public domain (scheme in (**a**) modified and simplified from the original))

$$\frac{G}{H} = aD^b \qquad (7.11)$$

where H is the depth of unperturbed water in the basin, $D = V/WH^2$ is the dimensionless volume of the falling block (displacement number according to Vischer), V is the volume of the block, W is the block width, while a, b, are fitting constants. It is remarkable that G is roughly independent of the Froude number of the block, and thus on block velocity. The maximum run-up R above quiescent water level obtained by numerous tests on an artificial flume is reasonably well fitted by the following law

$$\frac{R}{H} = k(\beta)\left(\frac{G}{H}\right)^{1/3} \qquad (7.12)$$

where $k(\beta)$ is again a constant to be found experimentally, and which depends on the dipping angle β of the simulated shore hit by the waves. For the tested limits, Vischer and Hager (1998) has found that the wave height H depends strongly on the volume of the falling model landslide, increasing approximately with the square root of the volume, while the mass velocity is less important. For more recent experiments and theory (see e.g. Watts and Grilli 2003).

7.3 Landslides Traveling on Glaciers

7.3.1 General Considerations

We have already mentioned the devastating Nevados Huascaran landslide (←Chap. 1) that started on an inaccessible suspended glacier. The travel on a glacier seems to be the key to understand the high velocities reached and the enormous devastation. The Sherman landslide is another famous example of landslide fallen on a glacier. Traveling completely on the flat surface of the Sherman glacier in an uninhabited area, it did not cause any damage.

One of the effects of global warming is the retreat of glaciers and the reduction of the permafrost layer in the mountain environment. The rocky layers adjacent to the glacier basins may undergo reduction of buttressing and become gravitationally unstable. Considering that earthquakes are fairly common along some of the highest and steepest mountain chains (initial rock collapse in both Nevados Huascaran and Sherman was caused by a strong earthquake) it must be concluded that the threat of landslides traveling on glaciers might have been underestimated. Figure 7.13 shows a recent landslide on a glacier: the Thurwieser rock avalanche in the Italian central Alps. Table 7.3 shows a partial list of some landslides fallen on glaciers. Cases have been presented among the others by Barla et al. (2000), Bottino et al. (2002), Evans et al. (1989), Evans and Clague (1988), Fahnestock (1978), Hewitt (2009), Sosio et al. (2008).

There are also examples of ancient landslides fallen onto glaciers. In northern Italy, the "Marocche" are characteristic deposits of landslides that, after being deposited on top of a glacier now extinct, were displaced by the movement of the glacial ice. Figure 7.14 shows another example of glacially lubricated landslide in Norway.

Fig. 7.13 The Thurwieser rock avalanche (Photograph courtesy of Giovanni B. Crosta)

The extraordinary mobility for a landslide of this limited volume (FB $= 0.27$ for $V = 5$ Mm^3) was probably due to the lubrication by ice. The dynamics of landslides traveling on glaciers is not well understood. In this section, some simple considerations and estimates are briefly discussed.

7.3 Landslides Traveling on Glaciers

Table 7.3 A short list of landslides traveling on glacier

Name of the landslide	Year	Volume (millions of cubic meters)	Locality	Measured ratio H/L In parenthesis the value predicted for a nonglacial slide based on the volume (Scheidegger 1973)	Reference
Sherman	1964	10	Alaska, U.S.A.	0.11 (0.332)	Shreve (1966); McSaveney (1978)
Pandemonium creek	1959	5 (only partly on glacier)	Canada	0.23 (0.371)	Evans et al. (1989)
Nevados Huascaran	1970	50 (only partly on glacier)	Peru	0.24 (0.256)	Plafker and Ericksen (1978)
Mount Rainier	1963	11	Washington, U.S.A.	0.27 (0.326)	Fahnestock (1978)
Mount Cook	1991	12	New Zealand	0.40 (0.322)	McSaveney (2002)
Brenva	1997	2	Monte Bianco (Italy)	0.42 (0.429)	Barla et al. (2000)
Triolet	1717	18	Val D'Aosta (Italy)	0.26 (0.302)	Bottino et al. (2002)
Thurwieser	2004	2.5	Italian central Alps	0.49	Sosio et al. (2008)

The table shows: the ratio FB $= H/L$ between the fall height and the runout length. The Sherman slide collapsed from Shattered Peak in Alaska following an earthquake, and traveled several kilometers entirely on a wide glacier (Shreve 1966; McSaveney 1978)

7.3.2 Dynamics of Landslides Traveling on Glaciers

7.3.2.1 Smooth Surfaces on Ice

The dynamics of a rock avalanche traveling on a glacier is poorly known. A simple observation is that skis, curling rocks, and sledges exhibit a low dynamic coefficient of friction. Values for different materials (e.g., ebonite or brass on ice, or ice on ice) measured at some m/s lie as low 0.02 in proximity of the ice melting point and increase by a factor of five when temperatures are decreased to $-80°C$ (see e.g. Weast 1989). The static friction coefficients are typically much greater than the dynamic ones, a result which has no equivalent in rocks.

Some measurements of friction on ice at low pressure also show that the friction between a solid body and ice does depend on the velocity, which is at variance with friction of rock against rock (←Chap. 2). Bowden (1953) has studied the behavior of aluminum against ice at a temperature $T = -10°C$ and different sliding velocities. He found a friction coefficient falling from 0.38 at very low speed to 0.04 at a relative speed of 5 m/s. Also rock against ice exhibits a similar behavior (Persson 2000).

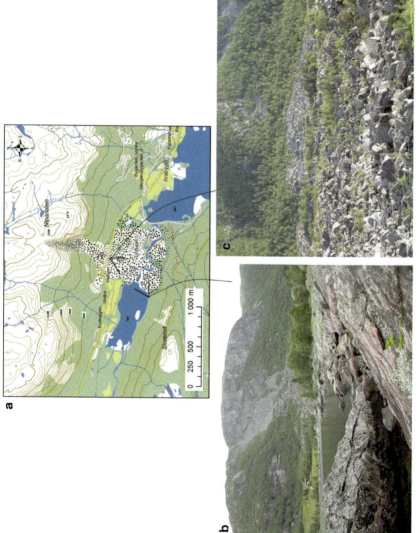

Fig. 7.14 Urdbø in Telemark, Norway is a relatively small rock avalanche (about 5 millions cubic meters) with a long runout. The front ends in a run-up at the opposite side of the valley. It is believed that the slide fell originally on the valley glacier. (**a**) General map. The landslide deposit is shown as the *dotted area*. (**b**) View of the detachment niche from the distal part. (**c**) View of the middle part with huge boulders, some "as large as a house" as reported by a tourist guide

7.3 Landslides Traveling on Glaciers

These experiments recreate the situation at the base of skates sliding on ice and skis on snow. In all these systems, lubrication is provided by a thin water film created by frictional heating (Colbeck 1995). A significant velocity dependence of the friction parameters can be explained in terms of heat diffusion. Penner (2001) finds that the friction coefficient in curling rocks behaves as $\mu_{lub} \propto U^{-1/2}$ where U is the sliding velocity. Because a greater mass dissipates more energy creating a thicker lubricating layer, the friction coefficient decreases with the load. This is also confirmed by the simple observation that sliding on skis is more effective with a heavier skier. We call this coefficient of friction $\mu_{lub}(M; U) \ll 1$, keeping in mind that the coefficient may depend on the mass and the velocity.

7.3.2.2 Granular Medium on Ice

A landslide is akin to a chaotic granular flow rather than a polite metal plate or the surface of smooth skis. The boulders at the base of the landslide behave like hard indenters into the soft icy surface. The situation is similar to the one of two metals in sliding contact, of which one is much harder than the other (Rabinowicz 1995). This suggests a contribution to friction due to plowing action of hard rock on ice. The friction coefficient for plowing is found to depend on the geometrical characteristics of the indenters, and in particular their average angle of plowing ϑ with the physical characteristics of rock or the size of the indenters playing only a minor role (Rabinowicz 1995).

$$\mu_{plowing} = \frac{\tan \vartheta}{\pi}. \tag{7.13}$$

For sandpaper, $\tan \vartheta \approx 0.2$ (Rabinowicz 1995); probably lower values should be used for rocks. The friction coefficient becomes so tentatively $\mu_{TOT} = \mu_{lub}(M; U) + \mu_{plowing} \approx \mu_{plowing} \approx 0.06$ or less. Such values of the friction coefficient are still lower than the ratio H/L exhibited by the Sherman landslide, but not completely at variance considering the crudeness of these estimates.

A contribution to friction will likely result from ice melting. Initially, ice is probably produced as a gouge of small particles, but the frictional and fragmentation energy is sufficient to melt a fraction of the gouge. An upper limit of the amount of water produced by frictional heat can be estimated equating the potential energy MgH where M is the landslide mass and g is as usual the gravity field to the energy necessary for melting a mass m of ice $\lambda m + c\Delta T$, where $\lambda = 3.3 \times 10^5$ J kg^{-1} is the latent heat, $c = 2.06 \times 10^3$ J kg^{-1} K^{-1} is the specific heat, and $-\Delta T$ is the initial temperature of ice (ΔT is positive). The thickness of water δ_W is thus found as

$$\delta_W \approx \frac{\rho}{\rho_W} \frac{g}{\lambda + c\Delta T} HD \tag{7.14}$$

where D is the landslide thickness. For Sherman ($\rho/\rho_W = 2.5$; $H = 500$ m; $D = 15$ m; $\Delta T = 10°$C) one finds about half meter of water. Water will probably mix with fine particles created by the partial disintegration of rock, forming a non-Newtonian fluid. The total resistance encountered by the landslide on the glacier will thus likely result from plowing and from the cohesion-viscosity of this fluid.

7.3.2.3 Origin of the Striations

Striations have been documented in many case studies of subaerial landslides (Dufresne and Davies 2009) and especially for the landslides falling on glaciers. They have been studied with particular detail for the Sherman landslide in Alaska (→Box 7.3). Probably the most important condition for their formation is the presence of a strong lubrication and a soft base. These allow the portions of the landslide to move driven by inertia, following the local gradient. Due to the earth pressure force, the rocky material has a tendency to spread on a broader area as it descends along the glacier. Thus, the landslide is torn apart along a series of lines. In contrast, a strong friction at the bottom reduces slippage. The effect of tearing of the landslide material transverse to the direction of motion is possible if the earth pressure force at the sides is greater than the resistance at the bottom, namely $\sigma_{zz} \tan \phi \ll \sigma_{xx}$. Considering that $\sigma_{xx} = k_A \sigma_{zz}$, where k_A is the earth pressure coefficient in the active state, it follows $\tan \phi \ll k_A$, which implies a low value of friction coefficient in comparison to the earth pressure coefficient.

It is also important that the base be soft in addition to slippery, else when meeting irregular terrain, the slide will fail along shear zones, perpendicular to the terrain. As a consequence, striations will be significantly disturbed in the long run. In the presence of a soft terrain, the landslide will cut into the ground, damping out vertical movements and shearing.

Figure 7.15 summarizes the conceptual model for the formation of striations in glacially lubricated landslides.

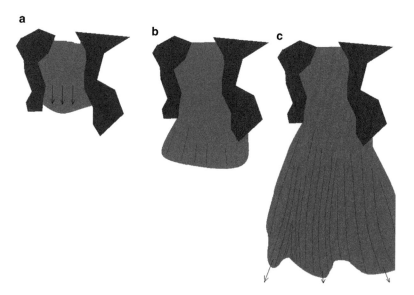

Fig. 7.15 Hypothetical mechanism for the formation of striations in a glacier-lubricated landslide like Sherman. (**a**) When the landslide is confined in a valley, there is no widening of the landslide body. (**b**) As the landslide travels past the natural confining walls, it partially spreads laterally responding to the lateral earth pressure force at the sides. The surface of the landslide, reduced to a fragmented granular medium, is torn apart along subparallel lines that subsequently develop into the striations. (**c**) As the landslide widens more, the old striations do not heal; new striations form to accommodate the space available. In the entire process, the presence of a well-lubricated and soft base is essential

7.3 Landslides Traveling on Glaciers

Box 7.3 Brief Case Study: Sherman and Glacier-Lubricated Landslides

Salient data
Name: Sherman landslide
Mount involved: Shattered Peak
Location: Alaska
Coordinates: 60 32° N; 145 06° W
Volume: Very large (10 Mm^3 according to McSaveney (1978); 30 Mm^3 according to Shreve (1966))
Runout: 6 km
Year: 1964
Kind: rock avalanche
Casualties: none
Cause: earthquake
Landslide material: greywacke and siltstone
Fahrböschung: 0.11
Special dynamic characteristics: it ran on a glacier and traveled lubricated, reaching high speed and long distance. It presents on its surface a series of enigmatic striations parallel to flow direction.

Southern Alaska is a seismic region. A particularly strong quake occurred in 1964. It triggered no less than 80 landslides, four of which fell onto the Sherman glacier. Figure 7.16 shows the deposit of the largest Sherman landslide. The crushed rock has initially traveled from the Shattered peak with direction N, partly constrained by the valley floors. After about 2 km of constrained flow, the slide has found no lateral impediment and opened widely in a fan increasing its width by about a factor 3. Part of the material has bent slightly the trajectory toward W and finally deposited toward NW, apparently continuing by inertia the substraight direction acquired in the initial acceleration phase. However, the largest part of the rock follows the steepest gradient, and curved toward W along the glacier's main axis. On the glacier, the average thickness of the deposit is 3–6 m.

Clasts, very variable in size from silt to huge blocks, often present the jigsaw-puzzle structures. Lichens on the surface of the largest blocks indicate the external wall of the peak as the source. Siltstone, a weaker rock than greywacke, tends to form long stripes made up of smaller clasts than the greywacke. The two different rock types do not appear to have mixed up thoroughly during flow.

The surface of the slide presents numerous striations longitudinal to the main flow (Shreve 1966) consisting of "V" shaped grooves about 8 m across and deep at most 2.5 m. The two walls of a groove often exhibit different rock composition. It has been suggested that striations represent the lines of divergence of the landslide material that widens during its motion along the glacier, or that they are the product of blocks (Shreve 1966).

(continued)

Box 7.3 (continued)

Fig. 7.16 The Sherman landslide (Photo by A. Post, August 25, 1965. Image USGS of public domain)

The movement of the slide material clearly resulted from a combination of inertial and gravitationally driven flow. Although this is obviously true for any landslide, frequently the protuberances of the valley floors together with the strong friction tend to redistribute the deposit masking the details of flow directions. In the case of Sherman, however, the trajectory of each single portion is made visible by the longitudinal striations.

(continued)

Box 7.3 (continued)

The fact that the underlying topography is reflected on the surface of the deposit makes Shreve (1966) to conclude: "Thus, the low "coefficient of friction" is due, not to low internal friction, but to low sliding friction".
Main references: Shreve (1966); McSaveney (1978).

7.4 Landslides in the Solar System

In the last few decades, the exploration of the Solar System with spacecrafts has revolutionized the field of planetary sciences. Among the most interesting discoveries are huge landslides in all the terrestrial planets (Mercury, Venus, and Mars) and on some satellites of the gaseous planets. Most of these planetary bodies lack recent tectonic activity and are scarcely affected by erosion processes. On Mars, water, wind, and ice erosion occur at a much slower pace than on Earth. Impact cratering plays a role in most planets and satellites, but except for the very large craters that are usually very old, impact cratering is incapable of erasing the features on the planetary bodies.

As a consequence, landslides deposits and scars represent faithful photographs from the past of the planets. They can help us greatly understand the conditions and environment in the past of the planets.

7.4.1 Landslides on Planets and Satellites, Except Mars

7.4.1.1 Venus

The thick Venusian atmosphere permits only radio waves to penetrate the planet's surface. The Magellano mission (NASA) has shown examples of many landslides along the slopes at high altitude: slumps, taluses, rock avalanches. Volcanic calderas gave origin to sector collapses, too. In particular, a sector collapse stretches for 50 km with a FB of only 0.07, a value that certainly requires some unidentified lubrication process. Being associated to a volcano, this landslide might represent the result of a lateral explosion similar to St. Helens.

7.4.1.2 Landslides on Dead Moons and Planets: Mercury, the Moon, and Iapetus

The Moon, Mercury, and the satellite of Saturn called Iapetus share lack of significant volcanic and tectonic activity. Devoid of an atmosphere, these planetary

Fig. 7.17 The crater Dawes on the Moon seen from Apollo 17. Coordinates: 17.2° N, 26.4° E. The crater diameter is approximately 18 km, the depth 2,300 m

bodies are not affected by erosion except thermal breakdown and meteoritic fall. Landslides usually evolve from the borders of craters, the locations where gradients are highest. The Fig. 7.17 shows the Dawes crater on the Moon.

The terraced structure in the crown of the crater indicates that the material has slid concentrically toward the centre. This has probably occurred in one single event, perhaps triggered by the fall of a nearby meteorite.

Similar localized landslides have been found in Mercury (e.g., Murray et al. 1981). On the Moon, there also appear bizarre rockfalls where boulders have apparently rolled for distances of several hundreds of meter. However, both the Moon and Mercury lack the spectacular character of Martian landslides.

Figure 7.18 shows an enormous crater failure in Iapetus. The small crater hosting the landslide deposit measures about 120 km across. The FB is about 0.25 and the volume is about 20,000–30,000 km^3.

7.4.1.3 Other Moons

On Callisto (Jupiter's satellite) several landslides are reminiscent of terrestrial mudflows. This is not surprising, as water is apparently abundant on Callisto. Io, another satellite of Jupiter famous for the astonishing volcanic activity, might hold the record for the largest landslide known in the Solar System. Schenk and Bulmer (1998) have suggested the existence of an enormous landslide deposit (50,000 km^3!) at the foot of Eubeoa Montes, with a FB of 0.1.

Fig. 7.18 A landslide on the border of a nearly equatorial crater of Iapetus, moon of Saturn

7.4.2 Landslides on Mars

Mars has a special place in the Solar system. It is the only planet other than the Earth with some probability of having hosted life. It is the second (and probably last) planetary body that will be targeted by manned missions. And it is one of the geologically most interesting planets. Landslides on Mars are also numerous and varied.

The close exploration of Mars started with the Mariner probes in the middle 1960s and early 1970s, followed by the Viking missions and then more recent probes like Mars Odyssey, Mars Reconnaissance Orbiter, and the mobile Rovers. Before the Mariners, the sole information came from observations from Earth. Maps were just an ensemble of blurred spots, while today the topographic data provided for example by the MOLA laser altimeter or the Themis camera, may be studied in detail. The most astounding characteristic of Mars is the global dichotomy, i.e., the division of the planet's surface in two very distinctive parts. The southern part of the planet is above average altitude. It consists mostly of mountainous areas, with often very old terrains of the Noachian and Hesperian ages (the two most ancient period of Mars). In marked contrast, the northern lowlands are on the average 5 km lower in altitude and comprise younger terrains of Amazonian age.

7.4.2.1 Valles Marineris

One of the most amazing characteristics of Mars is Valles Marineris, an equatorial valley discovered in 1971 by the mission Mariner 9. The figures of Valles Marineris are breathtaking: 4,000 km long, 200 km wide, and 7 km deep, it dwarfs the largest canyon on Earth, the Arizona Grand Canyon, that would appear as a mere tributary. Debate persists as to the nature of this impressive scar, which appears to be tectonic in origin. Valles Marineris is divided into a series of subcanyons called "Chasmata" (singular "Chasma"). The relatively stiff cliffs in Valles Marineris (of the order of 30°) have collapsed producing numerous spectacular landslides McEwen (1989). Mass wasting has been among the responsible processes for the enlarging of the Chasmata. Figure 7.19 shows a couple of landslides in Ganges Chasma. The landslide visible on the left is older, as it is covered by the one on the right. This is also indicated by the numerous craters on its surface. Note the alcoves of subcircular

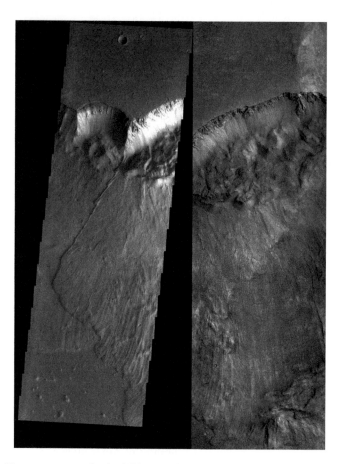

Fig. 7.19 The most spectacular landslides on Mars appear along the giant canyon of Valles Marineris and along its tributaries. The picture on the *left* shows two neighboring landslides in Ganges Chasma, one of the Chasmata of Valles Marineris. The picture on the *right* shows a detail of one of the previous landslides. These and other landslides in Valles Marineris share with Sherman and other landslides traveling on glaciers the impressive characteristic of striations parallel to the direction of the movement. (Images Themis PIA09057 and 20020401, coordinates −8.6° N, 315.7° E, reproduced with permission (Courtesy NASA/JPL/Space Science Institute).) Vertical length about 35 km

shape indicating the failing mechanism of a slump (←Chap. 1). Landslides have been probably fast-moving, with estimated speeds of at least 40 m/s (Lucchitta 1979). Note also the long runout of the landslide material, which extends for several tens of km, with a FB of 0.2 or even less. The volume of each landslide in the Chasmata is of the order 1,000–100,000 Mm^3.

Landslides data display a volume effect similar to the Earth, but with an interpolation line shifted vertically compared to terrestrial events (←Fig. 6.22). One basic question is: were these landslides more similar to cohesive debris flows or to rock avalanches?

The lower gravity field on Mars has also contributed to influence the size of the slumps. From (←Sect. 2.2.3), it follows that the radius of a slump composed of cohesive material goes approximately like

$$R \approx \frac{2\tau_y}{\rho g_{MARS} k}. \tag{7.15}$$

Because the gravity field is lower on Mars ($g_{MARS} \approx 3.7$ m s^{-2}), the radius necessary for instability will be greater for the same kind of rock. Using $R = 6$ km, the estimated value of the cohesion is $\tau_y \approx 20k$ MPa where k in Eq. 7.15 is of the order one. If the material is frictional in addition to cohesive, the calculation becomes more complicated. The mechanical instability of the walls in Valles Marineris has been the subject of some studies. Lucchitta et al. (1992) showed that with the friction angle and cohesion typical of basalt (about 30–35° and 40–80 MPa, respectively) the walls of Tithonium Chasma are stable. However, fractured basalt or volcanic tuff has lower friction angle and cohesion, and could develop instability in Valles Marineris.

The structure of the landslide deposit is especially interesting. At the foot of the slump (zone "A" in Fig. 7.20), the collapsed material is relatively thick and exhibits chaotic appearance. In the zone "B" there appears weak striations parallel to the direction of flow, which become neater in the zone "C". Striations appear to radiate from the centre of collapse. In the last part "D", the striations are curved. The collapse does not resemble a purely granular avalanche. The elongation of a relatively thin sheet of deposit that likely moved at very high speed reflects more the characteristics of a viscous fluid. The striations appear very similar to the ones exhibited by terrestrial landslides falling onto glaciers. The Sherman landslide (←Box 7.3) is perhaps the closest example coming to mind, and the explanation is probably similar to the one for Mars: the powerful lubrication at the base made inertial forces greater than the resistance at the bottom.

In the case of Sherman, the lubrication was provided by the glacier (Shreve 1966; McSaveney 1978; Dufresne and Davies 2009). The low ratio H/R, (<0.2) confirms a strong lubrication effect. There are many similar landslides on Mars, which probably indicates the important role played by some medium on the Martian surface that acted as a lubricant.

Could this imply that the landslides of Valles Marineris were ice-lubricated?

Figure 7.21 shows more landslides in Eos Chasma. At least seven landslides are visible in the field. The landslides probably fell in different times; based on crater

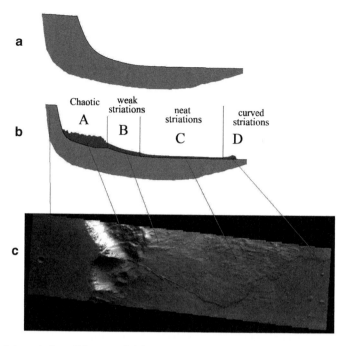

Fig. 7.20 Interpretation of the superficial structures of the Ganges Chasma. *Top*: prelandslide configuration (schematic) (Courtesy NASA/JPL/Space Science Institute)

counting, Quantin et al. (2004) estimated that the large landslide L2 visible in the bottom is about 3 Gy old, while the one on the top L1 is much more recent: about 200 My. Note a series of interesting features. Landslide L1, in colliding against the deposits of L2, has produced a compression ridge. Striations are very clear in L1 and well distinguishable in all the other landslides except L2, whose body has been subjected to erosion for a longer time. Note the lower alcove A2 from which the landslide L1 has taken place and an older scar A1 that probably gave origin to a smaller antecedent landslide blanketed by L1. The inset on the right shows the subdivision of the landslide L1 in four units as discussed in relation with Fig. 7.20.

7.4.2.2 Other Locations on Mars

Other two classes of landslides seen on Mars have been discovered with the Mars Orbiter Camera (MOC). The first type consists of small-scale dry avalanches observed especially in steep regions of the Olympus Mons aureole. These avalanches appear as dark strikes scars; they are still forming at the present time.

Debris flows similar to those observed in permafrost area in Greenland have also been described (Costard et al. 2007). These debris flows that exhibit the classical terrestrial morphology with an alcove, levees, and accumulation zone,

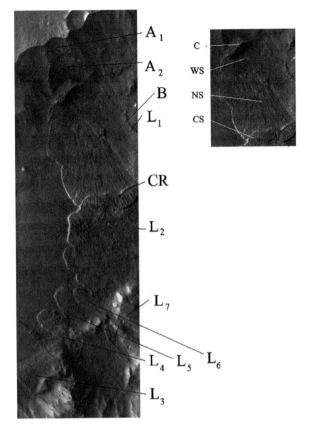

Fig. 7.21 Landslides in Eos Chasma. Coordinates: $-8.0°$ N, $318.6°$ E. Horizontal length 18.4 km. $L1$–$L7$ seven landslides visible in the field. $A1$ upper alcove of landslide 1, $A2$ lower alcove of landslide 1. B blocks often border the landslides. CR compression zone. For the landslide $L1$ the division in four parts according to the Fig. 7.20 is shown. C chaotic, WS weak striations, NS neat striations, CS curved striations (Image Themis V13283001. Courtesy NASA/JPL/Space Science Institute)

probably take origin from local melting of permafrost. This occurrence was probably triggered by changes in the local surface temperature (Costard et al. 2007). Figure 7.22 shows one series of debris flow in the interior of the East Gorgonium crater.

7.4.2.3 The Aureole of Olympus Mons

One of the most stupefying features of the entire Solar System is Olympus Mons, already identified as a white spot from Earth before the space missions. We know today that spots are clouds formed at the top of the Mons, which reaches an amazing height of 25 km. It is indeed the highest volcano in the Solar System. With its 750 km of diameter, it would partly be out of sight beyond the horizon to an observer standing on its surface. The slope of only $5°$ is compatible with tranquil basaltic eruptions like for the Hawaii islands. Figure 7.23 shows a Mola map of the

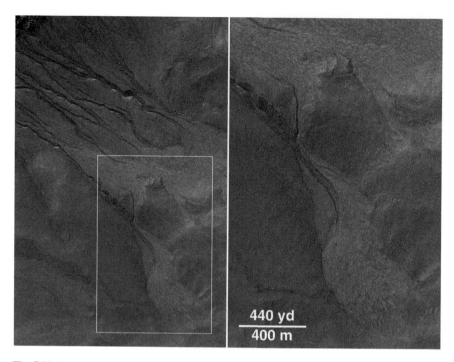

Fig. 7.22 Debris flows developing from the rim of the East Gorgonium crater. Coordinates 37.4° N; 168° W (Image MOC2-241. NASA/JPL/Malin Space Science System. Courtesy NASA/JPL/Space Science Institute)

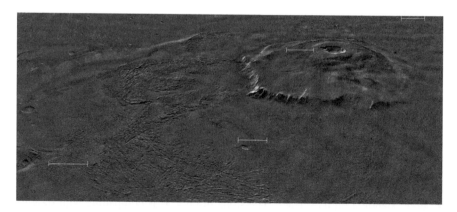

Fig. 7.23 The aureole of Olympus Mons. Each segment represents a length of 100 km (Courtesy NASA/JPL/Space Science Institute)

Olympus Mons. Notice the cliffs bordering the structure, about 8 km high and characterized by a much stiffer slope of 20°–30°.

The figure also shows the *Aureole,* a subcircular halo surrounding the whole edifice of Olympus Mons. The total estimated volume in the aureole is up to nearly one million cubic kilometers (Griswold et al. 2008). Among the explanations suggested so far to explain this enigmatic structure, some involve lava flows (McCauley et al. 1972), subglacial volcanic deposits (Hodges and Moore 1979), ash flows (Morris 1981), deep-seated deformations, and lateral spreading (Francis and Wadge 1983). The total volume deposited in the aureole is about equal to that of missing rock from the cliffs of Olympus Mons, which is compatible with the landslide hypothesis. Landslide might have been subaerial and catastrophic (Lopes et al. 1980; Harrison and Grimm 2003; McGovern et al. 2004), slow and lubricated by ice (Tanaka 1985). It is also possible that the aureole deposits are the product of subaqueous landslides similar to the rock avalanches of the Hawaii or Canarians (→Sect. 9.1; De Blasio submitted to Earth and Planetary Science Letters).

Chapter 8
Rockfalls, Talus Formation, and Hillslope Evolution

In contrast to a rock avalanche, which involves the flow of a large mass, a rock fall (or rockfall) is the movement of a single block. The product of rockfalls often appers as a rock heap, a talus, at the foot of mountain slopes. Although rock falls exhibit less mobility than a rock avalanche, they are far more common. Rock fall-prone areas are usually steep rock cliffs in mountain areas, especially if subject to frost–thaw activity in the rock joints. Rockfalls are a significant safety problem in mountainous areas, as single falling boulders may kill people, interrupt roads and railways, and demolish buildings. Moreover, the opportunity of building settlements, houses or hotels at the foot of mountains is often limited by rockfalls. It is thus advantageous to be able to predict the extent of these phenomena and the distance from the mountain foot where the terrain can be considered safe.

The trajectory of a single boulder is theoretically easier to predict than the travel of a whole rock avalanche, because the complications due to block–block interactions are missing. However, a rock fragment has irregular shape; it can both roll and bounce, and the terrain may have complicated geometry and variable properties. Even describing the process with simplified models for the block (e.g., a spherical block), the prediction of a block trajectory is a complicated task. The block may revolve around an axis not necessarily parallel to the ground, it may slide and roll. Bouncing and the transition to rolling is also an intricate process that requires simplified approaches. As usual, in this chapter we focus on the basic physics of the problem, often utilizing simple models.

The photograph besides shows a rock cliff threatening the road below. A system of net fences has been installed to mitigate the risk of rockfall. Via Mala, Val di Scalve, Northern Italy.

8.1 Introduction to the Problems and Examples

8.1.1 General

8.1.1.1 Block Trajectories

The processes producing rockfalls are mostly weathering (frost–thaw activity in rock joints, rain), seismic activity and human action (Turner and Schuster 1996). Figure 8.1 shows a multiple rock fall. Several boulders have fallen on a soft terrain, coming to a halt close to the boundary wall and houses. Figure 8.2 shows an example of boulder trajectories along a mountain incline. Two boulders (whose initial position is shown with red squares and trajectories are indicated by the lines) have detached from a limestone rock face and after about 1 km of runout (650 m in altitude difference) have come to rest in an area (gray) where future buildings are planned. To assess the safety of mountain locations, it is important to predict within an acceptable approximation the trajectory and velocity of such blocks. This also assists in planning the most efficient position for rockfall nets and ditches.

Fig. 8.1 An example of multiple rock fall in Val Venina, northern Italy. Several huge boulders have fallen from an overhanging cliff, plunging onto the grassed terrain. They left some grooves on the terrain; mud splash was observed against some of the house walls (Observations and original photograph courtesy of Giovanni Crosta)

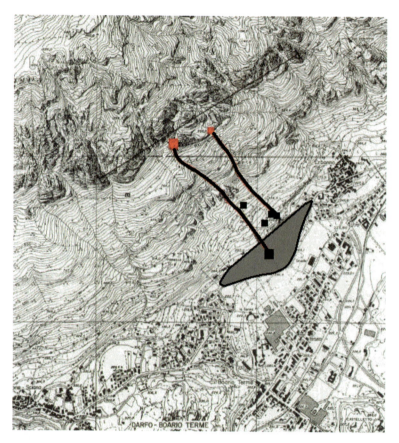

Fig. 8.2 Example of blocks trajectories along a mountain terrain near Erbanno, Lombardy (Italian Alps). An area intended for house building was affected by rockfall (*black lines*, with detachment and final points indicated with *red* and *black squares*, respectively), and countermeasures were taken to protect the area with nets. (From Cancelli et al. 1991, modified.) The fahrböschung of the block with the longest runout is about 0.64 (32.5°)

The trajectory of a block is the complicated result of the terrain geometry, of the block shape, and of the properties of the rock or soil involved including the presence of vegetation. The same terrain may exhibit different properties depending on the season. A frozen barren ground in winter favours rock mobility compared to soft terrain more typical of summer season. It is clearly impossible even in principle, to determine exactly the block trajectories. Various approximation methods have thus been proposed to predict the trajectory and velocity of blocky fragments, and adopted in numerous models of rock falls.

8.1.1.2 Return Period and Fahrböschung

In addition to the maximum distance reached by rock fragments, a complete safety assessment should also consider the return period. A relevant question is the probability that a boulder larger than a given volume will reach a certain distance from the source during a period of 100 years. The return period increases dramatically from the distance to the rock source. For example, in a Canadian example Evans and Hungr (1993) indicate a return period of 1 day at a distance of 200 m from the source. The period becomes 110 years at the toe of talus, approximately 300 m from the source, and 5,000 years at a distance of 400 m. People are particularly insensitive to geohazards when the return period is of the order of one human lifetime. As a consequence, building may be frequently shattered by rock falls in locations with large return periods. Remarkably, the boulders with the longest runout are also the largest, which makes rockfalls even more dangerous.

Rockfall blocks typically exhibit fahrböschungs between 32° and 45° (see Evans and Hungr 1993); they are thus less mobile than rock avalanches, but far more common.

8.1.2 Physical Processes During a Rock Fall

The detachment of a block and its movement down slope involve a series of different physical processes

1. Free fall, during which the boulder follows ballistic trajectory.
2. The impact against the terrain, in correspondence of which most of the dissipation is taking place, as well as significant changes in the boulder's velocity, spinning rate and direction of movement.
3. Rolling.
4. Sliding.

Cameras and computer simulations have aided elucidating the relative importance of each process, which depends on many factors like the slope angle, the block size and shape, and the velocity. Figure 8.3 shows filmed sequences of block impact (Fig. 8.3a) and rolling (Fig. 8.3b) along natural slopes. Note that on a hard terrain, the time of contact between the block and the terrain is very short. In contrast, on a soft terrain sliding may occur. Figure 8.4 shows schematically the dominant processes during the rock fall. The abscissa is the slope angle, while the ordinate scale represents a measure of the boulder sphericity. A block close to spherical (top of the figure) would typically roll below 30°, and bounce between 45° and 70°. For angles superior to 70° it will mostly propagate in ballistic flight, without much interaction with the terrain. For less spherical blocks, rolling becomes less significant at low angles, as the roughness of the block will make it jump more frequently above the terrain; for tabular blocks (lower part of the graph) the block will preferably slide, but only for angles above the friction angle, and will fall freely for angles superior to 70°. Obviously, there are important parameters not considered in such diagram. For example, a tabular block at very high speed might also roll around its main axis.

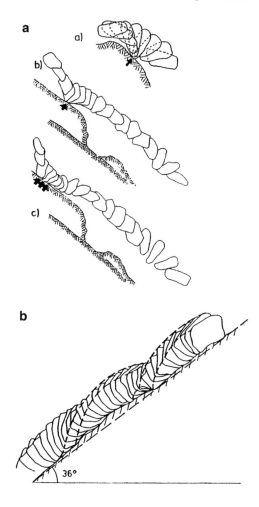

Fig. 8.3 Sequences from movies of a block traveling down a natural slope. (**a**) Impact with hard ground. In (**a**) and (**b**) the impact is on hard rock, and no sliding occurs. In (**c**), the impact on soft ground produces sliding. (**b**) Rolling of a block (Figures from Bozzolo and Pamini 1986)

8.2 Simple Models of a Simple Object Falling Down a Slope

8.2.1 Simple Models of Rolling, Bouncing, Gliding, and Falling

Here we consider the simplified motion of an object with regular shape, and in particular:

1. A sphere bouncing and rolling on a smooth terrain, and
2. A sliding slab.

The simplest model of terrain geometry is a slope dipping with constant angle β. The friction angle for sliding of the rock fragment is denoted by ϕ.

8.2 Simple Models of a Simple Object Falling Down a Slope

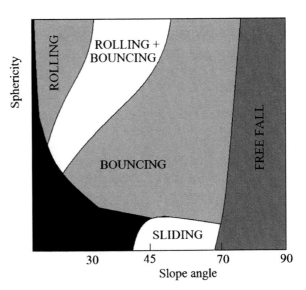

Fig. 8.4 Highly schematic illustration of the behavior of a falling block on hard terrain. The abscissa represents the slope angle and the ordinate scale the sphericity. Normally rolling prevails at small slope angles (<45°). At greater angles (between 45° and 70°) the block tends to bounce, and at still greater angles, it will fall freely for most of its trajectory

8.2.1.1 Rock Slab Moving Parallel to Slope (Sliding, Fig. 8.5a)

For the movement along the horizontal plane, equations are the ones previously introduced for the slab model of rock avalanches (←Chap. 6). If $\beta > \phi$, the acceleration and velocity of the fragment parallel to slope are so given as (6.4.1.2)

$$\ddot{l} = g \sin \beta \left[1 - \left(\frac{\tan \beta}{\tan \phi}\right)\right]; \quad \dot{l} = g \sin \beta \left[1 - \left(\frac{\tan \beta}{\tan \phi}\right)\right] t. \quad (8.1)$$

In order for the block to stop, the slope angle must be lower than the friction angle and thus the slope needs to flatten out. In case, the runout is then simply (←Sect. 6.4.2)

$$R = \frac{H}{\tan \phi}. \quad (8.2)$$

8.2.1.2 Cylinder or Sphere Rolling Down a Slope (Fig. 8.5b)

The coefficient of rolling friction μ_R is defined similar to the coefficient of sliding friction, with $\mu_R \ll \tan \phi$. Although it is often difficult to separate sliding from rolling in a rockfall as both processes may take place at the same time, here for simplicity we consider a purely rolling block. This is equivalent to imposing absence of slip at the contact with the bed.

Fig. 8.5 Analysis of the elementary processes during a rock fall

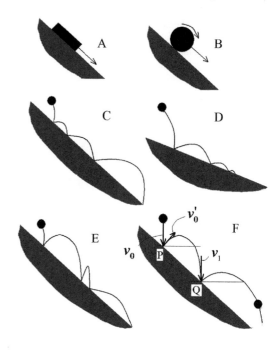

When a rigid body is spinning, the rotational energy sums up to the kinetic energy of the center of mass so that the total energy is given as (see also ←Sect. 8.2.1)

$$E = \frac{1}{2}mu^2 + \frac{1}{2}I\omega^2 \tag{8.3}$$

where the moment of inertia is

$$I = \vartheta mr^2 \tag{8.4}$$

and ϑ is a geometrical constant equal to 2/5 for a sphere spinning around an axis passing through the center, and ½ for a cylinder spinning around its main axis $\omega = 2\pi v$, v is the frequency of rotation (for a periodic rotation $\omega = 2\pi/T$, where T is the period of rotation of the sphere). Consider a cylinder of a sphere rolling down slope for a total length l measured along the slope and assume that μ_R is independent of the velocity and slope angle. The block acquires rotational energy which is absent with pure sliding, and that must be accounted for in the energy balance. The conservation of energy leads us to

$$mgl\sin\beta = \frac{1}{2}mu^2 + \frac{1}{2}I\omega^2 + mgl\,\mu_R\cos\beta \tag{8.5}$$

8.2 Simple Models of a Simple Object Falling Down a Slope

If the body does not slip, then

$$u = \omega r \tag{8.6}$$

and so the velocity can be found immediately as

$$u = \sqrt{\frac{2gl(\sin\beta - \mu_R \cos\beta)}{1 + \vartheta}}. \tag{8.7}$$

Note in Eq. 8.7 two main differences from the case of purely sliding object: (1) part of the potential energy of the falling block goes into the rotation; thus, the other conditions being the same, the velocity is smaller for a rotating than for a purely sliding block, as evident by the presence of the term ϑ in the denominator of Eq. 8.7. (2) Because the rolling friction coefficient is usually smaller than the coefficient of sliding friction, the numerator of Eq. 8.7 is normally higher for rolling than for sliding friction.

The expression for the runout is given again by Eq. 8.1, with the sliding friction coefficient substituted by the rolling coefficient. The dimensionless moment of inertia ϑ is thus absent from the final expression for the runout.

8.2.1.3 Bouncing of a Sphere on an Inclined Plane

Bouncing of a block against the terrain is frequently quantified in terms of the coefficients of restitution (←Chap. 4), which depend on the impact velocity, on the particle size, on the block geometry, and especially on the nature of the ground. The lowest values are typically measured for collisions on a granular bed or a loose soil, rather than hard rock. Because a large energy loss takes place during impact with the terrain, the coefficients of restitution are normally much smaller than unity. The numerical values of the coefficients of restitution for typical terrains are examined in more detail in →Sect. 8.4.

Let us consider a sphere bouncing down an incline like in Fig. 8.5c–f. If the terrain is smooth and the slope is sufficiently steep, the behavior will look like the one in Fig. 8.5c. The sphere bounces indefinitely, increasing with time the spacing between successive impacts with the terrain. This occurs if at each bounce the energy loss is lower than the energy gained by fall in the gravity field. Note that the angle of bouncing changes in general with successive bounces. By reducing the sloping angle (so decreasing the gain of gravitational energy), a transition to the behavior like the one shown in Fig. 8.5d takes place, where bounces become closer with distance. If the terrain is not smooth and we allow for the presence of unevenness, the outcome may resemble the one shown in Fig. 8.5e, showing that the sphere proceeds with irregularly spaced bounces.

8.2.1.4 Maximum Extension of a Bouncing Sphere Down an Incline

The previous discussion suggests the simple but instructive model shown in Fig. 8.5f. It is assumed that the sphere is not spinning, and that it always bounces with an inclination of 45° with respect to the horizontal. These assumptions result in an exaggeration of the sphere mobility, so that the result will show the maximum runout attainable.

After free fall from the rock overhang, the sphere hits the terrain in P. It bounces at 45° with respect to the horizontal, follows a ballistic trajectory, and hits the terrain again in Q. There, it bounces again at 45°, and so on to the next bounce.

We call V the velocity acquired by the sphere after free fall just prior to colliding in P, $v_0 = \sqrt{2gH}$ and $v'_0 = \varepsilon v_0$ the speed of the particle after the first bounce. The particle reaches the position Q with a velocity v_1 that can be calculated with simple kinematic considerations as $v_1^2 = v'_0{}^2 \Gamma(\beta)$ where

$$\Gamma(\beta) = 1 + 2\tan\beta(1 + \tan\beta). \tag{8.8}$$

and so after the second bounce the velocity is

$$v'_1{}^2 = \varepsilon^2 v_1^2 = v'_0{}^2 \varepsilon^2 \Gamma(\beta) \tag{8.9}$$

and after the third bounce

$$v'_2{}^2 = \left[\varepsilon^2 \Gamma(\beta)\right]^2 v'_0{}^2 \tag{8.10}$$

and after n bounces

$$v'_n{}^2 = \left[\varepsilon^2 \Gamma(\beta)\right]^n v'_0{}^2 = \left[\varepsilon^2 \Gamma(\beta)\right]^n \varepsilon^2 v_0^2 \tag{8.11}$$

The horizontal distance between two successive bounces is

$$\bar{x}_n = \frac{v'_n{}^2}{g}(1 + \tan\beta) \tag{8.12}$$

and thus the total runout after an infinite number of bounces is

$$R = \sum_{n=0}^{\infty} \bar{x}_n = \frac{(1 + \tan\beta)}{g} \sum_{n=0}^{\infty} v'_n{}^2 = \frac{(1 + \tan\beta)}{g} v_0^2 \sum_{n=0}^{\infty} \varepsilon^{2n+2} \Gamma^n(\beta)$$
$$= \frac{(1 + \tan\beta)}{g} v_0^2 \frac{\varepsilon^2}{1 - \varepsilon^2 \Gamma(\beta)} \tag{8.13}$$

Thus, if

$$\varepsilon^2 \Gamma(\beta) < 1 \qquad (8.14)$$

the sphere will stop at a horizontal distance given by Eq. 8.13, whereas for

$$\varepsilon^2 \Gamma(\beta) \geqslant 1 \qquad (8.15)$$

it will continue to bounce indefinitely. Obviously, there is a natural limit to the slope in a natural terrain that limits the total horizontal displacement. Moreover, a falling block is not a regular sphere and may also be captured by interstices in the ground, or collide against an obstacle. The calculation is, however, interesting in showing the importance of the coefficient of restitution in the runout of a bouncing sphere. According to Eqs. 8.14 and 8.15, the critical value of the coefficient of restitution giving an indefinite propagation at a slope angle of 45° is 0.45. It increases to 0.49, 0.59 and 0.71 at slope angles of 40°, 30°, and 20°, respectively.

The equation also shows the importance of the initial velocity of the clast. Because v_0^2 in (8.13) is proportional to the fall height, this equation predicts that the horizontal extension of a rockfall is proportional to the height of the overhang.

8.3 Simple Rockfall Models

In this section we examine some simple rockfall models. As already discussed in (←Sect. 5.3), we define two coefficients of restitution for the impact of a block: one parallel to the ground, $\varepsilon_=$, and one perpendicular to the ground, $\varepsilon_\perp$.

In some of these models (→Sect. 8.3.1) the block is considered as point-like, and the rotational degrees of freedom are suppressed. Other models aim at describing the rotation of rocks (→Sect. 8.3.2). In this case, we limit ourselves to consider simple symmetric bodies like disks, cylinders, or spheres.

8.3.1 A Simple Lumped Mass Model

In a lumped mass model, the block is described as a geometrical point. The block rotation is completely neglected, like in the model by Evans and Hungr (1993). Let us consider a block falling from a height H. Just prior to collision, the energy of the block is $E = mgH$. This energy is also equal to the block's kinetic energy

$$E = \frac{1}{2} m v^2 \qquad (8.16)$$

where $v = \sqrt{v_\parallel^2 + v_\perp^2}$ is the magnitude of the velocity. The ratio between the two velocity components evidently gives the angle of impact α with respect to the terrain, $\tan\alpha = v_\perp/v_\parallel$. The energy loss during an impact is

$$\begin{aligned}\Delta E &= \frac{1}{2}m\left[v'^2 - v^2\right] = \frac{1}{2}mv^2\left[\frac{v'^2}{v^2} - 1\right]\\ &= \frac{1}{2}mv^2\left[\frac{v_\parallel'^2 + v_\perp'^2}{v_\parallel^2 + v_\perp^2} - 1\right] = \frac{1}{2}mv^2\left[\frac{\varepsilon_\parallel^2 v_\parallel^2 + \varepsilon_\perp^2 v_\perp^2}{v_\parallel^2 + v_\perp^2} - 1\right] \\ &= \frac{1}{2}mv^2\left[\frac{\varepsilon_\parallel^2 + \varepsilon_\perp^2 v_\perp^2/v_\parallel^2}{1 + v_\perp^2/v_\parallel^2} - 1\right] = \frac{1}{2}mv^2\left[\frac{\varepsilon_\parallel^2 + \varepsilon_\perp^2 \tan^2\alpha}{1 + \tan^2\alpha} - 1\right].\end{aligned} \qquad (8.17)$$

Evans and Hungr (1993) define also the energy height as

$$e = E/mg = z + \frac{v^2}{2g} \qquad (8.18)$$

where z is the block height from a reference point. In the absence of impacts, the energy height remains constant in time (the weak effect of air drag is neglected). Each time the block collides with the terrain, there is an instantaneous plummet of the energy height by an amount

$$\frac{v^2}{2g}\left[\frac{\varepsilon_\parallel^2 + \varepsilon_\perp^2 \tan^2\alpha}{1 + \tan^2\alpha} - 1\right]. \qquad (8.19)$$

The trajectory between one impact and the next is calculated from the expression for ballistic flight. Figure 8.6 gives the trajectory profile, the velocity, and the energy height calculated with Eqs. 8.17–8.19. Evidently, if the expression in Eq. 8.17 is positive, the block will continue bouncing without end. If the expression is negative, the block travels to infinite distance with progressively shorter jumps. The latter condition leads to

$$\frac{1 - \varepsilon_\parallel^2}{1 - \varepsilon_\perp^2} < \tan^2\alpha. \qquad (8.20)$$

A rule for block stoppage must also be introduced in the model. In a natural slope, blocks will stop bouncing either because of collision with an obstacle, or following a transition to rolling. The problem of block stoppage and change to rolling is important to assess the runout of the block, but not of immediate solution, because at low energies even small perturbations may bring the block to a halt. As a consequence, models necessarily introduce some degree of arbitrariety.

In the model by Evans and Hungr, a transition to rolling occurs if after three consecutive bounces the ratio $\frac{\Delta E}{\Delta L}$ between the energy lost in each impact and the horizontal length ΔL of the corresponding bounce is greater than the rolling friction force

8.3 Simple Rockfall Models

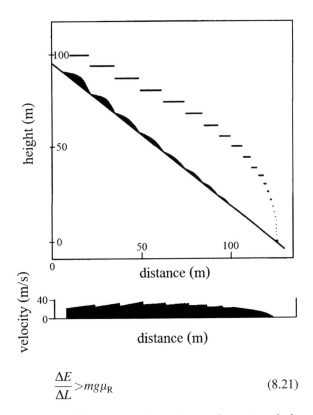

Fig. 8.6 A typical result from the model of Evans and Hungr (1992) (From Giani 1992, re-drawn)

$$\frac{\Delta E}{\Delta L} > mg\mu_R \tag{8.21}$$

In other models, the transition to rolling occurs when the post-impact angle is smaller than a certain critical angle. Stoppage may be also imposed if the velocity becomes sufficiently small that the block cannot shift the center of mass over its edges.

8.3.2 The CRSP Model

The Colorado Rockfall Simulation Program considers the impact of spheres or disks on a two-dimensional terrain (Pfeiffer and Bowen 1989). It thus represents an improvement from the simple models discussed earlier, in allowing for the finite size of a block and introducing block rotation. In addition, it addresses the roughness of the terrain.

The first basic equation of the model is based on a calculation of the energy dissipation during a collision (see also Fig. 5.8). The total energy of the disk is given by Eq. 8.3 that can be re-written as

$$E = \frac{1}{2}mu_\perp^2 + \frac{1}{2}mu_=^2 + \frac{1}{2}I\omega^2 \tag{8.22}$$

where the kinetic energy of the center of mass has been split in one part due to the velocity parallel to the terrain and a second one due to the velocity perpendicular to the terrain.

Pfeiffer and Bowen (1989) write a constraint on the total energy dissipation at impact, but considering only the part of the kinetic energy deriving from motion parallel to the terrain. Their equation reads

$$\left[\frac{1}{2}mu_\parallel^2 + \frac{1}{2}I\omega^2\right]f\,\mathrm{SF} = \left[\frac{1}{2}mu_\parallel'^2 + \frac{1}{2}I\omega'^2\right] \qquad (8.23)$$

where as usual the primed indices refer to the post-impact situation. The functions f and SF are given as

$$f = \varepsilon_= + \frac{1-\varepsilon_\parallel}{[(v_= - \omega R)/\sigma]^2 + \gamma}$$
$$\mathrm{SF} = \frac{\varepsilon_=}{(\delta v_\perp/\varepsilon_\perp)^2 + 1} \qquad (8.24)$$

where $\sigma = 20$ m/s; $\gamma = 1.2$; $\delta = 1/250$ are empirical parameters. The functions f and SF in conjunction account for the energy dissipation due to frictional force between the sphere and the terrain. The function f peaks when $v_= \approx \omega R$, i.e., when the lower part of the block approaches the terrain with zero relative velocity. In this case, the energy dissipation will be reduced to a minimum, because no sliding between the sphere and the terrain will occur. However, when $v_=$ differs much from ωR, f becomes small and little of the post-impact combination of the rotational + translational energy parallel to ground will be conserved. Similarly, the function SF accounts for the fact that frictional energy dissipation increases with the velocity normal to the ground. This is because the frictional force is proportional to the normal force, which in turn is given by the momentum change perpendicular to the terrain, and so to the normal velocity. A further equation gives the behavior of the normal velocity

$$u_\perp' = \frac{\varepsilon_\perp}{1 + (u_\perp/u_{\mathrm{REF}})^2} u_\perp \qquad (8.25)$$

where $u_{\mathrm{REF}} \approx 9$ m/s. This equation accounts for the inelastic character of the impact.

The roughness of the terrain is introduced in the following way. The angle of impact α of the block is calculated from a combination of the local average slope β and a random component, representing the roughness. The angle Θ is thus chosen as a random number with uniform distribution between 0 and an upper value

$$\Theta_{\mathrm{MAX}} = \tan^{-1}(S/R). \qquad (8.26)$$

where S stands for a scale length of roughness (Fig. 8.7). The post-impact velocity and angular speed $u_='$, $u_\perp'$, ω' are calculated with Eqs. 8.23–8.26 supplemented with the no-slipping condition after impact

$$u_=' = \omega' R. \qquad (8.27)$$

Fig. 8.7 Impact angles in the CRSP model

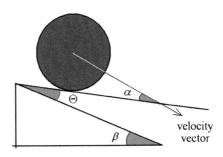

8.3.3 Three-Dimensional Programs

Modern programs handle three-dimensional simulations by dividing the terrain into a certain number of cells; they can introduce non-spherical blocks, and more realistic models for the terrain roughness. It is not the purpose of this book to review such models (see e.g. Dorren 2003; Guzzetti et al. (2002)). Figure 8.8 shows an example of calculation of several block trajectories on synthetic profile: the slope has inclination of 45° in the upper part of the trajectory (dark gray) and zero slope angle in the lower part (light gray). The steep terrain has a horizontal length of 500 m. Terrain unevenness in this particular calculation is introduced stochastically. The maximum velocities (dark red) of the order 70 m/s are of the order of 70 m/s. Models like this can predict in a statistical sense the velocities and runout of blocks.

8.4 The Impact with the Terrain

8.4.1 The Physical Process of Impact Against Hard and Soft Ground

So far, the impact with the terrain has been modeled in terms of the coefficients of restitution. As anticipated in Fig. 8.3, however, the impact is a complicated process. Even in the case of perfectly spherical block impacting against a smooth rock surface, the coefficient of restitution will depend on the angle of attack and on the speed; the sphere will change its rotational state and slippage may occur (←Sect. 5.3.1). In addition, if the block fractures, more energy will be used up in the collision, which further reduces the coefficient of restitution.

Considering more realistically a block as an irregular object, we see that complications can only increase. The block will hit the terrain with a large and uneven impact area, which will result in an unpredictable post-impact spinning. The terrain itself may have very different properties. A soft cohesive soil will respond very differently compared to a talus made up of irregular grains.

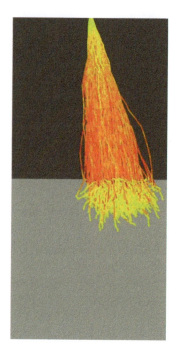

Fig. 8.8 Example of calculated block trajectories on an artificial slope with roughness. The color intensity gives the velocity (see text) (From Crosta and Agliardi 2004. Reproduced with permission)

Figure 8.9 shows a high-speed filming sequence of impact with a soft granular terrain. Note the formation of a crater, a process that dissipates a large amount of energy and leaves the impacting block with little residual energy. We can thus expect that the impact onto a granular medium with average particle diameter smaller than that of the impacting block will result in a low coefficient of restitution.

Keeping in mind the complexity of a real impact, it is however useful and practical to use the consolidated approach based on the coefficients of restitution. In the following, we briefly examine some of the values for the coefficients of restitution used in the simulations with different models.

8.4.2 Coefficients of Restitution and Friction

The coefficients of restitution may be determined based on filming or on back-analysis of calculated trajectories. Table 8.1 shows some suggested values. For hard rock the coefficient is higher than for loose detritus or soil, a result that appears reasonable. It is also evident that concerning the impact on loose detritus or talus, the coefficient of restitution depends on the size of the falling block relative to the talus blocks (Fig. 8.10). If the talus blocks are larger than the impacting block

8.4 The Impact with the Terrain

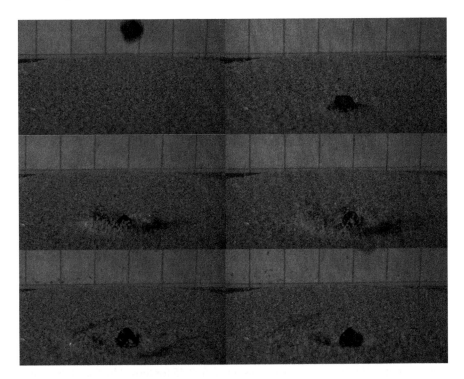

Fig. 8.9 Sequence of impact of a pebble (diameter about 5 mm) falling on a layer of coarse sand (diameter between 0.5 and 1 mm) from a height of 60 cm

Table 8.1 Normal and parallel coefficients of restitution

Kind of terrain	$\varepsilon_\parallel$	$\varepsilon_\perp$	Reference
Rocky outcrop	0.65–0.75	0.8–0.9	Piteau e Clayton
Rocky outcrop	0.33–0.37	0.83–0.87	Pfeiffer and Bowen (1989)
Rocky outcrop	0.75	0.50	Crosta and Agliardi (2004)
Rocky outcrop	0.99	0.53	Hoek (1987)
Rocky outcrop, forested	0.70	0.50	Crosta and Agliardi (2004)
Talus with large blocks, devoid of vegetation	0.82–0.85	0.3–0.33	Pfeiffer and Bowen (1989)
Talus	0.70	0.30	Lan et al. (2009)
Talus devoid of vegetation	0.82	0.32	Hoek (1987)
Talus with vegetation	0.80–0.83	0.28–0.30	Pfeiffer and Bowen (1989)
Talus with vegetation	0.80	0.32	Hoek (1987)
Talus, forested	0.60	0.30	Crosta and Agliardi (2004)
Cemented talus	0.70	0.50	Crosta and Agliardi (2004)
Muddy terrain with bushes	0.30	0.15	Lan et al. (2009)
Glacial deposit	0.50	0.25	Crosta and Agliardi (2004)
Glacial deposit, forested	0.40	0.20	Crosta and Agliardi (2004)
Colluvial deposit	0.65	0.20	Crosta and Agliardi (2004)
Colluvial deposit, forested	0.50	0.25	Crosta and Agliardi (2004)

Source: From Giani (1992) and other sources

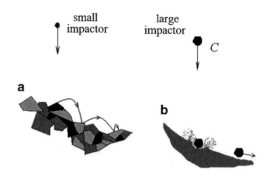

Fig. 8.10 Impact on a talus of a small (**a**) and of a large block (**b**)

(Fig. 8.10a), the latter bounces as if collision occurred on a hard ground. The randomly orientated talus blocks thus behave as random scatterers. If the talus blocks are smaller, the impact will be like the one shown in Fig. 8.8 and the coefficient of restitution will be small.

The normal coefficient of restitution also depends on the velocity (see also the discussion ←Box 5.1). A practical interpolation is provided by the relation given by Pfeiffer and Bowen (1989), Eq. 8.25. According to this formula, at a speed of 20 m/s the coefficient of restitution is reduced to less than 20% from its value at small velocity.

Rolling friction coefficients are typically of the order $\mu_R = 0.4$ for rock (Azzoni and De Freitas 1995). They increase to some 0.6–0.75 for loose blocks, while values of the order 0.55 have been measured on compact blocky terrain. Values of the order 0.4–0.5 have been found in experiments with glass spheres on fine sand (De Blasio and Sæter 2009a); μ_R also depends on the size of the sphere (smaller spheres have higher friction coefficient) and on the slope angle.

8.4.2.1 Impact on Ditches and Nets

Different kinds of artificial structures have been devised to mitigate the effect of rockfalls. Ditches are indicated especially when the block falls sub-vertically and the impact occurs with the maximum transfer of energy. Design charts for ditches are useful to plan the best shape in terms of width and depth (see, e.g., Giani 1992) and a layer of loose granular medium is often shed at the base to further diminish the coefficient of restitution.

Nets or catch fences (Fig. 8.11) have different design depending on the energy they need to absorb. Blocks of very low energy (less than 10 kJ corresponding to a 40 kg block falling from a height of 25 m) may be rigid. Blocks of energy lower that 250 kJ (corresponding to the free fall energy of a block of 1,000 kg from a height of 25 m) require mobile parts and special energy dissipators based on plastic deformation or friction of the base and of the ring bundles. Energy of the order 5,000 kJ are considered very high but still can be managed with modern nets provided by specialized companies.

Fig. 8.11 Rockfall nets

8.4.3 Block Disintegration and Extremely Energetic Rockfalls

Blocks may disintegrate during impact with the terrain. The energy released by a block of mass $m = \rho V$ falling from a height H is $\rho g H V$. While for energy of the order of 100 J/kg a block may break into few pieces, for much greater energies of the order of some *MJ* the block disintegrates into a myriad of smaller grains, often including powder-size broken rock (Fig. 8.12). The energy released may be sufficient to create a weak shock wave around the point of impact, and fell trees several hundreds of meters ahead of the impact point. Seismic waves are produced in the zone of impact (Deparis et al. 2008).

Such events happened during the initial failure of the Nevados Huascaran, when the initial free fall of ice and rock on the glacier 511 created concentric stripes on the icy surface. Perhaps the violence of the impact contributed to the high energy acquired by the rock avalanche on the glacier.

On July 10, 1996, two huge blocks in the Yosemite National Park, California, plummeted from a height of 550 m in free fall, disintegrating at once and producing a shock wave and a wind at a speed of 110 m/s that felled about 1,000 trees. Tree trunks were also scoured by the abrasive broken rock, while the suspension of thin particles produced a temporary darkness. Table 8.2 summarizes some of these events with the relevant literature.

Fig. 8.12 Some blocks fall from height sufficient for complete disintegration. Western Norway

Table 8.2 Data on extremely energetic rockfalls

Evento	Volume (Mm³)	Fall height (m)	Velocity (m/s)	Energy (MJ)	Notes
Nevados Huascaran (1970)	50–100	600	78–200	400–5,400	Shock wave followed by abrasion on the trees
Happy Isles, Yosemite National Park, California (1996)	0.03	550	110–120	0.53	110 m/s wind and felled trees. Abrasion on tree trunks.
Paretone, Gran Sasso (Italy 2006)	0.03	1,500	<170	1.2	Felled trees. Abrasion on tree trunks.

Source: From Wieczorek et al. (2000) and Bianchi Fasani (2007)

8.5 Talus Formation and Evolution

Often the rate of rock fall at the foot of the mountain cliffs may be so high that blocks form heaps of detritus, called *taluses* or *screes*. In a talus, the fallen blocks determine in the long term the path for the succeeding ones. It has been suggested that this may lead to a kind of self-organization of the talus, whereby blocks tend to stop in a position proper to their size, which results in a stripe-like pattern with small blocks distributed at the top of the talus and boulders at the terminus.

8.5.1 Kinds of Talus and Their Structure

Figure 8.13 shows some examples of taluses. Lengths range between some meters to hundreds of meters. Taluses are often associated to steep rock headwalls like along gorges and canyons rapidly excavated by mountain streams (Fig. 8.13c).

Taluses are often cut by transportation facilities. Because the deposit is poorly consolidated, increasing artificially the angle of repose of a talus often leads to instability; this causes immediate hazard along the new cut and in the long term a progressive creep and sliding of the talus material. In addition, single boulders detached from the mountain side may damage properties at the foot of the talus and threat human life. Usually the largest blocks on a talus reach longer runout distance, which makes construction at talus foots particularly unsafe (Turner and Schuster 1996).

The most evident properties of a talus are the following.

1. Often a longitudinal grading is present. Fine grains are so more abundant at the top of the talus, and boulders at the base (see also Fig. 8.13).
2. Although taluses appear superficially as chaotic heaps, geophysical methods have demonstrated that they are often stratified (e.g., Sass 2006; Van Stejin et al. 1995). For example, Sass (2006) found low electrical resistivity at the top of the talus due to the presence of large voids, followed by lower resistivity at greater depths due to more compact granular arrangement where voids are filled with smaller blocks. At greater depths the electrical resistivity may increase again.
3. Angles in the highest part of the talus fall typically between 30° and 40°; they decrease to some 30–35° midslope, and less than 20° at the foot. Hence, talus slopes usually consist of a straight upper slope and a concave lower slope segment (Fig. 8.14).

8.5.2 Physical Processes on Top of Taluses

8.5.2.1 Observations

Pěrez (1998) observing rock falls on talus slopes in the Californian Cascades noticed that most blocks stopped in the region of first impact. However, larger blocks were seen to roll down, traveling 450 m from the cliff. Trunks of pine trees hit by blocks some tens of cm across demonstrate the commonness of block bouncing on top of the talus.

Marking some of the blocks, Gardner (1969) showed that about half of the superficial blocks on talus slopes adjacent to Lake Louise in the Canadian Rocky

Fig. 8.13 Some examples of rock talus. (**a**) A talus in southern Norway with large boulders threatening the houses below. (**b**) Another talus in Southern Norway shows a marked longitudinal increase of block size

Fig. 8.13 (continued) (**c**) Vast taluses are often formed in correspondence of fast-excavated gorges. In the example shown, the erosion was produced by the catastrophic water flow that followed the drain of a glacial lake during early Holocene. Jotulhogget, Rondane (Norway)

Fig. 8.13 (continued) (**d**) Taluses with flat blocks may exhibit higher angle of repose. (**e**) A small-scale talus (10 m in length) shows nicely the longitudinal grading

Mountains creep erratically with displacement up to 70 m during the 2-years observation time. Creep on talus surfaces may also be revealed by bent tree trunks. Collapses of sectors of a talus occur especially on the talus apex (Pĕrez 1998; Sass and Krautblatter (2007)). These processes may lead to a partial re-distribution of blocks so that small blocks are carried in the mid-slope like in Fig. 8.14a.

8.5.2.2 Theoretical and Analogic Models

In the simplest view, the talus formation is related to the energy input by fall of single blocks. When the energy of a block is lost during the travel down slope, and the loss of energy is not counterbalanced by the gain in energy due to fall in the gravity field, the block will stop and contribute to steepening the slope. In the opposite case it will accelerate and decrease the slope angle. The small fragments, which predominantly bounce on the irregular surfaces of the talus, tend to be trapped in the interstices between ambient blocks, stopping early. In contrast, larger fragments travel longer partly because they avoid being trapped by the roughness of the surface, and perhaps also because the rolling friction coefficient decreases with size. These simple considerations seem to explain at least the longitudinal sorting observed in many taluses.

To investigate the propagation of grains on top of a talus, Statham (1972, 1976) conducted a series of experiments where a grain of radius R moved on a bed where grains of size R_0 were glued to a tiltable table. Increasing the tilting angle, the mobile grain began to move at a certain angle ϑ. Performing the measurement of the critical number for a statistically significant number of experiments, it was determined that

$$\tan \vartheta = \tan \phi + k \frac{R_0}{R} \tag{8.28}$$

8.5 Talus Formation and Evolution

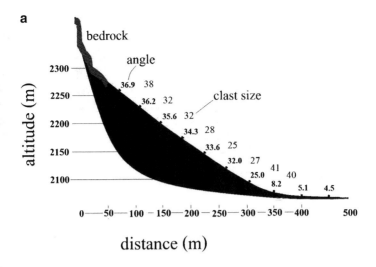

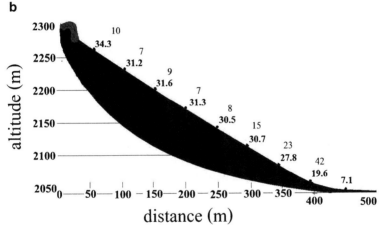

Fig. 8.14 Two typical sections of talus slopes with indicated angle and clast size. (From Pěrez 1985, modified.) Note that in the talus at the top (**a**) the clast size is nearly uniform along slope, while the one at the bottom (**b**) exhibits a marked longitudinal grading

where k is a constant ($0.17 < k < 0.26$), and ϕ is close to the internal friction angle. Thus, when the movable grain is very large in comparison to the glued grains, the friction angle is at the smallest. However, when the falling movable grain is smaller, it can hardly propagate without falling into holes, and its apparent friction coefficient increases.

In one recent analogical experiment, sand grains were dropped on a small board covered with non-glued grains (Sæter 2008). Figure 8.15 summarizes some results. Because small particles falling on large ones essentially bounce randomly with predominance for the direction along slope (Fig. 8.9), they tend to stop in the fall zone with an exponential tail of particle abundance as a function of the distance

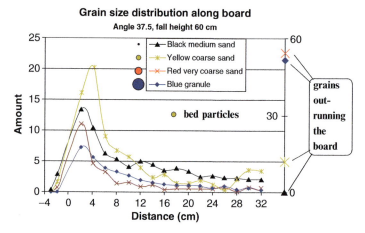

Fig. 8.15 Distribution of particles along an experimental board inclined at an angle of 37.5°, (close to the angle of repose) and covered with a thick layer of *yellow* sand grains of sieve diameter 0.5–1 mm. Sand grains of different size are dropped onto the board from a height of 60 cm. *Black* grains of diameter 0.25–0.5 mm are smaller than the yellow bed grains, while the *red* (diameter 1–2 mm) and the *blue* (diameter 2–4 mm) are larger. The figure shows the percentage of grains for a given distance from the point of impact. While the small black grains tend to stop at the point of impact and decrease in number as a function of the distance, larger grains (*red* and *blue*) either stop early within the first few cm from the impact point, or roll all the way down the table, outrunning the board. Note that yellow grains (same size as those resting on the board) are those stopping more efficiently in the zone of impact

from the impact point. In contrast, large particles lose most of their energy in the first impact (Fig. 8.9) and then may either stop at the point of impact, or start rolling down slope. In the latter case, the particles often roll to long distance, overrunning the board. As a result, large particles are abundantly distributed at the apex and at the end of the experimental board, and absent in the middle (Fig. 8.15).

In another experiment, grains of different size were cast at the same time to study the collective behavior at the surface of the granular heap (De Blasio and Sæter 2009b). At the beginning of the experiment, the small grains tend to stop at the apex, whereas the largest ones roll down to the slope break in accordance with the results of Fig. 8.15. However, in a later stage a superficial creep is observed, followed by a series of avalanches. Avalanching is preceded by vertical segregation similar to the "Brazil nuts effect", which brings small particles to the bottom (←Sect. 5.5.2). Because smaller grains have a lower angle of repose, the inverse grading gives rise to instability and to the observed avalanches. It is tempting to interpret the layered structures observed inside taluses as relics of periodic avalanching. Also the puzzling zones of low electrical conductivity observed by Sass (2006) could be the remnants of old collapses of the talus surfaces as indicated by the experiments, even though some caution should be used when extrapolating the results from a small board to the field (Fig. 8.16).

8.5 Talus Formation and Evolution

Fig. 8.16 An experiment in which grains of different sizes dropped on an experimental board initially exhibits longitudinal sorting (**a**) and (**b**) and then avalanches (**c**)

Snow residing along the rockfall chute for a long fraction of the year may prevent accumulation at the foot of the cliff. Blocks travel by sliding and rolling, and significantly lubricated by snow may reach long distance where they form a ridge known as a protalus rampart (Fig. 8.17). Ridges are characteristically some m

Fig. 8.17 Blocks with unusual runout, probably falling on snow, have contributed to the building of this talus in the central Italian Alps. Beloved daughter standing for scale in the center

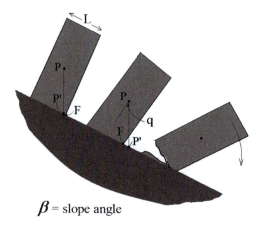

Fig. 8.18 Scheme for the analysis of the toppling failure

high, but some as high as 20 m or more have been exceptionally reported. The smallest blocks are more likely trapped in the soft snow, which explains the scarcity of small blocks in the ridges.

8.5 Talus Formation and Evolution

Fig. 8.19 (a) Topple failure in Norway (Telemark) resulting in strongly comminuted material. The fragmented material in the background derives from the disintegration of a columnar cliff about 80 m high

Fig. 8.19 (continued) (**b**) The same seen from far distance evidences the deposits deriving from the disintegration of the rock columns. (**c**) Multiple topple failure due to strong erosion at the base of a more erodible layer. Near Carboneras, southern Spain

8.6 Topple

Topple is a columnar rock fall pivoted at the base. The rock normally outcrops in vertical independent cliffs leaning downhill. The most advanced cliff becomes more unstable with time, and suddenly fails (Fig. 8.18, Dikau et al. 1996). Normally the mass falls forward, but backward movements are also possible.

A prismatic column of height H is statically stable if the projection of the center of mass P' of the prism on the terrain falls within the most forward point F of the prism foot, F (Fig. 8.18). As the base weakens, P' will lean toward a more advanced position until it crosses F. In this situation a torque sets in that tends to rotate the mass. If we assume that a certain cohesion C bonds the base of the prism at the terrain, the condition of instability becomes

$$\frac{gH^2 \rho \sin q}{L} > C. \qquad (8.29)$$

where q is the angle indicated in the figure and L is the prism length. If q increases with time, collapse will sooner or later occur. To find the maximum angular velocity $\dot{\Omega}$ reached by the falling columnar rock, we equate the potential energy due to the change in the position of the center of mass $Mg\frac{H}{2}(1 + \sin \beta)$ to the rotational energy $\frac{1}{2}I\dot{\Omega}^2$ where $I = \frac{1}{3}MH^2$ is the moment of inertia, and find

$$\dot{\Omega} = \sqrt{\frac{3g}{H}(1 + \sin \beta)} \qquad (8.30)$$

From which we also obtain the velocity of impact of the center of the column as $U_C = \frac{H}{2}\dot{\Omega} = \sqrt{\frac{3gH}{4}(1 + \sin \beta)}$ and the velocity of the top point as $U_C = H\dot{\Omega} = \sqrt{3gH(1 + \sin \beta)}$.

The impact with the ground is often capable to produce blocks of disintegrated material (Fig. 8.19a, b).

The fall and catastrophic disintegration consequent to toppling undoubtedly represent a hazard, even though the disintegrated material usually remains locally and will not flow as a rock avalanche (Fig. 8.19a, b). Erismann and Abele (2001) have also suggested as further hazard the possibility that the upper cap of the toppling columns, if loose, may be cast at higher speed and distance.

Another kind of toppling failure is shown in Fig. 8.19c. In this case, a relatively hard rock layer rests on much more erodible sediment. When the latter becomes worn and buttressing diminishes, a sudden topple of the harder cap may result. However, as also the figure shows, usually the block stops locally after a short sliding.

General References Chapter 8: Giani (1992); Turner and Schuster (1996).

Chapter 9
Subaqueous Landslides

On July 17, 1998 an earthquake shook the north-eastern shores of Papua New Guinea. The earthquake itself was not a particularly powerful one, and did not elicit much concern among the population, used to live on a seismic land. But after some minutes the sea level began to fall rapidly, unveiling the sea bottom along the shore. After short time, a first pulse of a series of three tsunami waves was travelling landward. Water towered 15 m above normal level, killing about 3,000 persons.

There is more awareness of the hazard represented by tsunamis after the Indian Ocean catastrophe of December 2004. The enormous devastation was also a consequence of the high intensity of the earthquake. How could then a big tsunami like the one of Papua New Guinea be triggered by a comparably weak earthquake? It is now believed that the tsunami was not caused directly by the earthquake itself, but by a catastrophic landslide mobilized by the earthquake. Understanding the triggering and dynamical behavior of subaqueous landslides is important not only for tsunami generation, but also to appreciate the continuous evolution of the ocean floor.

Subaqueous gravity flows may be really huge; the deposits of the largest landslides could blanket an area of the size of a European country. The material feeding underwater landslides is often made of clastic sediments, such as fine-grained glacial or fluvial material dumped into the sea basin. Whereas on land very large accumulation are rare, volumes of sediments available for subaqueous mass wasting reach thousands of cubic kilometers. The reduced apparent gravity in the sea due to Archimedean buoyancy also results in different mechanical balance of the forces. Together with the gentle gradients in the sea, this allows for large accumulations of sediments before instability occurs. Rock avalanches, which are very common in mountain areas, in the subaqueous setting are usually limited to steep areas, like the sides of volcanic islands.

Submarine landslides are more mobile than the subaerial counterparts, reaching fahrböschungs as low as 0.01 or even less. This is astonishing, considering that water exerts a much greater drag resistance than air, and that the effective gravity pull is lower underwater. Nevertheless, the understanding of the dynamics of subaqueous mass wasting is still at the beginning; many basic problems are still open.

The figure below shows the illuminated swath bathymetry of the Hinlopen-Yermak landslide. The color code in the lower panel shows the water depth. Note the scar and the hummocky surface in the distal part (original figure courtesy of Maarten Venneste).

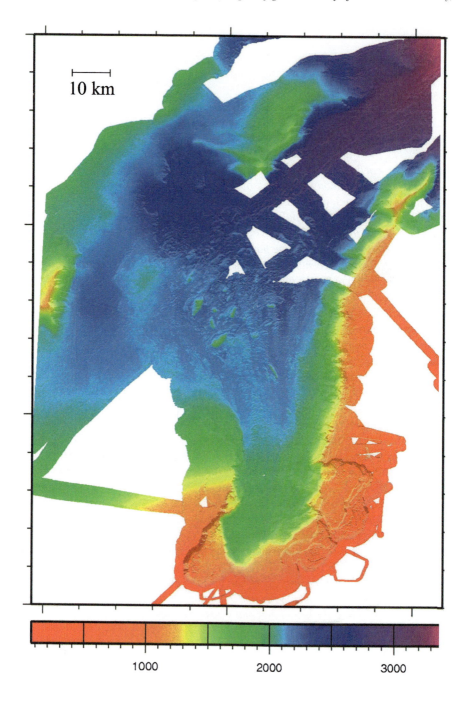

9.1 Introduction and Examples

If we could remove the cap of sea water from the globe, the ocean floor would appear very different from the scenery on the Earth's surface. The first thing one could notice is a much gentler landscape than on land. The ocean would appear as a flat area only locally punctuated by slopes greater than 5°. The surface would also reveal predominance of clastic sediments. The science of marine geology has been modernized by the introduction of new technologies such as the side-sonar mapping, swath bathymetry, and high-resolution seismic surveying. Box 9.1 is devoted to a short introduction to the subaqueous environment.

9.1.1 Some Examples in Brief

9.1.1.1 The Bear Island Debris Flows

Figure 9.1 shows the deposit of parallel debris flows in the Arctic off the Bear Island (latitude between 72°N and 74°N). The sediment that gave origin to the mass movement is mostly glacial clay dumped by glaciers along the continental margin, where it accumulates in huge masses. In this case, the debris flow has formed a series of independent lobes, each about 30 m thick and 10 km wide. The front of the debris flow lobes reaches the outstanding distance of more than 150 km despite the average slope of only 0.7°! No debris flow on land is capable of reaching this mobility, corresponding to a FB = 0.012. One of the most mobile subaerial mudflows is the long-runout lahar of the Nevado del Ruiz which, due to high water content in the pyroclastic material, has gained a FB of about 0.05 for an estimated volume of one hundredth of cubic kilometer. The Bear Island debris flows have been four to five times more mobile.

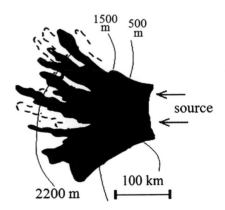

Fig. 9.1 The outline of the debris flows off the Bear Island, Arctic ocean

9.1.1.2 The El Hierro Rock Avalanche, Canarian and Hawaiian Aureoles

Figure 9.2 shows an example of a subaqueous rock avalanche of the Canarian island of El Hierro (Urgeles et al. 2003). Similar to the Hawaii, the Canarian Islands are a typical chain of oceanic islands created by a hotspot in the Earth's mantle. The magma initially accumulates at the sea bottom, until the volcanic edifice reaches the water level. In these conditions, the edifice remains relatively steep,

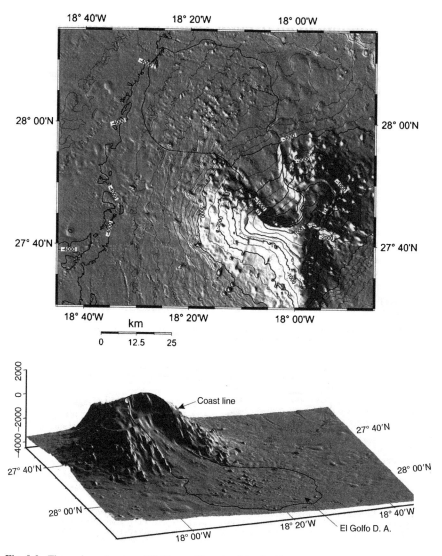

Fig. 9.2 The rock avalanche of El Hierro, Canarian Islands. (Images courtesy of Roger Urgeles)

because the magma cools and stops early in contact with water. The resulting islands have steep flanks compared to other oceanic features, and can fail catastrophically. The figure shows the scarp, partly subaerial and partly subaqueous, of a massive rock avalanche 150–180 km^3 corresponding to 3% of the island mass). The scar is linked to a 200 m thick submerged deposit, which develops a wide fan. The slide is about 70 km long; with a fall height of 6,500 m, the resulting FB results about 0.093, certainly low but still greater than that of the Bear Island debris flows. The mass failure took place about 15 ka ago. The islands themselves are young: subaerial rocks have been dated to slightly more than 1 Ma. In the fan area large blocks have been detected (1–2 km in diameter and 300 m high).

Similar landslide deposits form an aureole around the Hawaii, of area much greater than the surface of the islands themselves. The island of Molokai is missing of the main crater, slid in the ocean. Nowadays the scars of the landslides appear as cliffs parallel to the shore. The tsunamis related to these landslides must have been huge, as demonstrated by deposits of corals found well above the present shoreline. Also the southern part of the Kilauea is slowly sliding, and it is not known whether it will come to a halt or collapse in a catastrophic landslide.

9.1.1.3 The Hinlopen-Yermak Landslide

The remains of this old landslide in the Arctic are shown in the opening figure of the chapter Vanneste et al. (2010). The landslide developed from siliciclastic sediments during the Last Glacial Maximum, a period of low sea level, as ice was sequestered in the polar caps. Combined with increased ice volume, this resulted in intense glacial-tectonic stress. The landslide has volume of 1,350 km^3 and runout distance exceeding 300 km, for a total affected area of 10,000 km^2; the FB is only 0.01. Like other landslides in the boreal region of Europe (see also Storegga, →Box 9.3), this one was retrogressive. The deposit exhibits enormous tabular blocks with steep walls, several kilometers across and 100 m high. Blocks, which are partially buried by the landslide material, were rafted by the landslide itself. The slide probably generated a high tsunami.

Box 9.1 One Step Back: The Ocean Environment

The continental margins

Figure 9.3 shows the ideal section of a typical passive continental margin. The shelf, which is typically 50–100 km long, terminates abruptly in correspondence of the continental slope, where angles increase to some 5°. The length of the continental slope varies between 50 and 100 km. The foot of the continental slope gently develops into the ocean floor; the transition zone is also called the continental rise. The continental slope is often cut by huge deep submarine canyons similar to the gullies incised on land by mountain torrents, probably produced by turbidity currents (→Sect. 10.5).

(continued)

Box 9.1 (continued)

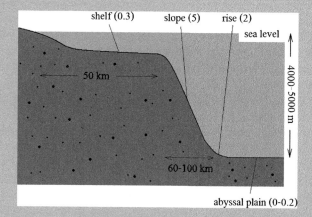

Fig. 9.3 An ideal section through a passive continental margin. The margin can be sub-divided into the continental shelf, the continental slope, and the continental rise, which gradually terminates into the ocean floor. In *parenthesis* the typical slope angles in degrees

The ocean floor
The ocean floor lies almost everywhere at a water depth between 4,000 and 5,500 m. Although usually very flat (it is also called the abyssal plain), it is sometimes punctuated by seamounts, trenches, and island arcs.

The abyssal plain starts at the end of the continental rise, where angles rarely exceed one part in 1,000. The abyssal plain is a universal feature of all the oceans, and aerially the most extended landscape on Earth. The reason for the perfect flatness is probably the efficient spread and uniform sedimentation of fine sediments, initially carried in the depths by turbidity currents and subaqueous landslides. Other structures worth mentioning are mid-ocean canyons, viz., wide channels up to 8–10 km across meandering along extensive sections of the abyssal plains.

Seamounts
Seamounts rise some thousands of meters above the sea floor. Most seamounts are extinct volcanoes. Seamounts may be the source of important and large landslides, such as in the Hawai'i and Canary Islands.

The mid-oceanic ridges
In the Atlantic ocean, the mid-Atlantic ridge is up to 2,500 km wide and 2,700 m high. Running almost exactly in the middle of the ocean, it witnesses the separation of African and Southern American continents. Ridges are formed by subaqueous eruptions of basaltic magmas in correspondence of the boundaries of tectonic plates, but are seldom the center of landslide movements.

Trenches and Island Arcs
With depths reaching 10 km or more, trenches are the deepest zones in the ocean and correspond to the geologically active zone regions of lithosphere

(continued)

> **Box 9.1** (continued)
>
> subduction. The sink of lithosphere is associated to strong earthquakes, relatively steep angles, and building of substantial deposits of clastic sediments. Because of the combinations of all these factors, numerous submarine landslides may occur along these regions. Trenches are often linked to arcs of parallel islands. Particularly striking is the ring of trenches surrounding the Pacific ocean.
>
> Most landslides occur in locations of high sediment budget and relatively steep slopes. These include the continental rise, the seamounts and island, and deep-sea trenches.
>
> The oceanic sediments are shortly illustrated in the (→GeoApp).

9.2 Peculiarities of Subaqueous Landslides

9.2.1 Types of Subaqueous Landslides

There exists a variety of subaqueous landslides.

- Rock avalanches derive from the collapse of rock, usually basalts of oceanic islands but also carbonate rocks. Perhaps a rock avalanche may change in a debris flow as it disintegrates and absorbs ambient water.
- Debris flows result from coarse clastic sediments with perhaps some percentage of marine clay. Examples include the Storegga (→Box 9.3) and Hinlopen-Yermak (←Sect. 9.1) landslides. There are also examples of sandy debris flows.
- Glacial clay-rich mudflows (up to 70% of clay and silt) are common along passive margins at high latitudes, where they are associated to glacial deposits. Volumes of each individual debris flow range from less than 1 to about 50 km^3 of sediments. Examples include the Bear Island mudflows (←Sect. 9.1).
- Rock falls also occur in the subaqueous environment. Boulders may fall from a coastal cliff and continue travelling underwater.
- Outrunner blocks are small blocks occasionally observed at the front of submarine landslides. Their distinctive trait is that they appear to have travelled for several kilometers with very little impediment ahead of the main slide. Although volumetrically small, they may be useful in understanding the dynamics of subaqueous landslides.
- Landslides may also occur in limited water basins like lakes and fjords (←Sect. 7.2), Fig. 9.4.

9.2.2 Differences Between Subaerial and Subaqueous Landslides

To appreciate the differences between subaqueous and subaerial landslides, it is important to keep in mind that the dissimilar environment influences the landslide products for a series of reasons.

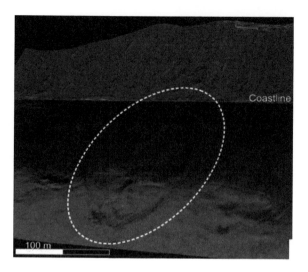

Fig. 9.4 Landslides in lakes are usually much smaller than those in the sea. A landslide in the Albano lake, Italy. Many of these debris flows tend to develop ring shape morphologies (enclosed in the *dotted white ellipses*) probably also because of the sudden slope break that characterizes the topography of closed basins and fjords (From Mazzanti and De Blasio 2009)

9.2.2.1 The Slope Angle

As we have seen earlier, the average slope angles are lower in the subaqueous environment. Thus the gravity pull, proportional to $\sin \beta$ is accordingly smaller.

9.2.2.2 The Volume Available

On land, the volume of material available for landsliding is limited for a variety of reasons. First, because slopes are steeper than under the sea, the probability that thick amounts of loose sediments will build up thick units is scarce. In contrast, the gentler angles present in the sea makes sediments more stable. Second, on land the erosion globally prevails over sedimentation. Locally, powerful deposition may take place due, for example, to rivers or glaciers, but the associated deposits are relatively thin. In contrast, large amounts of sediments are deposited underwater, for example at the deltas of large rivers. The continuous sedimentation coupled to the small but nonzero slope angle may lead to enormous depositional systems of clastic sediments, often poorly consolidated and under pressure due to rapid sedimentation, like in the Ganges fan. At high latitudes, like around the Arctic ocean, glacial systems may substitute rivers in the process of producing huge deposits in shallow waters. The failure of these enormous masses may lead to the largest landslides known on Earth, in the range of thousand of km^3.

9.2.2.3 The Material (See Also →GeoApp3)

In the subaerial environment, volcaniclastic deposits, morains, loess, colluvium, and talus are relatively common clastic sediments. In the ocean, some of these types may be present but are not volumetrically significant. Clays and sands of various origin, transported by rivers offshore to long distances, are more common.

The rocks involved in submarine rock failures are usually basalts building up seamounts and ocean islands. Rock failures may also occur along the continental shore and involve a larger variety or rock types like crystalline or carbonate rocks. In contrast, there is no limitation to the type of rock available for sliding in subaerial environment, even though the weakest rock types are obviously more likely to fail.

9.2.2.4 The Presence of the Water as Ambient Fluid

Water alters in a substantial manner the behavior of a slide, not only at failure (the presence of pore water under pressure may reduce stability) but also during the flow. Water exerts a much greater drag force than air; in addition, the buoyancy decreases the effective gravity acting on the sliding material. However, water seems to enhance the flowing of landslides (→Sect. 9.2.3).

9.2.3 The H/R-Volume Diagram for Submarine Landslides

Figure 9.5 illustrates the *H/R* diagram for a series of subaqueous landslides: subaqueous rock avalanches, debris flows, and outrunner blocks. For comparison,

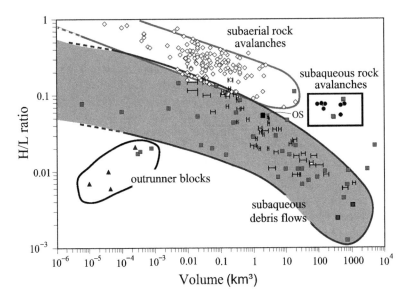

Fig. 9.5 *H/R* (FB) plot for subaqueous landslides. Debris flows are indicated with *grey squares*; the lobes of the Storegga sliding area are shown with *horizontal segment lines*, with the length indicating the uncertainty in the volume; the Storegga data provide one of the best sources of information. "OS" shows the data point for the subaerial Osceola debris flow. Data for subaqueous rock avalanches (*dots*) and outrunner (*triangles*) blocks are also shown, together with subaerial rock avalanches (*diamonds*). Similar data plots have been previously explored by Locat and Lee (2002) and Elverhøi et al. (2002), see also Edgers and Karlsrud (1982) (From De Blasio et al. 2006, slightly modified)

the data for subaerial rock avalanches (←Sect. 6.2.2) are also reported. Debris flows exhibit runout ratio as low as 0.01; in the most extreme cases the FB drops to 0.001. This implies that a fall of 10 m in the gravity field is capable of driving the debris flow horizontally for a distance of 10 km! The Osceola debris flow, one of the most mobile lahars on land (←Box 4.1, indicated with "OS" in Fig. 9.5), falls much higher on the plot than most of the subaqueous debris flows. Subaqueous and subaerial rock avalanches, on the other hand, fall approximately on the same fitting line. It appears that subaqueous rock avalanches reach lower FB because of their greater volume, but they are not intrinsically more mobile than the subaerial ones. Outrunner blocks (bottom left corner), reach a runout ratio of the order 0.005 or less, an exceptionally small value.

The conclusion from a diagram like that of Fig. 9.4 is unavoidable. The presence of water enhances mobility instead of acting as a hindrance as one would expect based on the presence of the drag force and the gentler slopes. The possible explanations for the extraordinary mobility of subaqueous landslides are examined in (→Sect. 9.6.2).

9.3 Triggering of Subaqueous Landslides (Especially Submarine)

While earthquakes are potential landslide triggers both on land and in subaqueous environment, subaqueous landslides may be activated by a variety of causes that have no counterpart on land.

1. Water seeping through rock fractures is an important source of instability in subaerial rock and soil masses. It can be inferred that a loose subaqueous mass, which is completely submerged, must also be significantly affected by water. If the resistance force at the bottom is purely frictional, the condition of stability Eq. 2.12 is unaffected by Archimedean buoyancy. This is because although buoyancy does affect the gravity pull, it also changes the component of gravity perpendicular to slope by the same ratio.

   However, the stability of a Coulomb-frictional material could be affected by excess pore water pressure. Let us consider again the example in Fig. 3.3 where a certain amount of sand is suspended in water. Before sand settles, pore pressure P_W and total pressure σ are comparable. Because the effective friction resistance force is of the form (←Sect. 2.2.2)

$$F_{res} = (\sigma - P_W) S \cos \beta \tan \phi \qquad (9.1)$$

   where S is the surface, ϕ is the friction angle and β is the slope angle, it follows that the sliding resistance is lower than when sand is deposited. A similar situation may occur as a consequence of fast sedimentation rate. In clastic

submarine sediments, water filling the pores may be present over normal hydrostatic pressure if the sedimentation rate has been so fast that water has not been completely expelled from the pores.
2. Gas hydrates are a mixture of ice and gas of low molecular weight such as methane (CH_4). They are solid and burn easily; their abundance makes them a possible energy source. The ideal places for the formation of gas hydrates are submarine environments dominated by high pressure and low temperature. Therefore clastic deep sediments are just right, especially at high latitudes.

   If the water pressure decreases because of eustatic changes of the sea level, isostatic uplift, or increase in local or global temperature, the gas hydrates can sublimate, turning into gaseous state. The gas in the pores consequently destabilizes the sediments. This effect might become relevant if the tendency to global warming will continue in the future. Box 9.2 examines how the massive collapse of landslides consequent to discharge of gas hydrates might have played a major role in an episode of extreme warming at the boundary between the Paleocene and the Eocene, with important consequences for the evolution of life.
3. During the ice ages, glaciers at high latitudes periodically bulldozed clastic sediments around the continental margins, causing them to slide. For example, the glacial sediments offshore western Norway have slid periodically during the Quaternary as the Scandinavian ice sheet reformed during the glacial episodes. Although the ultimate cause for the slides might have been related to earthquakes, the sediments were destabilized by mechanical action of glaciers.
4. Volcanic islands host the largest rock avalanches of the planet. Because molten basalts have very low viscosity (→Insert 10.1), single basaltic eruptions may reach distances of tens to hundreds or even thousands of kilometers. However, in contact with water, lava cools rapidly and stops. As a consequence, subaqueous eruptions rather than redistributing lava far from the source tend to produce a steep volcanic edifice, which may become gravitationally unstable. In addition, phreatic explosions and the presence of layers of pelagic sediments may promote instability on volcanic islands.
5. Waves close to the shore change periodically the pore pressure in loose sediments beneath the sea level due to the oscillation in the water height. The process, which is particularly accentuated during storms, may favor instability of sediments deposited under the water surface. Instability may also follow from the shear stress of the alternating motion of waves along the shore. Tides are much slower but can produce instability, too. Because the water column may change by several meters between low and high tides, pressure in the pore of loose sediment builds up if the permeability is too low to adjust to the actual pressure with time. Instability along cliffs may also result from the direct erosion of high-energy waves battering the foot of the cliff.

Box 9.2 External Link: Slope Instability, Landslides, Global Warming, and Biological Turnover in the Early Tertiary

The analysis of the relative abundance of the oxygen isotopes ^{16}O and ^{18}O in fossil animals and plants is a consolidated technique to assess paleotemperatures. It has been established that the period in which we live is one of the coldest of the last 65 million years, the time span including Tertiary and Quaternary eras. The highest temperatures, well above 10° higher than the present ones, are recorded for the Eocene period, when palm trees and tropical fish populated areas now at temperate latitudes. Here we focus on one particular event at the boundary between the Paleocene and Eocene periods about 54–55 Ma ago, termed PETM (Paleocene–Eocene thermal maximum). Figure 9.6 shows the thermal history across this boundary reconstructed with the analysis of the skeletons of microfossils. A sharp spike, taking place in less than 10,000 years, signals an increase in the temperature by some 5–10° (these values are also debated; some scholars favor lower excursions along the spike). Note that the organisms used for isotopic studies lived at diverse latitudes, from the tropic to subarctic regions; PETM thus represents a global event. Like many issues in paleoclimatology, the origin of this spike is still debated. However, one of the most promising explanations stems from slope instability and landsliding in the continental margins.

As a first clue, notice that the spike was preceded by a steady increase in the temperature at the end of the Paleocene. Another hint is the negative excursion $\delta^{13}C$ of the isotope ^{13}C of Carbon along the same spike, indicative of major release of carbon in the atmosphere. The possibility that the temperature spike is due to increase of greenhouse gas CO_2 appears very likely. But what caused the release of carbon in the first place? The amount of carbon estimated is so high that only one obvious source can be invoked: gas hydrates.

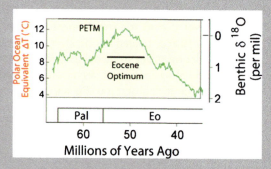

Fig. 9.6 The *PETM* from oxygen isotopic data (Modified from Wikipedia in agreement with the license GFDL)

(continued)

Box 9.2 (continued)

According to the model by Dickens et al. (1955), the high temperature of the late Paleocene released gas hydrates (methane, CH_4) sequestrated in the ocean surrounding the continental margins. The release of methane induced an increase in pore pressure of the sediment and promoted instability. Massive debris flows detached from the continental margin, probably at an unprecedented rate for the whole Phanerozoic. Travelling down-slope, landslides liberated the methane in their interior, increasing the amount of methane even more. It is well known that bacteria convert CH_4 into carbon dioxide according to the reaction

$$CH_4 + 2O_2 \rightarrow 2H_2O + CO_2. \tag{9.2}$$

Carbon dioxide would have increased the temperature due to its greenhouse potential, which is a possible explanation for the spike. In addition, in the marine realm the increase of CO_2 decreased the stability of calcium carbonate, lowering the lysocline (i.e., the level above which calcium carbonate is stable; below the lysocline the carbonate dissolves in water). This had a negative impact on organisms whose skeletons is made of calcium carbonate. Indeed, major biotic changes occurred in correspondence of the PETM, such as a marked decline of foraminifera and plants. However, as often occurred in the history of life, the event was not merely destructive, but promoted the radiation of some groups of organisms, such as the mammal orders of artiodactyls and perissodactyls on land.

Box 9.3 Brief Case Study: Storegga and the Inordinate Mobility of Large Submarine Landslides

Salient data
Name: Storegga landslide
Location: offshore mid Norway
Coordinates: 63 32′ N; 5′ 35′ E (the scarp)
Volume: about 3,000–3,500 km^3
Age: 8.2 ka
Kind: probably partly retrogressive
Landslide material: glacial clay, marine clay, ooze
Vertical fall (m): 3,000
Runout: 450 km
FB: 0.007
Volume: 3,000–3,500 km^3

(continued)

Box 9.3 (continued)

Dynamical characteristics: It is the largest landslide known on Earth; it exhibits a very long runout; it has generated a large tsunami well studied in the sedimentological records; it is the submarine landslide for which the best information is available.

The Storegga slide has been the subject of numerous recent studies. Figure 9.7 shows the general setting of the Storegga slide. The sea along the western coastline of Norway shows thick units of clastic sediments forming a very extensive plateau. The units comprise basal layers of biogenic ooze surmounted by quaternary sediments. The quaternary units are partly of glacial origin; however, during the interglacial periods, marine clay was deposited, so determining a layered structure of the plateau. The marine

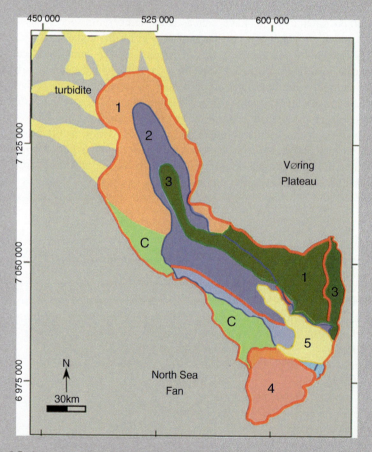

Fig. 9.7 General setting of the Storegga slide. The numbers *1–5* represent successive phases of the landslide. "C" represents a compression zone. Note also the turbidites (Reproduced from Elverhøi et al. 2010)

(continued)

Box 9.3 (continued)

sediments are much richer in clay (50–60% against the 30–40% of the glacial clays). The presence of marine clay determined an excess pressure due to the rapid sedimentation during peak glaciations compared to the expulsion rate of water; excess pressure of the order 20% higher than the hydrostatic has been measured in regions near the landslide. Considering also the high sensitivity of the marine clays, it follows that the layers of marine clay are units with decreased stability. The cyclic character of glaciations determined a periodic thrust of the sediment by the thick glacier. In this unstable situation, the Storegga landslide might have been triggered by an earthquake, also because seismic activity was greater during ice melting due to fast isostatic rebound. The failure occurred along one of the weak layers of marine clays.

The dynamics of the Storegga slide poses several problems. First, it is unclear whether the slide was retrogressive, even though it seems that it occurred in several phases, the first of which reached the maximum distance of about 400–450 km ("phase 1", Haflidason et al. 2005). Phase 1 was followed by a swarm of smaller lobes, notably those in the Ormen Lange region (area "3" in Fig. 9.7). The smaller debris flow lobes exhibit a runout length of 10–15 km. Thus landslides in the area were very different in terms of scale. Thick deposits of glacial clays such as those of the Storegga slide may achieve high shear strength (up to a fraction of MPa) due to compaction. With such values of the shear strength, no motion of the slide is possible at such gentle slopes (←Chap. 4). It thus remains to be explained how the phase 1 landslide could travel to a FB of only 0.008.

9.4 Forces on a Body Moving in a Fluid

9.4.1 General Considerations

When dealing with the dynamics of submarine landslides, it is important to account for the forces that the water exerts on the moving sediment. These are the same kind of forces acting on an aircraft and are thus well known from aeronautics.

The resistance encountered by an object moving against a fluid is called the drag force (←Sect. 3.3.5). It is the kind of force a person experiences walking against wind on a stormy day. If the medium is static, the drag force acts along the direction of movement. As an example, one can think of the force exerted by air against a hand leaning out of the window of a moving car. The air drag against subaerial landslides has been introduced earlier (←Sect. 6.4.3), and it was found to be negligible. It seems appropriate to re-introduce here the subject with more rigor, as it turns out that in water the drag force on landslides becomes essential.

A second force, acting perpendicular to the movement of the object, is the lift force. In the previous example of the hand, imagine bending the hand to an angle of attack α with respect to the horizontal. A vertical lift force will arise, which tends to lift up or pull down the hand depending on the angle of attack. The lift force is of fundamental importance in aerodynamics; it is the one permitting taking off of airplanes, while the drag force has a negative effect on efficiency as it counteracts the thrust of the engine.

The absolute values of the drag and lift forces are typically written in the following way

$$F_D = \frac{1}{2}\rho_F C_D U^2 S \tag{9.3}$$

$$L = \frac{1}{2}\rho_F C_L U^2 S \tag{9.4}$$

where S is a characteristic surface of the body, usually considered as the maximum cross section perpendicular to the direction of movement. Note that while in Chap. 3 the fluid density is indicated with ρ, here we use the symbol ρ_F since ρ stands for the density of the solid. The coefficients C_D and C_L, called, respectively, the drag and the lift coefficients, do not vary dramatically with the velocity. Thus, in a first approximation, it can be assumed that the drag and lift forces are proportional to the velocity squared. Of these two forces, the drag force is considered to be the most important for subaqueous landslides, while the effect of the lift is hardly ever considered. However, more recent work suggests an important role of the lift force especially for small slides. In the following, we consider only the drag force; the discussion about the lift is deferred to ($\rightarrow$Sect. 9.6).

9.4.2 Drag Force

9.4.2.1 Drag Force on a Sphere

Let us consider for sake of example a sphere of radius R and diameter $D = 2R$ immersed in a water flow (see also $\leftarrow$Sect. 3.3.5). The Reynolds number is

$$Re = \frac{DU\rho_F}{\mu} = \frac{2RU\rho_F}{\mu} \tag{9.5}$$

where μ is the viscosity of water. The drag force of the sphere at variable Reynolds number (from $Re \ll 1$ to $Re \approx 10^8$) is well known. When the Reynolds number is less than about 1, the flow is laminar. The drag force is in this case given by the Stokes formula

$$C_D = \frac{24}{Re}. \tag{9.6}$$

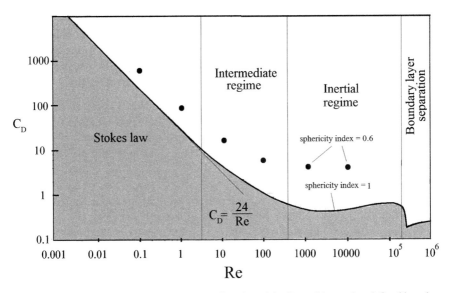

Fig. 9.8 The drag coefficient on a sphere as a function of the Reynolds number defined based on the diameter of the sphere. For the case of small particles it is customary to define the sphericity index as the ratio between the surface area of a sphere of volume equal to the one of the particle, and the surface area of the particle. For example, for a cube the index is 0.806. For a sphere it is evidently 1

This result can also be demonstrated theoretically starting from the Navier–Stokes equation. Figure 9.8 reports the drag coefficient as a function of the Reynolds number. The value 9.6 is referred to as the "Stokes law" in Fig. 9.8. Substitution into (9.3) gives an expression for the drag force

$$F_D = 6\pi R \mu U. \tag{9.7}$$

With increasing Reynolds number, vortices begin to form behind the sphere. As a consequence, the resistance increases and the drag coefficient as a function of Re flattens out (Fig. 9.8). The behavior remains stationary up to $Re \approx 10^4$ (inertial regime) after which the drag coefficient rises again reaching a maximum. Then C_D decreases by a factor three as a consequence of the boundary layer separation (Fig. 9.8). Note also that nonspherical bodies ("sphericity index" < 1) show higher drag than a sphere for the same Reynolds number.

9.4.2.2 More Asymmetric Bodies

The drag force on a body of whatsoever geometry may be difficult to establish because the flow pattern around the body that determines the drag force is in general complex. Let us consider a specific example: if the body is skewed with respect to the fluid flow (e.g., a cylinder whose axis is not perpendicular to the velocity vector) a torque will tend to line up the cylinder axis normal to the velocity at low Reynolds numbers. The drag coefficient will also depend on the orientation of the cylinder.

In the following, we examine some examples of drag coefficient that can be useful for the study of subaqueous landslides. We restrict ourselves to $Re > 1{,}000$.

- As a first example, let us consider a plate of height H and length W. The flow is normal to the surface (Fig. 9.9b). The drag coefficient is found to vary from 1 to 2 depending on the aspect ratio of the plate. Table 9.1 gathers some values.
- For a circular plate the coefficient is $C_D = 1.17$, close to the square plate when $H = W$.
- For a circular cylinder (Fig. 9.9c) the drag coefficient ranges between 0.63 when the cylinder long axis has the same length of the diameter to a value of about 1.2 when the length-to diameter ratio tends to infinity (Table 9.2).
- The drag coefficient for a square cylinder of infinite length is about two when the flow is perpendicular to one of the faces.
- A cone provides an example of axisymmetric body (Fig. 9.9d). When the flow is directed against the tip of the cone, the drag coefficient at $Re > 2{,}000$ ranges between 0.64 and 0.2 depending on the apex angle of the cone (Table 9.3).

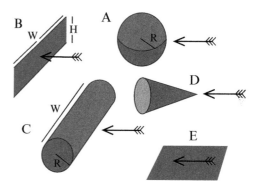

Fig. 9.9 Elementary forms for the drag coefficient

Table 9.1 The drag coefficient for a plate of sides W and H at Reynolds number greater than 1,000 for different aspect ratios of the plate. The flow is perpendicular to the plate

$W/H \rightarrow$	1	5	10	20	30	∞
C_D	1.18	1.2	1.3	1.5	1.6	1.95

Table 9.2 The drag coefficient for a cylinder of axis length W and radius R at Reynolds number greater than 1,000 for different aspect ratios. The flow is perpendicular to the axis of the cylinder

$W/(2R) \rightarrow$	1	5	10	20	30	∞
C_D	0.63	0.80	0.83	0.93	1.00	1.20

9.4 Forces on a Body Moving in a Fluid

Table 9.3 The drag coefficient for a cone of apex angle ϕ. The flow is directed against the tip of the cone

$\phi \rightarrow$	120	90	60	30
C_D	0.64	0.52	0.38	0.20

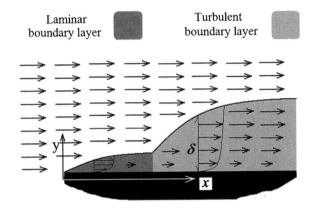

Fig. 9.10 Illustration of the laminar and turbulent boundary layers

9.4.3 Skin Friction

Skin friction is the drag exerted when the relative movement of the fluid occurs parallel to the object (Fig. 9.9e).

Let us consider a fluid flow parallel to a plate of finite length L (Fig. 9.10) with the x-axis and y-axis, respectively, parallel and perpendicular to the plate. The fluid in direct contact with the plate is at rest, while at large distance the fluid flows with a certain velocity U. Therefore, there is a velocity gradient along the direction perpendicular to the flow, which gives origin to a shear force on the plate contributing to the drag force. This form of the drag force is called skin friction.

The flow pattern along the plate depends on the horizontal distance x from the edge of the plate (Fig. 9.10). The flow is affected by the presence of the plate for a thickness δ from the plate called the boundary layer. In the boundary layer, the velocity depends on the distance y from the plate. Outside the boundary layer, the velocity becomes constant and equal to U.

Close to the leading edge, the flow regime in the boundary layer is laminar and the velocity of the fluid is found to be (Schlichting 1960)

$$U(y) = U \left[\frac{3}{2} y_* - \frac{1}{2} y_*^3 \right] \tag{9.8}$$

where $y_* = y/\delta(x)$ is the dimensionless height, and

$$\delta(x) \propto \sqrt{x} \tag{9.9}$$

is the thickness of the boundary layer as a function of the distance from the leading edge. Beyond a certain length, the boundary layer becomes turbulent (light grey in Fig. 9.10) and the velocity profile becomes

$$U(y) = U y_*^{1/7} \tag{9.10}$$

whereas the thickness of the boundary layer increases as $\delta(x) \propto x^{4/5}$. The force acting locally on the plate at the distance from the leading edge x is calculated as

$$D(x) = W \rho_F \int_0^{\delta(x)} [U - U(y)] U(y) dy \tag{9.11}$$

where W is the width of the plate (perpendicular to the figure). It is evident that because $U(y) \propto U$, a quadratic dependence of the skin friction term on the velocity follows, $D(x) \propto U^2$. The formulas are more complicated for the turbulent boundary layer. We do not show these profiles in detail, also because the interesting question is the drag friction exerted by the whole plate. We limit ourselves to provide the most widely used formula for the smooth plate in the form of the so-called Prandtl–Schlichting equation, which also accounts for the initial laminar behavior

$$C_{SF} = 0.455 [\log Re_L]^{-2.58} - A Re_L^{-1} \tag{9.12}$$

where $Re_L = VL/\nu$ is the Reynolds number based on the plate length L, and A is a coefficient ranging between 1,050 at low Re_L up to 8,700 at $Re_L \approx 3 \times 10^9$ (Schlichting 1960). At such high Reynolds numbers, the first term of (9.12) is numerically more significant. Typical values for the skin friction on the lateral face of a submarine landslide with L of the order 100–500 m of length becomes of the order

$$C_{SF} \approx 10^{-3} \Rightarrow 2 \times 10^{-3}. \tag{9.13}$$

The drag coefficient for a rough plate changes radically if the roughness affects significantly the boundary layer. The formula for the skin friction coefficient for a rough plate is

$$C_{SF} \approx [1.89 + 1.62 \log(L/k)]^{-2.5}; \quad 100 < L/k < 10^6 \tag{9.14}$$

where k represents the roughness length.

9.4.4 Added Mass Coefficient

When a body is accelerating in a dense medium like water, an extra force arises due to the change in the velocity field of the fluid in contact with the body. Because the force is proportional to the body acceleration, this force can be formally expressed

9.4 Forces on a Body Moving in a Fluid

as if the mass of the body changed, hence the name of added (or virtual) mass. It must be emphasized that this force, which is present also in an ideal (i.e., nonviscous) fluid, has nothing to do with drag force, which is due to viscous dissipation in the fluid. This is also clear from the fact that the drag force is approximately proportional to the square of the velocity (and to the velocity in the Stokes regime), whereas the virtual mass accounts for a force proportional to the acceleration. The equation of motion of an object moving in a fluid can so be written as

$$(m + m_v) \frac{dU}{dt} = F \tag{9.15}$$

where F are the external forces and m_v the virtual mass that can be calculated as explained in Box 9.4 for objects of different shapes. The sum of bare and virtual mass is also called the effective mass.

Box 9.4 Calculation of the Added Mass Coefficient

Let us examine the problem in more detail and work out a general expression for the virtual mass. Figure 9.11 shows the velocity field of a viscous fluid generated by a cylinder of infinite length moving with a certain speed. Because the velocity of the fluid has to satisfy the no-slip boundary condition with the surface of the body, the velocity of the fluid at the surface of the body must be equal to the velocity of the body itself. This implies that the velocity field of the fluid is determined by the velocity of the body.

At a certain time t, the total kinetic energy of the fluid is

$$T = \frac{1}{2} \int d^3r \rho_F V^2(x, y, z) \tag{9.16}$$

where $V^2(x, y, z) = u^2(x, y, z) + v^2(x, y, z) + w^2(x, y, z)$ is the velocity field of the fluid. Let us suppose that the cylinder is accelerating. At an instant of time $t + dt$, the velocity of the cylinder has increased and so also the velocity field of the fluid and the kinetic energy (9.15). A certain amount of energy has thus been used up by the cylinder to accelerate the water around it.

In the absence of dissipation, the force acting on a body can be calculated as the negative gradient of the potential energy. For example, the potential energy of a body in the gravity field is mgy. The gradient of this potential energy gives the force as $-mg\hat{k}$, where $\hat{k}$ is the vertical versor. Likewise, the force acting on the cylinder can be calculated as the space derivative of the kinetic energy with respect to the coordinate x along the direction of motion of the cylinder. Assuming a perfect fluid (the drag contribution if present can be summed separately)

$$F = -\frac{dT}{dx} = -\frac{d}{dx} \frac{1}{2} \int d^3r \rho_F V^2(x, y, z) \tag{9.17}$$

(continued)

Box 9.4 (continued)

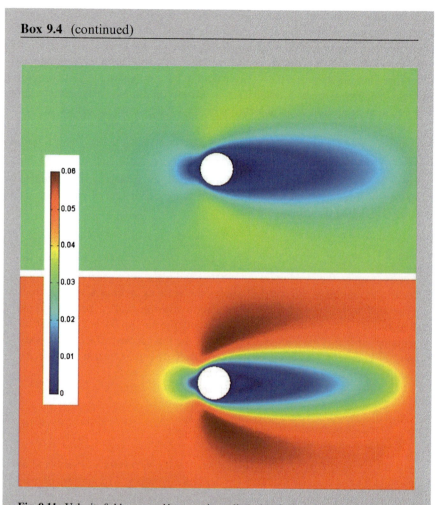

Fig. 9.11 Velocity field generated by a moving cylinder in a fluid, seen by a frame attached to the cylinder. *Upper*: the cylinder is moving at a speed of 0.03 m/s. *Lower*: the cylinder is moving at a speed of 0.05 m/s. The viscosity of the fluid is 1 Pa s. The change in the field velocity at different cylinder speeds determines different total kinetic energies. If the body is accelerating, this results in a modification of the effective mass of the body, the virtual mass. The velocity is in m/s. The horizontal length is 5 m. Figures generated by finite element computation

We assume that the velocity field is proportional to the velocity of the body itself $V(x, y, z) = U f(x, y, z)$ where $f(x, y, z)$ is a field dependent on the geometry, which in the present approximation is independent of the velocity U. This approximation is valid for potential flow. Writing the derivative $\frac{d}{dx} = \frac{1}{U}\frac{d}{dt}$, Eq. 9.17 can be written as

(continued)

Box 9.4 (continued)

$$F = -\frac{d}{dt}\frac{1}{2}\int d^3r\rho_F f^2(x,y,z) = -\frac{1}{U}\frac{d}{dt}\left[\frac{1}{2}\int d^3r\rho_F U^2 f^2(x,y,z)\right]$$
$$= -\frac{dU}{dt}\int d^3r\rho_F f^2(x,y,z) \qquad (9.18)$$

and remembering the definition of the virtual mass m_V (we can set the bare mass to zero without loss of generality)

$$F = -m_V \frac{dU}{dt}. \qquad (9.19)$$

By comparison of (9.18) and (9.19) we get

$$m_V = \int d^3r\rho_F f^2(x,y,z) = \frac{1}{2}\int d^3r\rho_F \left[\frac{V(x,y,z)}{U}\right]^2. \qquad (9.20)$$

As an example, let us calculate m_v for an ideal fluid.

For the ideal fluid it is possible to calculate analytically the velocity field as a solution of Laplace's equation. This technique is beyond the scope of the book, and we make use of the expression of the velocity field without demonstration. Adopting cylindrical coordinates (→MathApp), the velocity field around a non-spinning cylinder of infinite length and radius R is

$$v_r = U\frac{R^2}{r^2}\cos\theta \quad v_\theta = U\frac{R^2}{r^2}\sin\theta \qquad (9.21)$$

and the square of the velocity is so

$$v_r^2 + v_\theta^2 = U^2\frac{R^4}{r^4}\cos^2\theta + U^2\frac{R^4}{r^4}\sin^2\theta = U^2\frac{R^4}{r^4}. \qquad (9.22)$$

From the expression of the volume element in polar coordinates (→MathApps)

$$d^3r = r\,dr\,d\theta\,dz \qquad (9.23)$$

it is found that

$$m_V = \int d^3r\rho_F \left[\frac{V}{U}\right]^2 = \frac{1}{U^2}\rho_F \int_{\theta=0}^{2\pi} d\theta \int_R^\infty dr\, r \int_{-\infty}^\infty dz \left[U^2\frac{R^4}{r^4}\right]$$
$$= \pi\rho_F L R^2. \qquad (9.24)$$

The result can also be written as

$$m_V = M_{\text{WAT}} \qquad (9.25)$$

indicating that the virtual mass of the cylinder is equal to the mass M_{WAT} of the liquid contained in the same cylinder volume, which we call the equivalent water.

(continued)

Box 9.4 (continued)

For a sphere of radius R the analytical expression for the velocity is

$$v_r = -\frac{UR^3 \cos\theta}{r^3} \quad v_\theta = -\frac{UR^3 \sin\theta}{2r^3} \quad (9.26)$$

and the element of volume in polar coordinates is

$$d^3r = r^2 \sin\theta \, dr \, d\theta \, d\varphi. \quad (9.27)$$

A direct calculation with (9.20) yields

$$m_v = \frac{2}{3}\rho_F \pi a^3 = \frac{1}{2}M_{WAT} \quad (9.28)$$

i.e., for a sphere the virtual mass is half the mass of the equivalent water. For bodies that are not rotationally invariant, the virtual mass becomes a second rank Cartesian tensor and thus must be specified with two indices $m_{v,ij}$. For the purposes of the present book, the example of the elliptical cylinder of length L will suffice. If the motion occurs parallel to one of the axes of the ellipses, the virtual mass coefficient per unit cylinder length is

$$\frac{m_v}{L} = \pi\rho_F a^2; \quad \frac{m_v}{L} = \pi\rho_F b^2. \quad (9.29)$$

depending on whether the motion is, respectively, parallel to the minor axis of the ellipse b, or to the major axis a (Fig. 9.12). Figure 9.12 shows also the values of the virtual mass for a square cylinder of infinite length and for a flat plate.

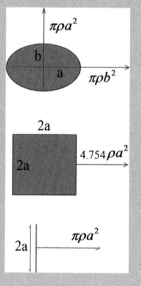

Fig. 9.12 The virtual mass per unit length for an elliptical and square cylinder, and for a flat plate. The direction of motion is indicated with an *arrow*

9.5 Block Model for Subaqueous Landslides

9.5.1 Equation of Motion for the Block

It is often impossible to assess the rheology of subaqueous landslides. Not only is the material often inaccessible and the composition imprecisely known; water may also alter significantly the density of the material and its rheological behavior. Many submarine landslides composed of fine-grained material have probably rheology akin to a subaerial debris flows. Due to the intrinsic cohesion of clay, it is likely that subaqueous mudflow will not be much diluted when flowing, especially if composed of strongly reactive clay. Fine materials will probably result also from the fragmentation of originally intact rock in rock avalanches, and become wetted by ambient water. Thus, both cohesion and Coulomb friction may play a role in subaqueous rock avalanches.

In the present section, the slab model examined earlier for subaerial slides ($\leftarrow$Sect. 6.4.3) is introduced as a simple model for subaqueous landslides. Similar to the subaerial case, the slab is subjected to frictional resistance at the bottom. Finite pore pressure, which decreases the effective friction at the base, the drag force, and the virtual mass are the most significant implementations needed to model the subaqueous landslides with the block model.

9.5.1.1 Drag Force

The results obtained in ($\leftarrow$Sect. 9.4) for the drag coefficients can now be applied to the slab model. Let us start assuming a submarine landslide as a parallelepiped of length L, width W, and thickness D. When travelling in water, the landslide is subjected to the front drag acting on the front face, and skin friction of the three parallel sides. We neglect drag force on the rear face of the block. The total drag force is so written as

$$F_\mathrm{D} = \frac{1}{2}\rho_\mathrm{F} U^2 [C_\mathrm{D} WD + 2C_\mathrm{SF} LD + C_\mathrm{SF} WL]. \tag{9.30}$$

To determine which of the three contributions in the square bracket of (9.30) is dominant, let us notice that usually submarine landslides are wide and flat, so that the term $2LD$ can be neglected in comparison to WL. Hence, if

$$\frac{C_\mathrm{D}}{C_\mathrm{SF}} > \frac{L}{D} \tag{9.31}$$

the resistance at the front is greater than that on the top. Because typically $C_\mathrm{F}/C_\mathrm{D} \approx 10^{-2} - 10^{-3}$, it follows that landslides with aspect ratio D/L smaller than 10^{-3} are chiefly affected by the friction with the upper surface, while landslides with aspect ratio $H/L > 0.01$ are more affected by the drag force at the front. For aspect ratios between these two limits, both front drag and skin friction are important in this simple model.

9.5.1.2 Virtual Mass

For the calculation of the virtual mass, is it reasonable to assume a streamlined shape like an ellipse. The virtual mass can be written as

$$m_v = \Gamma \rho_F H^2 W \tag{9.32}$$

where the coefficient Γ lies somewhere between $\pi/4$ and 1.2. In the block model for subaerial landslides ($\leftarrow$Sect. 6.4.3), the bare mass term on left hand side should then be replaced by the virtual mass

$$(M + m_v)\frac{dU}{dt} = M\left(1 + \frac{m_v}{M}\right)\frac{dU}{dt} \tag{9.33}$$

and from (9.32) it follows

$$M\left(1 + \frac{m_v}{M}\right)\frac{dU}{dt} = M\left(1 + \alpha \frac{\rho_F}{\rho}\right)\frac{dU}{dt} \tag{9.34}$$

where

$$\alpha = \frac{\Gamma D}{W}. \tag{9.35}$$

(Note that If the landslide is higher than wider, a more realistic coefficient would be $\alpha = \Gamma D/L$, but this geometry is uncommon).

9.5.1.3 Equation of Motion for the Block

Using Eqs. 9.34 and 6.31, the equation of motion for the submerged block becomes

$$\begin{aligned}\left(1 + \alpha\frac{\rho_F}{\rho}\right)\frac{dU}{dt} &= g\frac{\Delta\rho}{\rho}\sin\beta\left[1 - (1-r)\frac{\tan\phi}{\tan\beta}\right] \\ &- \frac{1}{2}\frac{\rho_F}{\rho}\left[\frac{C_D}{L} + \frac{C_S}{D} + 2\frac{C_S}{W}\right]U^2.\end{aligned} \tag{9.36}$$

A comparison with the corresponding equation for the subaerial case shows three novel terms in Eq. 9.36:

9.5 Block Model for Subaqueous Landslides

1. The virtual mass coefficient α introduced in (9.35);
2. The relative pore pressure $r = \frac{P_W}{\sigma}$ given as the ratio between the pore water pressure P_W and the total pressure has been explicitly added in the expression (9.36)
3. The Archimedean term $\frac{\Delta \rho}{\rho} \equiv \frac{\rho - \rho_F}{\rho} = 1 - \frac{\rho_F}{\rho}$ that multiplies the gravity acceleration. The effective gravity g' is so reduced from the subaerial case

$$g' = g \frac{\Delta \rho}{\rho} \qquad (9.37)$$

4. The last term on the right hand side of (9.36) accounts for the drag force. Figure 9.13 shows the meaning of the various terms in Eq. 9.36.

9.5.2 Application of the Block Model to Ideal Cases

Before applying the slab model to the case of Grand Banks, it is instructive to consider the results of typical calculations with the slab model for idealized topographies. The calculations for a 700 m long landslide are performed for different friction coefficients, which are reported in the figure. Because submarine landslides normally travel with low effective friction coefficient compared to subaerial landslides (←Sect. 9.2), the values adopted for the calculation are comparatively smaller.

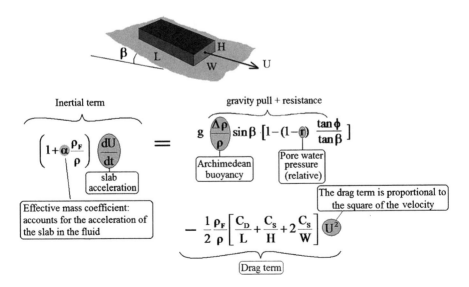

Fig. 9.13 Various terms in the equation of motion for the slab model

The results are reported in Fig. 9.14. Note the high value of the maximum velocities reaching nearly 250 km/h. To assess the role of water in damping the speed of the slide, Fig. 9.15 shows the calculations with the same geometry, but without water. Water greatly diminishes the velocities and the runout for the same friction coefficient.

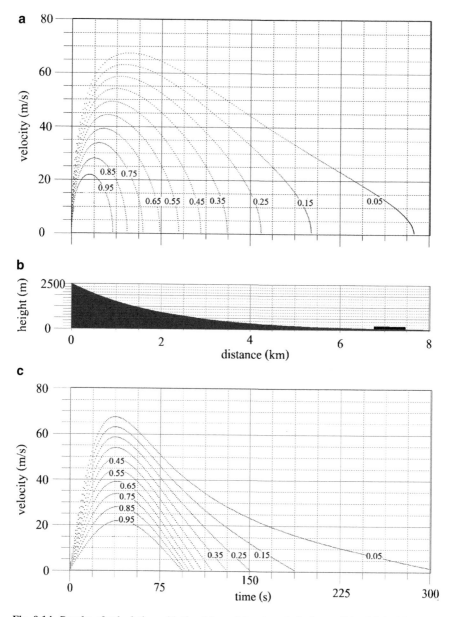

Fig. 9.14 Results of calculation with the slab model and a relatively small landslide. The block has the following dimensions: $L = 700$ m; $H = 30$ m; $W = 300$ m. (**a**) velocity as a function of the position of the front point for different (indicated) values of the friction coefficient; (**b**) the profile of the artificial topography; (**c**) velocity as a function of time

9.5 Block Model for Subaqueous Landslides

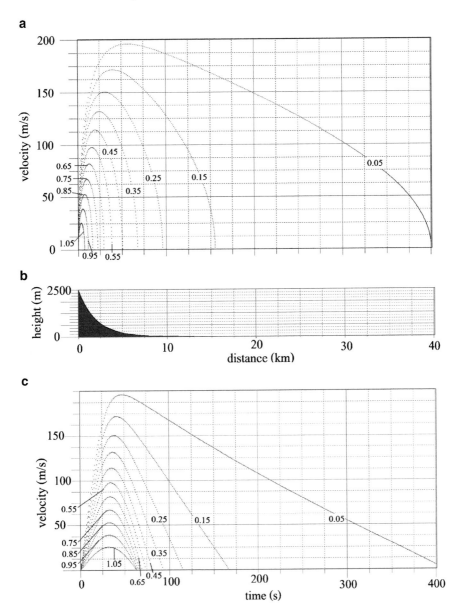

Fig. 9.15 Results of calculation with the slab model without water. The parameters are the same as in Fig. 9.14. (**a**) Velocity as a function of the position for different (indicated) values of the friction coefficient; (**b**) the profile of the artificial topography (identical to the one of Fig. 9.14 but extended to longer distance); (**c**) velocity as a function of time. Notice the much higher velocities and the longer distances reached especially with the lowest friction coefficients

Figure 9.16 presents the case where the virtual mass coefficient is set to zero. The results appear to be different from the calculation in Fig. 9.14a, albeit not dramatically.

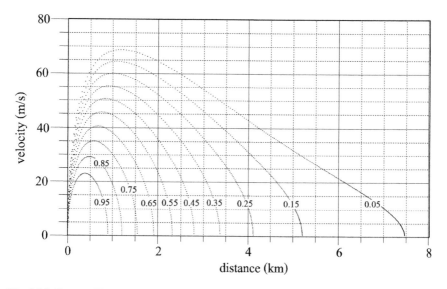

Fig. 9.16 Same as Fig 9.14a with zero virtual mass

9.6 Tsunamis

9.6.1 Introduction

The word *tsunami* (from the word "tsu" meaning "harbor" in ancient Japanese and "nami," or "wave") has become infamous after the Indian Ocean catastrophe of December 2004. The name is appropriate: invisible in deep water, a tsunami becomes perceptible only in the vicinity of the coast, where it may suddenly grow to devastating heights. A tsunami is often confused with tidal waves or rough waves, which in reality have little to do with a tsunami. A tsunami is generated by impulsive displacement of huge water masses in the ocean. Lifting slowly the hand immersed in a bath tub, no significant displacement of the water surface will be observed. This is because the water in front and around the hand has time to move sideways. However, if the hand is lifted fast, a bump of water will form on the free surface, directly on top of the hand. Alternatively, waves can be created with sudden downward movement, in which case the water free surface will be sucked down. A stone cast into a pond is another example of process generating impulsive waves in water.

In nature, these impulsive waves can be created by a series of phenomena:

- Earthquakes. During strong earthquakes, hundreds of km long fault lines may be displaced instantaneously. Depending on the magnitude of the earthquake and on the length of the fault, a huge perturbation may be generated on the water surface, creating a tsunami. The 2004 Indian Ocean tsunami was generated by an earthquake.
- Volcanic eruptions.

- Meteoritic impacts.
- Landslides.

The dreadful fact about tsunamis is that they do not discharge their destructive power only in the vicinity of the source. Once generated, tsunami waves travel fast and are almost invisible. They may pass through a quarter of the globe in matters of 5–7 h, transporting a vast amount of energy while water, which moves only vertically upon the passage of the waves, is not transported. When reaching the shore of a distant land, waves grow higher and start moving toward shore, carrying enormous energy and momentum, and at that point leaving little time for warning.

In this section tsunamis are introduced from a physical viewpoint. The present chapter focuses on general aspects of tsunamis. The next chapter then considers on tsunamis generated by landslides.

9.6.1.1 Basic Definitions of Wave Phenomena

It is worth recalling a few basilar facts about waves. The wave length λ of a wave of infinite length is the distance between successive crests. The wave number is defined as $k = 2\pi/\lambda$. The period T is the time taken for a wave to pass a given point. The frequency $\nu = 1/T$ is the number of waves that pass a given point in 1 s. The pulsation is defined as $\omega = 2\pi/T$. The wave phase velocity is the speed at which a wave front passes an observer at rest. It is given as

$$C = \frac{\lambda}{T} = \frac{\omega}{k} \qquad (9.38)$$

9.6.1.2 Triggering of a Tsunami

For simplicity, in the following discussion we consider a wave with a well-defined wave number k and wavelength $\lambda = 2\pi/k$. The typical wavelength of a tsunami is related to the dimension of the source. In the case of earthquake-triggered tsunamis, the size of the source can reach several hundred of kms, which implies long wavelengths of tsunami waves. A tsunami generated by a landslide will normally have shorter wavelength, but typically still in the range of tens to hundreds of km. A simple small-scale test illustrates the relationship between wavelength of surface waves and size of the source. If a small stone 1–2 cm across is dropped from a small height into a poll, it creates a concentric pattern of short centimetric waves. Waves are comparably higher and longer using a much larger stone.

The initial height of the wave is linked to both the time needed to produce the perturbation, and to its amplitude. Whereas the movement of a fault is in practice instantaneous, this is not so for a landslide. Figure 9.17 shows schematically how a landslide-generated tsunami consists of a region where the water level momentarily rises, and a region where it is depressed from the unperturbed level.

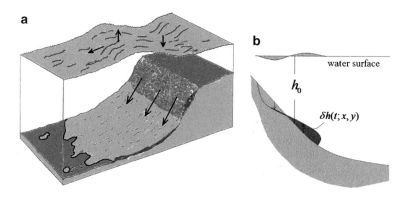

Fig. 9.17 (a) Scheme of generation of a tsunami by a submarine landslide. (b) The perturbation consists of a bulge

---

Box 9.5 One Step Forward: Physics of Surface Waves – A Classic Calculation

The present box is more technical, and requires knowledge of some topics of fluid mechanics not covered in Chap. 3.

Let us consider a water mass contained in a basin of depth H (Fig. 9.18). The vertical coordinate y is measured from the free surface of water at rest; the horizontal coordinate parallel to the water free surface is x. The motion along the coordinate z perpendicular to the line of sight is neglected. The upper surface of the liquid, which changes with time, is denoted by $\delta(x,t)$, or simply δ.

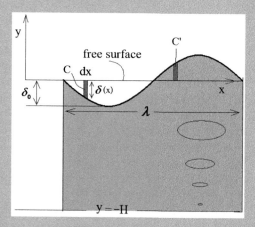

Fig. 9.18 Analysis of the surface waves

(continued)

Box 9.5 (continued)

Because the viscous forces are small compared to water inertia, we can use Bernoulli equation in the analysis of the problem (←Sect. 3.3.3)

$$\frac{1}{2}u^2 + gy + \frac{p}{\rho_F} = 0 \qquad (9.39)$$

which is valid in every part of the fluid, from the bottom of the basin to the upper surface. We can use this equation also when the free surface is perturbed by a wave. Applied to the upper surface of the oscillating fluid, this equation gives a boundary condition of the form

$$\frac{1}{2}\left[u^2(\delta) + v^2(\delta)\right] + g\delta + \frac{p_{\text{ATM}}}{\rho_F} = 0 \qquad (9.40)$$

where $u(\delta)$, $v(\delta)$ are the velocities along x and y, respectively, and p_{ATM} is the atmospheric pressure. Without loss of generality, we can also put $p_{\text{ATM}} = 0$ in Eq. 9.40, as the pressure term is constant.

In the bottom of the basin the velocity of water is zero because of the no-slip condition. The boundary condition is

$$\begin{aligned} u(-H) &= 0 \\ v(-H) &= 0 \end{aligned} \qquad (9.41)$$

From the Navier–Stokes equation without the viscous term

$$\rho_F \left[\frac{\partial \vec{u}}{\partial t} + \vec{u}\cdot\nabla\right]\vec{u} = \rho_F \vec{g} - \nabla P \qquad (9.42)$$

it can be noticed that the magnitude of the term $\vec{u}\cdot\nabla\vec{u}$ on the left hand side is much smaller than the first term $\frac{\partial \vec{u}}{\partial t}$. This can be appreciated estimating the ratio $\left|\frac{\vec{u}\cdot\nabla\vec{u}}{\partial \vec{u}/\partial t}\right| \approx \frac{u^2/\lambda}{u/(\delta/u)} = \frac{\delta}{\lambda} \ll 1$. Equation 9.42 is so reduced to

$$\frac{\partial \vec{u}}{\partial t} = \vec{g} - \frac{1}{\rho_F}\nabla P. \qquad (9.43)$$

Because the fluid is perfect, it is possible to write the velocity as the gradient of a potential function

$$\vec{u} = \nabla \psi. \qquad (9.44)$$

(continued)

Box 9.5 (continued)

and the above equation becomes

$$\nabla\left[\frac{\partial \psi}{\partial t} + \frac{P}{\rho_F} + gy\right] = 0. \tag{9.45}$$

where $\vec{g} = -\nabla(gy)$ expresses the gravity field in terms of the gradient of the gravity potential. The argument of the gradient is thus a constant that can be set equal to zero by a re-definition of the potential

$$\frac{\partial \psi}{\partial t} + \frac{P}{\rho_F} + gy = 0 \tag{9.46}$$

Notice that the vertical component of the water surface velocity is $\frac{D\delta}{Dt} = v(\delta)$. Let us recall the relationship between Lagrangian and Eulerian derivatives can be written as (←Sect. 3.6.1) $\frac{D\delta}{Dt} = \frac{\partial \delta}{\partial t} + u\frac{\partial \delta}{\partial x} + v\frac{\partial \delta}{\partial y}$ and notice that the second and third terms on the right hand side are of the order $\frac{u\delta}{\lambda}$ and v, respectively; they are thus negligible compared to the first term. This implies that the Lagrangian and Eulerian derivatives are nearly equal, $\frac{D\delta}{Dt} \approx \frac{\partial \delta}{\partial t}$. Therefore the velocity of the free surface is

$$v(\delta) = \frac{\partial \delta}{\partial t} \tag{9.47}$$

Substituting Eq. 9.44 in the continuity equation for an incompressible fluid $\nabla \cdot \vec{u} = 0$ (←Sect. 3.5.2), it is found that the potential satisfies Laplace's equation

$$\nabla^2 \psi = 0. \tag{9.48}$$

Because we seek solutions in the form of plane waves propagating along the x direction, a solution must contain a term of the form $\cos(kx - \omega t)$. We thus write the potential as

$$\psi(x, y; t) = G(y)\cos(kx - \omega t) \tag{9.49}$$

(continued)

Box 9.5 (continued)

and from (9.48) we find that G must satisfy the equation

$$\frac{\partial^2 G}{\partial y^2} = k^2 G \qquad (9.50)$$

which has exponential solutions of the form

$$G(y) = A\exp(ky) + B\exp(-ky) \qquad (9.51)$$

where A and B are integration constants.

The velocity at $y = -H$ must be zero. Thus, one of the constants is found imposing the condition $G(-H) = 0$, from which after some manipulation it is found that $B = A\exp(-2kH)$ and

$$\begin{aligned} G(y) &= A[\exp(ky) + \exp(-k(y+H))] \\ &= 2A\exp(-kH)\cosh[k(y+H)] \end{aligned} \qquad (9.52)$$

and so the potential becomes

$$\begin{aligned} \psi(x,y;t) &= G(y)\cos(kx - \omega t) \\ &= 2A\cosh[k(y+H)]\cos(kx - \omega t)\exp(-kH) \end{aligned} \qquad (9.53)$$

where "$\cosh(\ldots)$" denotes the hyperbolic cosine ($\rightarrow$MathApp). From the Eq. 9.44, the y component of the velocity is $v = \partial \psi / \partial y$. Using Eq. 9.53 it is found

$$v = \frac{\partial \psi(x,y;t)}{\partial y} = 2Ak\sinh[k(y+H)]\cos(kx - \omega t)\exp(-kH) \qquad (9.54)$$

and at the free surface the velocity is

$$v(\delta) = \left.\frac{\partial \psi}{\partial y}\right|_{y=\delta} = 2Ak\sinh[k(\delta + H)]\cos(kx - \omega t)\exp(-kH) \qquad (9.55)$$

whereas in a generic point below the water surface the velocities are

$$\begin{aligned} u(x,y) &= \frac{\partial \psi(x,y)}{\partial x} = -2Ak\cosh[k(y+H)]\sin(kx - \omega t)\exp(-kH) \\ v(x,y) &= \frac{\partial \psi(x,y)}{\partial y} = 2Ak\sinh[k(y+H)]\cos(kx - \omega t)\exp(-kH). \end{aligned} \qquad (9.56)$$

(continued)

Box 9.5 (continued)

The time integration gives the trajectory of the water particles

$$X(x,y;t) = \int_0^t u(x,y)dt = -\frac{2Ak}{\omega}\cosh[k(y+H)]\cos(kx-\omega t)\exp(-kH)$$

$$Y(x,y;t) = \int_0^t u(x,y)dt = -\frac{2Ak}{\omega}\sinh[k(y+H)]\sin(kx-\omega t)\exp(-kH)$$

(9.57)

where x,y represent the initial point.

Notice that squaring the two above equations

$$\left\{\frac{X(x,y;t)}{\frac{2Ak}{\omega}\cosh[k(y+H)]\exp(-kH)}\right\}^2 = \cos^2(kx-\omega t)$$

$$\left\{\frac{Y(x,y;t)}{\frac{2Ak}{\omega}\sinh[k(y+H)]\exp(-kH)}\right\}^2 = \sin^2(kx-\omega t)$$

(9.58)

and using the identity $\sin^2(x) + \cos^2(x) = 1$, we obtain

$$\left\{\frac{X(x,y;t)}{\frac{2Ak}{\omega}\cosh[k(y+H)]\exp(-kH)}\right\}^2 + \left\{\frac{Y(x,y;t)}{\frac{2Ak}{\omega}\sinh[k(y+H)]\exp(-kH)}\right\}^2 = 1.$$

(9.59)

The equation

$$\left\{\frac{X}{a}\right\}^2 + \left\{\frac{Y}{b}\right\}^2 = 1 \qquad (9.60)$$

is the locus of an ellipse of semi-axes a and b. Hence, Eq. 9.59 shows that the movement of water follows elliptical trajectories. The semi-axes a and b are not constant but depend on the height y. It is found that

$$a = \frac{2Ak}{\omega}\cosh[k(y+H)]\exp(-kH)$$

$$b = \frac{2Ak}{\omega}\sinh[k(y+H)]\exp(-kH)$$

(9.61)

(continued)

Box 9.5 (continued)

Thus, the shape of the ellipses changes with the height (see Fig. 9.18). In particular, the eccentricity of the ellipse, given as the ratio of the minor to major semi-axes

$$e = \frac{b}{a} = \tanh[k(y+H)] \qquad (9.62)$$

takes on the value zero at the ocean bottom. If the depth of the basin is much greater than the wave length, then $kH \gg 1$ and at the free water surface the hyperbolic tangent becomes $\tanh[kH] \approx 1$. Hence, water moves in a nearly circular fashion near the free surface, while the movement appears more rectilinear in the bottom of the ocean. In contrast, if the basin depth is smaller than the wave length and $kH \ll 1$, the hyperbolic tangent has the limit $\tanh[kH] \approx kH$ and the eccentricity at the top becomes

$$e = \frac{b}{a} = kH \ll 1 \qquad (9.63)$$

meaning that the water trajectories are flat ellipses.

It is now interesting to work out the dispersion relation for the waves, i.e., the relationship between pulsation and wave number. Considering again Bernoulli equation at the free surface Eq. 9.46 $\frac{\partial \psi}{\partial t}\Big|_{y=0} + g\delta = 0$

$$\delta = -\frac{1}{g}\frac{\partial \psi(x,y;t)}{\partial t}\Big|_{y=0} = \frac{2A\omega}{g}\cosh[kH]\sin(kx - \omega t)\exp(-kH) \qquad (9.64)$$

and using the expression for the potential from Eq. 9.53 one obtains

$$\omega = \sqrt{gk \tanh[kH]}. \qquad (9.65)$$

which is the dispersion relation.

9.6.2 Propagation of Tsunami Waves in the Ocean

9.6.2.1 Velocity

The tsunami waves spread from the source region with a phase velocity given as the ratio

$$C = \frac{\omega}{k} = \sqrt{\frac{g}{k}\tanh[kD]} = \sqrt{\frac{g\lambda}{2\pi}\tanh\frac{2\pi H}{\lambda}} \qquad (9.66)$$

while the group velocity is

$$U_G = \frac{d\omega}{dk} = \frac{1}{2}\left[\frac{\omega}{k} + \frac{kHg}{\omega}\right]\sqrt{gk\tanh[kH]}. \tag{9.67}$$

Notice that both the group and the phase velocities depend on the wave number $k = 2\pi/\lambda$, a property called dispersion. This is a consequence of the nonlinear relationship between pulsation and wave number in the dispersion relation (9.65), and implies that a group of waves of different wave length and frequency have different phase velocity. This is different from other forms of wave motion like light in vacuum, whose velocity is independent of the wave number.

As discussed earlier, the source of tsunamis typically measures tens to hundreds of km depending on the size of the source, and is thus much wider than the depth of the ocean basins (4–5 km, Fig. 9.3). Setting $kH \gg 1$ in (9.66), the dispersion relation simplifies to a linear relationship between pulsation and wave number $\omega = \sqrt{gH}k$.

The group and phase and velocities become equal and also independent of the wave length

$$U_G(\lambda/H \to \infty) = C(\lambda/H \to \infty) = \sqrt{gH}. \tag{9.68}$$

This is also called the shallow water approximation for the wave velocity. From Eq. 9.68, it follows that a tsunami wave typically travels at velocities of the order 800–900 km/h, comparable to the travel speed of a jumbo jet. Thus, it takes several hours for a tsunami to travel across a basin of the size of an ocean. The tsunami of December 26, 2004, originated in Sumatra, reached Thailand after 1 h, southern India after 2 h, the Maldives after 4 h; after passing through the whole Indian Ocean, it reached Somalia 7 h after the shock. It takes a longer time to cross the whole Pacific Ocean. Where a tsunami warning system is installed, this travel time may be sufficient to evacuate the coasts.

Note that the approximation of deep basin breaks down for shorter wavelengths (higher wave-numbers), as both the phase and the group velocities decrease from the limit value $U_G(\lambda/H \to \infty)$ and $U_P(\lambda/H \to \infty)$ with decreasing wave length. Some numbers from Eqs. 9.66 and 9.67: for waves of wavelength 10 km travelling in a 5.5 km deep ocean, the velocities are $C(\lambda = 10 \text{ km}) = 0.6C(\lambda \to \infty)$ and $U_G(\lambda = 10 \text{ km}) = 0.3U_G(\lambda \to \infty)$. Hence, waves of shorter wavelength are significantly delayed.

9.6.2.2 Energy Carried by a Wave

The energy carried by a wave is calculated as a sum of the potential and the kinetic contributions. Let us consider again Fig. 9.18. The volume is limited by one wavelength along the x coordinate. To calculate the increase in the potential energy from the unperturbed case, let us consider two rectangles C and C' where the wave has opposite phase. Let x be the abscissa of C. C and C' have infinitesimal base length dx. A column of water of height $\delta(x)$ and base area S has a potential energy with respect to the ground state given as $(1/2)\rho_F \delta(x)^2 S g$ ($\leftarrow$Sect. 3.1). The potential energy per unit area of these two elements is thus

$$dP = \frac{1}{2}\rho_F \delta^2(x) g \, dx + \frac{1}{2}\rho_F \delta^2(x) g \, dx = \delta^2(x) \rho_F g \, dx. \tag{9.69}$$

The total potential energy of the wave is obtained integrating Eq. 9.69 and using sinusoidal shape $\delta(x) = \delta_0 \sin kx$:

$$P = 2\rho_F g \delta_0^2 \frac{1}{2} \int_0^{\lambda/2} (\sin kx)^2 dx = \frac{1}{2k} \rho_F g \delta_0^2. \tag{9.70}$$

The kinetic energy per unit length perpendicular to the figure is obtained integrating in space the kinetic energy density

$$T = \frac{1}{2}\rho_F \int_0^H dy \int_0^\lambda dx \left[u^2(x,y) + v^2(x,y) \right]$$

$$= 2\rho_F A^2 k^2 \exp(-2kH) \int_0^H dy \int_0^\lambda dx \tag{9.71}$$

$$\left\{ \cosh^2[k(y+H)]\sin^2(kx-\omega t) + \sinh^2[k(y+H)]\cos^2(kx-\omega t) \right\} = \frac{1}{2k}\rho_F g \delta_0^2$$

Thus, kinetic and potential energy per unit length are equal. The total energy per unit length is then

$$E = P + T = \frac{1}{k}\rho_F g \delta_0^2 = \frac{\lambda}{2\pi}\rho_F g \delta_0^2 \tag{9.72}$$

This is the total energy carried by a single wave. Note the linear relationship with the wavelength and quadratic with wave height. The calculation of the energy carried by a whole tsunami is more complicated (e.g., Levin and Nosov 2009).

9.6.2.3 Behavior of a Tsunami Approaching the Coast and Refraction of Waves in Shallow Water

In deep water, a perturbation some hundreds of km long and some meters high will go unnoticed. However, as the tsunami reaches the coast, the waves will grow to considerable height and water will acquire a high velocity component landward.

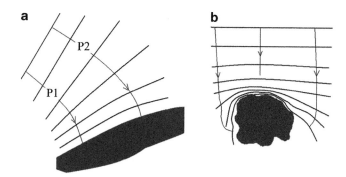

Fig. 9.19 Wave refraction. (**a**) waves approaching the shore. (**b**) waves embedding an island

Due to refraction, waves in shallow water may change direction of propagation depending on the local changes of the water depth. Consider a wave train approaching a coastline (Fig. 9.19a). Because the wave fronts that are travelling along the direction "P1" meet shallower water than those in "P2", their velocity is lower according to Eqs. 9.66–9.68. Thus, the wave directions bend as shown in the figure. This is also the reason why waves along a beach have fronts nearly parallel to the shore. If a wave train approaches an island like in Fig. 9.19b, it will embed a large portion of the island beach. The local submerged topography is thus critical in realistic simulations of tsunami propagation. As a wave approaches the shore, the front meets shallower water and as a consequence the wavelength decreases. The conservation of energy density implies from Eq. 9.72 a growth of the wave amplitude, which for a tsunami wave can reach several meters.

9.6.3 Tsunamis Generated by Submarine Landslides

9.6.3.1 Tsunami Generation by a Landslide

We have already mentioned in the beginning of the Chapter the Papua New Guinea earthquake and related landslide-generated tsunami. In 1953, an earthquake in the Fiji islands generated numerous small submarine landslides. The epicenter of the earthquake was far from the coast. Yet the associated tsunami, which reached a height of 15 m in the town of Suva, was created immediately after the shock and reached the coast in matter of seconds, leaving no time for escape. This is because the tsunami was created by a small landslide very close to the shore. Its thickness was only 30 m for a volume of 350 Mm^3, but moved at high speed owing to the steep slope characteristic of active margins.

The best way to assess the tsunami hazard along the coastlines is to carry out time-dependent numerical simulations of tsunami generation and propagation. To this purpose, usage is made of the equations of fluid dynamics for shallow water

9.6 Tsunamis

waves, namely the continuity equation for a incompressible medium (←Chap. 3). Integrating Eq. 3.54 vertically leads to

$$\int_0^D dy \frac{\partial u}{\partial x} + \int_0^D dy \frac{\partial v}{\partial y} + \int_0^D dy \frac{\partial w}{\partial z} = 0 \qquad (9.73)$$

and using Leibnitz's rule (MathApp) the three addenda in (9.73) can be written as

$$\int_0^D dy \frac{\partial u}{\partial x} = \frac{\partial}{\partial x} \int_0^D u\, dy - u(D)\frac{\partial D}{\partial x} = \frac{\partial}{\partial x}(<u>D) - u(D)\frac{\partial D}{\partial x}$$

$$\int_0^D dy \frac{\partial w}{\partial z} = \frac{\partial}{\partial z} \int_0^D w\, dy - w(D)\frac{\partial D}{\partial z} = \frac{\partial}{\partial z}(<w>D) - w(D)\frac{\partial D}{\partial z} \qquad (9.74)$$

$$\int_0^D dy \frac{\partial v}{\partial y} = v(D) - v(0)$$

where $<u>$, $<w>$ are the depth-integrated velocities. Noting that $v(D) = \frac{\partial \delta}{\partial t}$; $v(0) = \frac{\partial \delta h(t)}{\partial t}$ it is found (a few mathematical passages are skipped for brevity)

$$\frac{\partial \delta}{\partial t} = -\nabla \cdot [(h_0 - \delta h(t))<\vec{u}>] + \frac{\partial (\delta h(t))}{\partial t} \qquad (9.75)$$

where $\delta h(t)$ gives the change of the sea bottom level as a function of time caused by the landslide (Fig. 9.17b). Note that the first and second terms on the right hand side of Eq. 9.75 are of the order CD/λ and UD/λ, respectively, where U is the landslide velocity. The Navier–Stokes equation (Eq. 3.81) with zero viscosity gives

$$\frac{\partial \vec{u}}{\partial t} = -g\nabla \delta - (\vec{u} \cdot \nabla)\vec{u}. \qquad (9.76)$$

The inertial term $(\vec{u} \cdot \nabla)\vec{u}$ of Eq. 9.76 is also often dropped.

To simulate the tsunami with a computer, the perturbation of the sea level caused by a landslide is introduced by providing $\delta h(t)$ as a function of time. This requires information on the shape of the landslide body and its dynamics. Simulations may be performed with either simplified geometry and dynamics (e.g., a landslide with the shape of a rounded box with a predetermined movement, for example, in the form of a sinusoid), or simulating the landslide progression more accurately with a dynamical model and then using the results as an input to the tsunami model.

From the computational viewpoint, simulating an earthquake-generated tsunami is easier because it can be accomplished using an instantaneous rise of a sea bottom block.

9.6.3.2 Differences Between Earthquake-Generated Tsunamis and Landslides-Generated Tsunamis

A first major difference between earthquake and landslide-provoked tsunamis is that the maximum extension of submerged faults is greater than the area of a submarine landslide. Hence, the energy and the wavelength involved in earthquake-generated tsunamis are greater than in a landslide tsunami. Moreover, an earthquake generates the perturbation of water instantly, while a landslide builds up a perturbation during a time span of several seconds to minutes. Additionally, the perturbation to the sea surface induced by a slide is of the dipole type. In this kind of motion, shown in Fig. 9.17, the frontal part of the landslide thrusts the water surface upward while the rear end sucks the water downward. The perturbations of opposite sign tend so to interfere negatively, especially in the far field. That means that tsunami waves generated by a landslide tend to damp out more quickly than those due to an earthquake.

9.6.3.3 An Example of Landslide-Generated Tsunami

Although the equations for shallow water theory (9.75 and 9.76) are a simplified version of the Navier–Stokes and continuity equation, they are sufficiently complex to require numerical solution. Figure 9.20 shows the results of a simulation of the Hinlopen-Yermak landslide (opening figure of the chapter). The dynamics of the landslide is provided by a Bingham-like simulation. The red and blue colors represent the elevation over and below the undisturbed sea level, respectively. Note how in Fig. 9.20a and b the superelevation is heading the tsunami, while a lower water level appears in the rear end (the landslide is moving from south to north). The maximum height of the perturbed water layer is about 10 m.

9.6.3.4 The Importance of the Froude Number

The Froude number is the ratio

$$Fr = \frac{U}{C} = \frac{U}{\sqrt{gH}} \qquad (9.77)$$

between the landslide velocity U and the phase velocity of tsunami waves in shallow water approximation. Consider a deep and relatively slow landslide perturbing the sea surface. The Froude number will be low because the phase velocity

9.6 Tsunamis

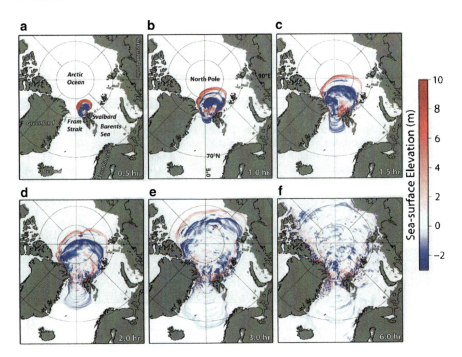

Fig. 9.20 Tsunami generated by the Hinlopen-Yermak landslide (see opening figure of the chapter; from Vanneste et al. 2010). The landslide is moving from south to north (Simulation by F. Løvholdt and C. Harbitz)

of the waves in the denominator of Eq. 9.77 is large compared to U. In this situation, the landslide cannot catch up with the waves it creates. Because the amplitude of the perturbation is enhanced by the persistence of the source (as follows from the superposition principle), the waves will leave the source before the landslide can build a strong perturbation, and the resulting tsunami will be weak. If, on the other hand, the landslide is fast and shallow, the ratio (9.77) may approach unity. The landslide will build up perturbation on the same wave packet and the tsunami will be stronger. This shows the importance of the landslide speed and seafloor topography in the generation of tsunamis, and the need for a better understanding of the landslide dynamics in the ocean (see Harbitz et al. 2006).

9.6.3.5 Ancient Tsunami Deposits

Tsunami deposits of ancient or more recent landslides have been described in many locations (Dawson and Stewart 2007; Bourgeois 2009). They allow reconstructing the run-up reached by the tsunami on shore. Figure 9.21 shows a tsunami deposit in northern Norway due to the Storegga landslide (←Box 9.2). Similar deposits, found also in other locations of Norway and Shetland, allow for a test of the run-up calculated by modeling (Harbitz 1992; Bondevik et al. 2005).

|————————————————————————————|
tsunami
deposit

Fig. 9.21 Tsunami deposits found in Lyngen (Tromsø, northern Norway) related to the Storegga event. The tsunami deposit is the chaotic unit in the middle, about 80 cm long. The deposit has been found 22 m above sea level. At the time the tsunami took place, this level was a few meters below sea (Original photograph and explanations courtesy of Stein Bondevik. Photographed by Henrik Rasmussen)

Box 9.6 Brief Case Study: The Grand Banks Landslide and the Velocity of Subaqueous Landslides

Salient data
Name: Grand Banks landslide
Location: offshore Newfoundland (Canada)
Coordinates: 44°30′ N; 57°15′ W
Volume: between 100 and 550 km^3, only 30 km^3 according to other authors
Year: 1929 (18th November)
Kind: Initially a rock, sand and clay slide turned into a debris flow and probably a turbidity current
Casualties: 30 caused by the tsunami
Landslide material: rock
Vertical fall (m): 2,000
Runout: 600–700 km
FB: 0.005
Cause: earthquake
Dynamical characteristics: it has generated a large tsunami; it has travelled with extremely low friction; the landslide transformed first into a debris flow, and then perhaps into a turbidity current; among the largest submarine landslides, it is the one for which best velocity data is available

The Grand Bank subaqueous landslide represents a unique event in the studies on landslides; it provoked a large tsunami, and also provided the first detailed information on the velocity of a subaqueous landslide. At that time, the telegraph cables between Europe and North America stretched south of Newfoundland, some 100 km apart. On 18th November 1929, a 7.2 Richter scale Earthquake shook the region of Grand Banks, southern Newfoundland. Because the epicenter was well far from the shore, little direct damage was caused by the quake. However, a tsunami killed 30 people some hours later. The telegraph cables broke in sequence from north toward south, showing

(continued)

Box 9.6 (continued)

that some kind of mass had travelled down slope along the submarine basin of Grand Banks, for a total length of some 700 km. Controversy aroused as to the nature of this mass.

Figure 9.22 shows the ramp of the Grand Banks. The material probably detached from the epicenter area of the quake, at a slope of less than half a degree. Another possibility is that the slide commenced at slightly higher slope. At a distance of 130 km from the epicenter, the velocity was measured from timing of cable breaks: 28 m/s. At greater distances the velocity decreased. At a distance of 700 km something was still moving hours after the shock at a velocity of 6 m/s, moderate but still twice the speed of a running man. What was exactly the nature of the moving mass? Was it a rockslide, a debris flow, or turbidity current? This question is also considered in the next section.

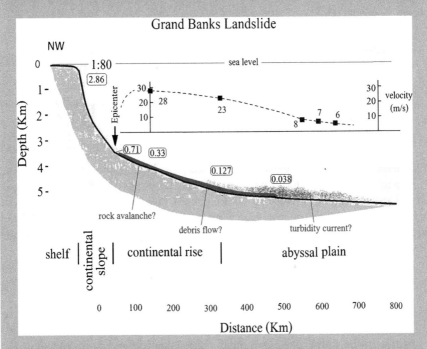

Fig. 9.22 Profile of the Grand Banks submarine setting and landslide. In *rectangles* the slope angles in degrees

The tsunami was observed around over 30 stations along Canada, the USA, and the Bermudas. Interestingly, no anomalous waves were reported

(continued)

Box 9.6 (continued)

in Africa and Europe, except for Portugal. This is consistent with the fast damping of tsunamis caused by landslides. The water height registered with a mareograph on the 18th November at Halifax, Canada, showed the long-period waves due to the tides superimposed to the short-period oscillations some meters high due to the Grand Banks tsunami.
References: Hsü (2002), Fine et al. (2005).

9.6.3.6 Possible Interpretation of the Grand Banks Event

If the Grand Banks event was a turbidity current, its velocity can be predicted by results in (→Sect. 10.4), and in particular Eq. 10.29 for the velocity of a turbidity current (→Sect. 10.1.3):

$$U \approx \sqrt{\frac{8}{f} g \frac{\Delta \rho}{\rho} \sin \beta H} = C \sqrt{\frac{\Delta \rho}{\rho} \sin \beta H} \quad 40 \text{ m}^{1/2} \text{ s}^{-1} < C < 88 \text{ m}^{1/2} \text{ s}^{-1} \quad (9.78)$$

Using an average value of $C = 60 \text{ m}^{\frac{1}{2}} \text{ s}^{-1}$, it follows that to reach a velocity of 20 m/s the current should be some kilometers thick and have a density difference of $\frac{\Delta \rho}{\rho} = 0.01$ with $\beta = 0.1°$. Alternatively, the turbidity current of Grand Banks might have been denser. With $\frac{\Delta \rho}{\rho} = 0.1$, a still implausible thickness of some hundreds of meters results. It is more likely that the flowing material was much denser, which indicates that it was a debris flow or a landslide rather than a turbidity current. Because the source material is composed of sand, clay, and larger blocks, the mass has probably travelled with density similar to a debris flow on land.

As a second model for Grand Banks let us consider the slab model (←Sect. 9.5) with a landslide of volume 300 km³. Figure 9.23 shows the results. One simulation has been performed with constant friction coefficient of 0.0003, height of 20 m, width of 100 km and length of 150 km. The friction coefficient has been chosen to give a value of the runout of about 650 km. It turns out that the simulated velocities in the initial phase are much higher than observed. In a second run, the friction coefficient is increased to 0.0013; the height is 10 m, the width is 300 km, and the length is 100 km. Velocities are now reasonable in the early phase of the Grand Banks, but decrease fast and the landslide stops too early. In another simulation, the friction coefficient is taken again equal to 0.0013. The landslide starts with a height of 10 m and a length of 100 km, stretching during the flow but maintaining the same product *LH*. In this case, a better fit with data is

9.6 Tsunamis

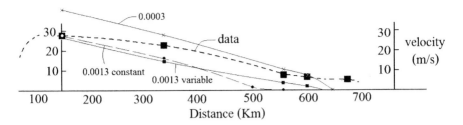

Fig. 9.23 Grand Banks landslide: results with the slab model. The values of the friction coefficients are indicated

obtained ("0.0013 variable" in Fig. 9.23). The main outcome of the simulations is that very low friction coefficients must be used to reproduce the velocity and runout of the Grand Banks landslide within a Coulomb frictional model. The reason for the low hindrance met by submarine landslides is addressed in (→Sect. 9.6.2).

It is likely that the Grand Banks landslide started off as a slide and transformed into a debris flow, and possibly only after 600–700 km continued as a turbidity current.

Box 9.7 Experiments with Subaqueous Debris Flows

Small-scale experiments with artificial subaqueous debris flows have become commonplace since the early experiments by Kuenen. The density of the medium, the local velocity field of tracer particles and the pressures can so be measured under strict experimental conditions. Figure 9.24 shows some images of an artificial debris flow obtained with 15% clay, 25% water and 60% fine sand along a 5° inclination flume at the St. Anthony Falls Laboratories in Minneapolis (Minnesota). In this case the flow is constrained between the two transparent flume walls 20 cm apart. The first figure shows fully developed hydroplaning (→Sect. 9.6.2) revealed by the thin water layer underneath the front of the debris flow. The same experiments with more sand and less clay do not show signature of hydroplaning. This demonstrates that the presence of a sufficient amount of clay is necessary to provide a low-permeability matrix necessary for hydroplaning. In the second picture, the artificial debris flow develops a small turbidity current on the top. Experiments like this suffer the problem of scaling. For example, the pull of gravity increases with the mass of the debris flow, while the velocity, turbulence, and permeability have more complex scaling behavior. Nevertheless, they are very instructive and indispensable tools for better understanding of the submarine debris flows.

(continued)

Box 9.7 (continued)

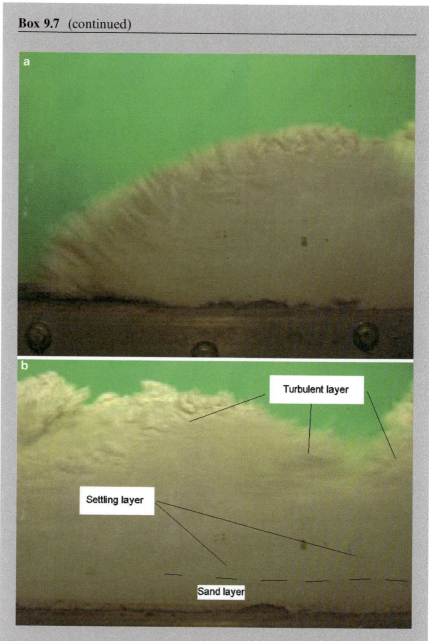

Fig. 9.24 Snapshots from the experiment at 15% clay content. (**a**) the head of the debris flow. (**b**) the body of the debris flow with the *turbulent layer*, *settling layer* and *sand layer* identified. Experiment performed at St. Anthony Falls laboratories (Pictures courtesy of A. Elverhøi)

9.7 More Dynamical Problems

9.7.1 Outrunner Blocks

Outrunner blocks are intact pieces of debris that detach from the front of a submarine debris flow and travel ahead of the rest of the material. There are numerous examples: in the Kitimat fjord area, Canada, in the Nigerian continental slope, in the Faeroe margin (Nissen et al. 1999; Prior et al. 1982; Kuijpers et al. 2001). The Box 9.8 reports the case of an outrunner block in the small Finneidfjord slide in Norway, and Table 9.4 gathers some salient data for the mentioned landslide areas.

In the Faeroe, blocks have created a groove, which appear deeper at regular intervals λ along the flow direction. Blocks in themselves are small compared to the

Table 9.4 A compilation of four data sets for outrunner blocks

Slide (and setting)	Nigeria	Kitimat (fjord)	Far Øer	Finneidfjord (fjord)
Age of main event (s)		1952–1975	Late Quaternary	1996
Dimensions of the blocks (m) length × width × height	$100 \times 250 \times 10$	$<75 \times 125 \times 5$	$70 \times 100 \times 18$	$60 \times 110 \times 2$
Volume (m^3)	250,000	<47,000	37,800	10,000
Slope (deg)	1–1.4	0.37–0.56	0.6	0.43
Maximum runout (km)	Up to 12	Up to 1	Up to 25	1.6
Depth of the track (m)	5–10	Detected, but value not reported	4	10–20 cm
Width of the track (m)	10–100	Up to 70	90	60 m
Distance between erosion marks λ (m)	150–225	5	60–70	Not detectable
Material of the debris flow and of the sea/fjord floor		Coarse grained sand and silt from deltaic deposits. Late-Pleistocenic marine clays on the floor		Layered glaciomarine with dropstones superposed to undisturbed bioturbated sediments
Notes			Prevailing orientation of the blocks perpendicular to the tracks	Orientation of the longest runout block parallel to the tracks
Country	Nigeria	Canada	Denmark	Norway
References	Nissen at al. (1999)	Prior et al. (1982)	Kuijpers et al. (2001)	Longva et al. (2003)

main body of the debris flow. For example, the front of the Nigerian debris flow is several tens of kilometers long, while the corresponding outrunner block is about 100 m wide. Hence, blocks are volumetrically of minor importance, and their tsunamigenic potential is limited. However, their study is attractive because the analysis of block movement may contribute significantly to an understanding of the dynamics of submarine landslides. Blocks are in fact extremely mobile, which contrasts with their small mass (Fig. 9.5).

The minimum energy per unit length required to create a groove like those observed in Nigeria and the Kitimat fjord in Canada is $\Delta E/\Delta x = (1/2)\Delta\rho g h^2 w$ where w and h are the width and depth of the groove, respectively. This gives for the Nigerian case about 5×10^7 J/m. The gravitational energy per unit length due to the fall of the block in the gravity field is $\Delta E'/\Delta x = \Delta\rho g V \sin\beta$ where V is the volume; imposing $\Delta E' \gg \Delta E$ one finds $V \gg \dfrac{h^2 w}{2\sin\beta} \approx 2 \times 10^5$ m^3, which for a cubic block would correspond to a length of 60 m, close to the volume of the Nigerian blocks. This simple analysis is confirmed by the fact that smaller blocks, such as those in Finneidfjord, do not excavate grooves.

Outrunner blocks are subjected to drag and lift forces. As a simple model, let us consider a rigid block shaped as a half-cylinder with its axis perpendicular to the velocity and travelling parallel to the sea bed. The ratio between the lift to the weight force is of the order ($\leftarrow$Sect. 3.3.3 and Eq. 3.26)

$$r \approx \frac{\rho_F}{\Delta\rho\cos\beta}\frac{U^2}{R} \tag{9.79}$$

and is thus inversely proportional to the radius R of the cylinder. This shows that lift forces are especially important for the smaller blocks, which explains their characteristic small size. A numerical and experimental analysis, in which small rectangular blocks are let to descend along a submerged board, shows that blocks become very mobile due to the effect of water and may oscillate in response to the lift and drag forces (Lars Engvik, private communication). If the front of the block is pushed up by the lift force, the drag force increases; the block slows down, and gets closer to the table. This decreases the drag, increases the velocity and the lift force. The block is lifted up again, and a new cycle starts. This could be an explanation for the regularly displaced erosion marks with length λ observed in some of the field cases (Table 9.4).

Some blocks travel embedded by the mobile mass of a large debris flow (examples of such blocks can be seen in the deposits of the Hinlopen-Yermak landslide in the opening figure to this chapter). The acceleration of a block of length L rafted by a debris flow travelling at a speed U_0 can be estimated considering the impact force of the debris flow against the block as the dominant force. The block acceleration is so

$$\frac{dU}{dt} \approx \frac{\rho_{DF}}{2L\rho}(U(t) - U_0)^2 \tag{9.80}$$

9.7 More Dynamical Problems

where U is the block velocity, ρ_{DF} and ρ are the debris flow and block densities, respectively. The solution to Eq. 9.80 is

$$V(t) = \frac{AU_0 t}{1 + At} \tag{9.81}$$

where $A = (\rho_{DF} U_0)/(2L\rho)$. The acceleration time for the block is thus of the order $\tau \approx \frac{18 L \rho}{\rho_{DF} U_0}$.

Box 9.8 Brief Case Study: The Finneidfjord Landslide and the Bizarre Outrunner Blocks

Salient data
Name: Finneidfjord landslide (from the name of a fjord)
Location: shoreline in northern Norway.
Coordinates: 66°10′ N; 13°48′ E
Volume: (1 Mm³)
Year: 1996 (June 20th).
Kind: debris flow, partly originated on land (10%)
Casualties: 4
Cause: liquefaction of sensitive clays
Landslide material: soft sensitive clays, quick clay, silt, sand
Vertical fall (m): 50
Fahrböschung: main slide: 0.05; outrunner block: 0.025

This submarine landslide illustrates well the potential for single isolated blocks to run ahead of the main landslide body and travel with almost no interaction with the sea floor even on very gentle slopes.

The collapse of the shoreline of the Finneidfjord on the coast of northern Norway partly affected buildings on land; it killed people in a car and in a house and swallowed a beach, only minutes earlier packed of people enjoying a summer outdoor party. Most of the slide material collapsed from the fjord. The failure started offshore, probably as a retrogressive quick clay slide, that only in the last phases of collapse affected the subaerial sediment.

The slide deposit can be divided into four different sectors (Fig 9.25). Zone A comprises the bulk of the sediment deposited along the fjord, and stretches for a distance of 1 km from the slide escarpment. Zone B starts from a water depth of 45–50 m, where the sea bottom slopes less than 1°. The sediment appears to be more scattered and a discrete structure made of harder blocks becomes discernible. Evidently, the material was not completely remolded.

The zone C appears devoid of any deposition, except for scant traces indicated by white lines in Fig. 9.25, probably the remains of glide blocks along the path of some more voluminous blocks appearing in zone D. The largest block, 100 m wide and 50 m long, is elongated transversely to the flow

(continued)

Box 9.8 (continued)

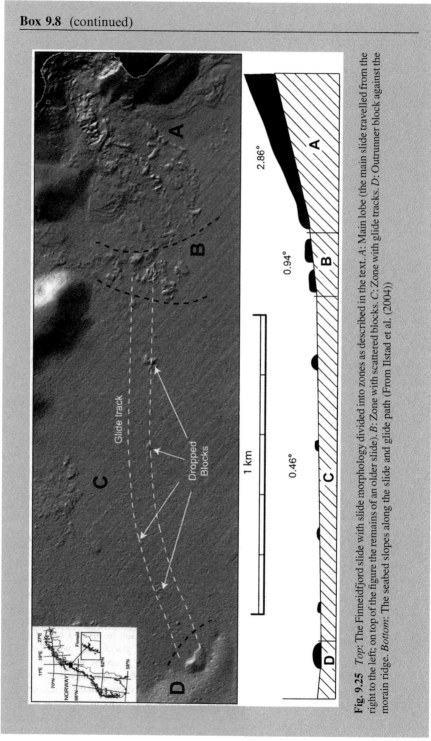

Fig. 9.25 *Top*: The Finneidfjord slide with slide morphology divided into zones as described in the text. *A*: Main lobe (the main slide travelled from the right to the left; on top of the figure the remains of an older slide). *B*: Zone with scattered blocks. *C*: Zone with glide tracks. *D*: Outrunner block against the moraine ridge. *Bottom*: The seabed slopes along the slide and glide path (From Ilstad et al. (2004))

(continued)

Box 9.8 (continued)

direction. With a thickness of only 2 m, the block is basically flat. It appears to have outrunned the main slide by at least 1.4 km, travelling ahead the main lobe on a slope of only 0.46° without major scouring of the sea floor. The block stopped when colliding with a submerged moraine. It would have likely travelled further. The coring analysis of the block shows maximum shear strength of 20 kPa, followed by a decrease to 10 kPa at contact with the seafloor. Interestingly, the original layering is conserved, which demonstrates that the block was not tilted, but travelled parallel to slope.

9.7.2 Debris Flows

It is clear that the slab model is just a simple approximation that holds only for frictional materials. For submarine debris flows, the equation (←Sect. 4.60) can be extended to

$$\frac{dU}{dt} \approx \frac{\Delta\rho}{\rho} g \sin\beta - \frac{\Delta\rho}{\rho} g \cos\beta \frac{\partial D}{\partial x} - \frac{1}{2} C_{SF} \frac{\rho_L}{\rho D} U^2 - \frac{1}{\rho D}\left(\tau_y + 2\mu_B \frac{U}{D}\right) \quad (9.82)$$

Note the presence of the Archimedean buoyancy in comparison to 4.60. One of the problems with the rheological models for submarine debris flows is that the rheological parameters will probably vary as ambient water is entrained in the debris.

Box 9.9 Numerical Simulation of Subaqueous Debris Flows

Although analytical calculations are useful in understanding the significance of flow equations, for a realistic prediction of debris flow behavior a numerical solution is necessary. The basic idea of a computer simulation of subaqueous debris flows (and more in general of every gravity mass flow) is to reduce the continuum into a discrete set of equations. This is accomplished adopting a discrete space lattice, and a discrete time. Once a solution is found for a certain time step, it is propagated in time by use of the momentum and continuity equations. The most used numerical methods for rheological flows are finite difference, finite elements and finite volume. The topic of numerical simulation of debris flow is a subject that would deserve a chapter in itself. Here, just as an example, some results from the depth integrated model BING are presented (Imran et al. 2001). BING is a well-known program that adopts the depth-integration technique (Huang and Garcia 1998; Savage and Hutter 1989), forming the basis of numerous other algorithms. BING is a model for the flow of a Bingham

(continued)

Box 9.9 (continued)

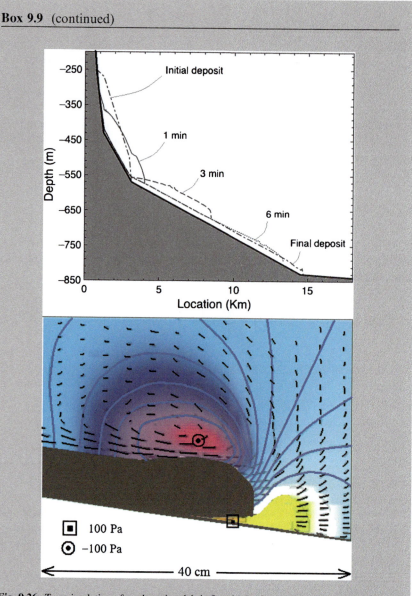

Fig. 9.26 *Top*: simulation of a submarine debris flow in the Ormen Lange area (headwall of the Storegga slide) with the BING model. The resulting debris flow profiles are shown after the indicated time. The debris flow is simulated with a Bingham rheological model with a shear strength of 12 kPa. (From De Blasio et al. 2005). *Bottom*: snapshot from a finite-volume numerical solution for a small-scale debris flow. High pressure builds up at the front (the stagnation pressure) whereas an area of low pressure develops at the top, resulting in a lift force on the debris flow, which lifts up the snout. This allows for ambient water to form a water corridor and so lubricate the flow. The process is termed hydroplaning (Calculation by P. Gauer From Gauer et al. 2006)

(continued)

Box 9.9 (continued)

Herschel-Bulkley or a bi-viscous fluid. In its basic formulation, BING is two dimensional. The equations of fluid mechanics are vertically integrated reducing the information along the vertical direction, but at the same time greatly simplifying the equations. The geometry of the debris flow is divided in several sectors and the acceleration of each sector is calculated according to the forces acting on it. With this technique one can obtain workable equations for the acceleration of the debris flow. MathApp reports more details about the BING model. Figure 9.26 top shows the dynamics of a large-scale debris flow of the Ormen Lange area (Storegga area) simulated with the BING model.

Figure 9.26 bottom shows a computation fluid dynamics (CFD) calculation, not based on depth integration. The result is a much more detailed result, but also a greater computational effort that often allows only for small-scale calculations (Gauer et al. 2005).

9.7.3 Theories for the Mobility of Submarine Landslides

9.7.3.1 Hydroplaning

It has been shown that subaqueous landslides and in particular debris flows exhibit extremely high mobility (Fig. 9.5). Experiments (←Box 9.7) indicate hydroplaning as a possible explanation especially when the artificial debris flow is rich in clay (Mohrig et al. 1998; Elverhøi et al. 2002). The process consists in the formation of a water corridor underneath the artificial mudflow (Mohrig et al. 1998). The water layer, typically some cm thick, causes the debris flow to travel with much reduced resistance at the bottom. In the absence of a water layer, the resistance at the base of a cohesive debris flow is of the order (←Sect. 4.7.2) $\tau \approx \tau_y + \mu_B U/D$ where a Bingham model is adopted for simplicity. The water layer modifies the shear resistance to a form of the skin-friction type (Eqs. 9.12–9.14). The ratio between resistance of the mud against water and that against the terrain defines a dimensionless ratio

$$M = \frac{\text{Friction against water layer}}{\text{Friction against terrain}} \approx \frac{(1/2)\rho_F U^2 C_{SF}}{\tau_y + 2\mu_B U/D} \approx \frac{1}{2} \frac{\rho_F U^2 C_{SF}}{\tau_y} \quad (9.83)$$

that in the experiments is of the order *0.01*. In a large-scale natural debris flow, shear strengths of the order of some kPa may be expected. Velocities will be 10–30 times higher, so that the ratio remains well below unity. Hydroplaning thus decreases substantially the resistance, promoting debris flow mobility.

The cause for the spontaneous formation of a water layer has to do with the Bernoulli principle and the stagnation pressure (←Sect. 3.3.3). The snout of the debris flow experiences a stagnation pressure given by (←Eqs. 3.24–3.26) $P_{STAG} \approx (1/2)\rho_F U^2$ (the pressure increase is also visible in the simulation of Fig. 9.26b). Whenever P_{STAG} exceeds the overburden pressure due to the weight of the material

$$P_{BASE} = \Delta \rho g D \cos \beta \tag{9.84}$$

the pressure will promote water seepage underneath the debris flow. The two pressures P_{STAG} and P_{BASE} need not be exactly equal, as the shape of the debris flow and the viscosity of water (which rules the time needed for seeping and is affected by suspended fines) will play a role in the ignition of hydroplaning. The condition for hydroplaning can be conveniently written in terms of the densimetric Froude number defined as

$$Fr = \frac{U}{\sqrt{\frac{gD\Delta\rho}{\rho_A}\cos\beta}}. \tag{9.85}$$

Hydroplaning starts whenever $Fr > Fr_C$. Experiments indicate that the critical Froude number with an experimental value for the critical Froude number $Fr_{crit} \approx 0.3$. Velocities should thus be greater than

$$U > Fr_{crit}\sqrt{\frac{2\Delta\rho g D \cos\beta}{\rho_W}} \tag{9.86}$$

for hydroplaning to start. With a slurry density of 1.7 kg m^{-3}, an artificial debris flow like the one of Fig. 9.24 starts hydroplaning when $U > 0.4$ m/s. For natural debris flows the velocities needed are relatively low (of the order of some meters per second) and are probably easily attained. Thus, hydroplaning is in all probability a likely process in cohesive submarine debris flows. Lubrication is more pronounced at the front, and this explains the frequently observed detachment of the front of the debris flow from the rest of the body. If the debris flow is not sufficiently cohesive and impermeable, the water layer will seep into the debris flow, obliterating the lubrication effect. This could explain the reason why subaqueous debris flows are more mobile than rock avalanches (Fig. 9.5).

There have been attempt to simulate hydroplaning on a computer. An implementation of the BING code called Water-BING is based on depth integration also for the water layer (De Blasio et al. 2004). Other more refined calculations based on finite volume algorithm show spontaneously the generation of the lubricating water layer (Fig. 9.26), but are computationally demanding and can be applied only to the laboratory scale.

Hydroplaning is likely a complicated effect involving a combination of drags and lift forces.

9.7.3.2 Increase in Pore Water Pressure

Pore water pressure may increase in submarine debris flows due to overloading, as suggested for the long runout (>400 km) of Saharan debris flow (volume 1,100 km^3 and runout ratio 0.0036, Gee et al. 2005; Masson et al. 1998), where the deposits of volcaniclastic debris flow were covered by a more recent pelagic debris flow, with consequent increase of pore pressure in the former.

9.7.3.3 Dilution and Remolding

It is natural to expect that depending on the composition, the landslide material may dilute during the flow due to the intense shear stress exerted by water. The decrease of both the yield stress and the viscosity with time is however very difficult to estimate from first principles. Once the material is diluted, its resistance with the bottom diminishes and the slide acquires mobility. This scenario has been suggested for the Storegga landslide (←Box 9.3) in Norway as an alternative to hydroplaning (Gauer et al. 2005; De Blasio et al. 2005). In order to reproduce the runout of the Storegga landslide without hydroplaning, a shear strength as low as 500 Pa is necessary (De Blasio et al. 2005). This requires an extreme dilution of sediments in water. It is likely that the process of remolding, if it occurs, is gradual.

The dilution of the whole debris flow is not strictly necessary for a dynamical decrease of the landslide resistance. A remolding at the base will be sufficient, because the center of mass acceleration is affected only by the forces at the boundary in addition to gravity (← see, e.g., the simple example in 6.6.1). It is natural to expect that most of the remolding will occur at the base of the landslide, where shear stresses and shear rates are highest.

General References Chapter 9: Hampton et al. (1996); Locat and Lee (2002); Hsü (2002).

Chapter 10
Other Forms of Gravity Mass Flows with Potentially Hazardous Effects

It is useful to briefly consider other gravity-driven mass flows, and their differences and similarities with landslides. In this chapter we limit ourselves in illustrating a few basic facts about lava streams, ice avalanches, flash floods due to dam breaks, snow avalanches, and turbidity currents. A brief discussion on slow landslides is also included.

The opening figure shows the relict of the Gleno dam in northern Italy, which broke up on 1 December 1923. The consequent water wave destroyed the Dezzo valley, claiming 400 lives.

10.1 Lava Streams

Low-viscosity lava streams and sheets can travel to very long distances. Enormous lava plateaus are the results of gigantic basaltic eruptions, such as the Deccan plateau in India (which spread on an area of 500,000 km^2 during a series of eruptions 66 Ma B.P.) or the Columbia plateau in Washington state (16 Ma B.P., 160,000 km^2). Lava plateaus of this extension are rare in the geological record. Smaller-scale lava flows are far more frequent and are a hazard problem in volcanic areas characterized by fluid lava.

> **Box 10.1** Physical Properties and Rheology of Lava
>
> Volcanic rocks can be roughly classified based on the content of free silica in the melt. When free silica is absent and all the silicon atoms are bound to iron and magnesium to form ferromagnesians, the resulting basaltic magma has a mafic composition. These melts are typical of oceanic ridges but may be also erupted on land. The content of silica may increase due to the contamination of the continental crust. Lavas rich in free silica are called felsic; between the felsic and mafic limits, melts have intermediate composition.
>
> Viscosities of silicate melts vary enormously according to the composition and temperature. Basaltic lavas, which achieve the highest temperatures, have viscosities between about 6 and 8 Pa s at 1,400 C up to 100 Pa s at 1,100 C. There is a sharp increase of viscosity in approaching the melting point. Intermediate and felsic magmas have far higher viscosities. This determines slow velocity of lava or even stagnation of thick lava plugs followed by explosions due to gas pressing from below. These kinds of magmas that may give rise to pyroclastic flows are not considered in this short illustration of lava flows.
>
> Several fitting formulas have been proposed for the viscosity as a function of the temperature. One of the most widely used formulas is the Arrhenius form (Spera 2000)
>
> $$\mu = \mu_0 \exp\left[\frac{a}{T}\right] \qquad (10.14)$$
>
> where the constants μ_0, a can be fitted to reproduce observed viscosities. Values of μ_0 for basalts are of the order $10^{-5} - 10^{-6}$ Pa s and $a \approx 15,000 - 30,000$ K. The presence of solid crystals increases the viscosity of the melt (←Sect. 4.2).

A simple model for basaltic lava flow has been suggested by Danes (1972). Here we follow his model, with some small differences. Assuming laminar flow and Newtonian behavior, the Navier–Stokes equation in Lagrangian form for the flow down a plane surface reads (←Sect. 3.6.4)

$$\frac{Du}{Dt} = -\frac{1}{\rho}\frac{\partial p}{\partial x} + g\sin\beta + \frac{\mu}{\rho}\frac{\partial^2 u}{\partial y^2} \tag{10.1}$$

where the usual symbols are used (←Sect. 3.6.4). The second of the Navier–Stokes equation leads to the condition of hydrostatic pressure $p = \rho g(h-y)\cos\beta$ where h is the thickness of the lava flow (←Sect. 3.7.2). Imposing absence of acceleration $Du/Dt = 0$, it is found

$$-\rho g\frac{\partial h}{\partial x}\cos\beta + \rho g\sin\beta + \mu\frac{\partial^2 u}{\partial y^2} = 0. \tag{10.2}$$

The first term is the usual lateral pressure force. Similar to a debris flow, in general, this term changes with time and position depending on the height of the lava sheet. We denote the derivative of the slope of the lava profile as $\partial h/\partial x \equiv h'$. The velocity profile as a function of the height becomes upon integration (←Sect. 3.7.2)

$$u(y) = \frac{1}{2\mu}\rho g(\sin\beta - h')(2h - y)y. \tag{10.3}$$

Debris flows may change the viscosity during flow as a consequence of water entrainment, water expulsion, and erosion (←Chap. 4). Also lava changes the viscosity with time as a consequence of cooling, as the traveling molten rock exchanges heat to the surroundings. This introduces simple concepts of thermodynamics in the problem. First, we need to know how the temperature decreases as a function of the distance from the source. The energy loss in 1 s by a body of surface S at a temperature T is partly due to emission of radiation, and partly due to convection, i.e., exchange of heat with air. The energy loss by radiation emission is given by the law of Stefan and Boltzmann

$$W = \sigma S\varepsilon T^4 \tag{10.4}$$

where $\sigma = 5.67 \times 10^{-8}$ W m^{-2} K^{-4} is the Stefan–Boltzmann constant, $\varepsilon \approx 0.95$ is the emissivity; convection follows Newton's law (one of the many laws named after him)

$$\frac{dT}{dt} = -h_C S(T - T_{\text{AMB}}) \tag{10.5}$$

with $h_C \approx 10 - 50$ Wm^{-2}K^{-1} is the convective heat transfer coefficient and T_{AMB} is the ambient temperature.

Assuming that the lava body changes uniformly the temperature in response to the energy lost from the surface, it is found that the temperature derivative as a function of time behaves as

10.1 Lava Streams

$$\frac{dT}{dt} = -\frac{\sigma\varepsilon}{\rho Ch}(T^4 - T^4_{AMB}) - \frac{h_C(T - T_{AMB})}{\rho Ch} \tag{10.6}$$

where C is the specific heat of lava. Convection is often small, so that in the following we neglect the second term in Eq. 10.7. Moreover, because $T \gg T_{AMB}$ we can put $T_{AMB} = 0$.

Equation 10.6 shows that cooling rate decreases with lava sheet thickness due to the presence of the thickness h in the denominator. Putting

$$dt = \frac{dx}{\bar{u}} \tag{10.7}$$

where

$$\bar{u} = \int_0^h u(y)dy = \frac{2}{3}u_{max} \tag{10.8}$$

is the average flow velocity, one finds by substitution in the previous equations

$$\frac{\exp(-a/T)}{T^4}dT = -\frac{3\sigma\mu_0}{C\rho^2 g(\sin\beta - h')(h_0 - h'x)^3}dx \tag{10.9}$$

where h_0 is the initial lava thickness (close to the vent); upon integration an implicit equation for the runout R of the lava flow can be found

$$\frac{\rho^2 Cg(\sin\beta - h')}{3\sigma a^3 \mu_0}\left|\left[(a/T+1)^2 + 1\right]\exp(-a/T)\right|_{T_0}^{T_F} = -\frac{h'R^2 + 2h_0R}{2h_0^4(1 - h'R/h_0)^2}. \tag{10.10}$$

Neglecting the term h' on the right hand side, Danes (1972) obtains for the example of $\sin\beta - h' = 0.001$ the following result: for an initial temperature of 1,400 C (1,673 K) a minimum thickness of about 11 m is necessary to produce a 300-km long lava flow, where the following values are used: $\mu_0 \approx 1.318 \times 10^{-6}$ Pa s and $a \approx 26,500$ K, $C=840\,\mathrm{J\,kg^{-1}\,K^{-1}}$; $\rho=3{,}000\,\mathrm{kg\,m^{-3}}$; $\sigma=5.7\times 10^{-8}\,\mathrm{W\,m^{-2}\,K^{-4}}$. For an initial temperature of only 1,250 C (1,473 K), a 15-m thickness is necessary to reach the same distance.

Retaining the terms on the left hand side of Eq. 10.10, an explicit expression for the runout can be in principle obtained. However, here for simplicity we evaluate the runout for constant lava thickness, $\partial h/\partial x \equiv h' = 0$. The above Eq. 10.10 becomes

$$R = -\frac{h_0^3 \rho^2 Cg\sin\beta}{3\sigma a^3 \mu_0}\left|\left[(a/T+1)^2 + 1\right]\exp(-a/T)\right|_{T_0}^{T_F} \tag{10.11}$$

Using the values suggested earlier

$$R = -\frac{9 \times 10^6 h_0^3 \times 840 \times 9.8 \times \sin\beta}{3 \times 5.67 \times 10^{-8} \times 26{,}500^3 \times 1.318 \times 10^{-6}} \\ \left|\left[(26{,}500/T + 1)^2 + 1\right]\exp(-26{,}500/T)\right|_{1{,}400}^{1{,}100} \quad (10.12)$$

it is found

$$R = 388 \times h_0^3 \times \sin\beta \quad (10.13)$$

showing the strong dependence of the runout on the cube of the thickness.

The cooling of lava flows and the problem of their maximum flow length is a complex one, and more approaches have been devised. Of particular importance is the distinction between volume-limited and cooling-limited lava flow. In the first case, the source ceases to emit lava when the lava flow is still fluid; the flow thus stops because of lack of feeding from the vent. In the second case, closer to the Danes model, the flow extension is limited by the cooling that after a long travel has led the lava to the solidification point. Attempts have been made to relate the length of lava flow to effusion rate, as field data indicate a strong correlation between the two.

10.2 Ice Avalanches

Glaciers result from snow accumulation turning to ice in cold regions of the Earth. Glaciers may be temperate if the temperature of the ice increases with depth due to heat flow from the interior of the Earth, or polar, if the temperature constantly decreases with depth. Temperate glaciers typically flow downslope with velocities ranging from some tens of meter per year to some tens of kilometer per year. Polar glaciers typically form extensive ice sheets like in Antarctica, lasting for millions of years without major recycling. At times, portions of a glacier may collapse forming an ice avalanche similar to a small sturzstrom. In the following, disasters associated with ice avalanches are briefly examined.

10.2.1 Fall of an Ice Avalanche

There may be various sources of instability in a glacier body leading to the detachment of large portions of ice. The Nevados Huascaran events of 1962 and 1970 have already been mentioned (←Chap. 1). Well-documented cases occurred in Switzerland, where ice falls and ice avalanche events have been studied since the time of Heim. On 11th September, 1895 five million cubic meters of ice collapsed

10.2 Ice Avalanches

from the top of the Altels, a mountain in the Bernese Alps, killing six people. On 30th August 1965, two million cubic meters belonging to the terminus of the glacier Allalingletscher (Valais) fell from a height of 400 m; continuing its descent along the bottom of the valley below, the ice killed 88 people. Even though most victims are killed by the impact with ice, soil, and rock entrained during the flow, the air pressure generated by the falling ice can also be harmful, as documented in an ice avalanche that occurred on 5 September 1996 in the Bernese Alps, when hikers were injured by the wind created by the moving mass.

The causes of detachment of large portions of a glacier seem to be related to increase in the glacier speed. Many ice falls occur at the front of hanging glaciers. Normally the glacier moves slowly, allowing for a partial melting of the glacier and to the collapse of small quantities of ice from the front. However, if the speed of the glacier increases, the ablation may be insufficient and the volume of the collapsed ice increases dramatically. Haefeli (1966) has distinguished two types of ice falls:

- Type I occur in the form of wedge failure; they have volumes of the order $10^3 - 10^5$ m^3 and typically start on gently inclined slopes followed by a cliff.
- Type II is a slab-shape failure of greater volumes ($10^5 - 10^6$ m^3) usually forming along a slope in the form of a ramp.

Saltzmann et al. (2004) further distinguish between temperate and cold-based type II ice falls according to the thermal regime of the glacier. In the cold-based kind, the ramp needs a steeper slope for ice to detach.

The runout of ice avalanches is difficult to quantify; the apparent friction coefficient of ice blocks ranges from 0.25 to 0.44 which, except for the Altels event, is lower than the friction coefficient for a rock avalanche of comparable volume. The effect of partial melting of ice due to energy accumulated during fall is not significant. A unit mass of ice falling from a height of H meters accumulates an energy gH, sufficient to melt a fraction $gH/\lambda \approx 3 \times 10^{-5} H$ of the mass, where $\lambda = 3.3 \times 10^5$ J kg^{-1} is the latent heat for melting and H is the drop height in meters. For a fall height of the order 300–400 m reported for the Allalingletscher event, this corresponds to 1% of the mass, in the hypothesis that all the energy available has been used up to melt the ice.

The best mitigation measure for ice avalanches is probably prediction. Attempts have been made to use the acceleration of the flow as a means to predict the final detachment, similar to the collapse of rock avalanches (←Box 6.2). Some researchers have shown that the ice creep leading to failure can be well described by a function of the kind

$$u(t) = A + \frac{a}{(t_\infty - t)^n} \qquad (10.15)$$

where $u(t)$ is the velocity at time t, t_∞ is the time at the moment of detachment, and the other parameters are constants. A best fit analysis with sufficient number of observations during the acceleration phase may allow to determine the empirical constants in Eq. 10.15 and thus find t_∞, the likely time of detachment.

10.2.2 Stability Condition of an Ice Avalanche

Let us consider the instability condition of the glacier front of type II in a simple model of the Altels-kind event. The glacier slides on a slope as in Fig. 10.1; we investigate the conditions of instability of the frontal portion. The dimensions of the front block are W, L, and D. The stability analysis is similar to that for the rocky slab (←Sect. 2.2.2). Neglecting ice cohesion, the pack of ice is unstable if

$$\tan \phi < \tan \beta - \frac{1}{\rho g \cos^2 \beta} \frac{C_T}{L}. \qquad (10.16)$$

Equation 10.16 shows that stability is enhanced by high values of the tensile strength. Instability increases with the length L, to a point where the ice may break along a crevasse. Water pressure may also play a role; for simplicity its role has not been included in Eq. 10.16. Tensile strengths for ice depend on many variables, primarily the temperature, the strain rate, and the ice grain size. For example, at 10° below zero, the tensile strength is about 0.8 MPa for grain diameters d of 7–8 mm, and increases to 1.2–1.3 MPa for diameters of 1 mm, following a relationship common also to metals known as the Hall–Petch law,

$$C_T = C_{T0} + k d^{-1/2} \qquad (10.17)$$

where k and C_{T0} are two constants (Petrovich 2003). It follows from Eq. 10.16 that with ice friction coefficient of 0.25, the tensile strength contributes in a decisive manner to stability for length L shorter than about 150 m. For longer lengths L, stability diminishes because the contribution of tensile strength becomes negligible. Small crystals determine a more stable configuration, as the tensile strength increases.

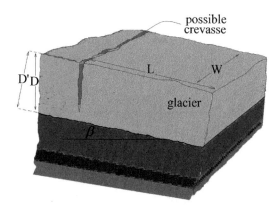

Fig. 10.1 Stability of an ice slab

10.3 Catastrophic Flood Waves

Catastrophic dam breaks have taken place in the past with poorly constructed dams (like with the Gleno catastrophe, see the opening figure to the chapter) or reduced rock quality and adverse geological setting (e.g. Malpasset in France, see Fig. 10.2). They may also occur as a consequence of breaching of natural deposits damming ephemeral lakes such as moraines, glaciers, or landslide deposits. One of the effects of global warming is the formation of extensive lakes on the upper surface of temperate glaciers. These lakes are temporarily dammed by ice. When the ice obstruction is removed, water is released as a flood wave termed glacial lake outburst flood (GLOF). Flood waves may also be caused by a landslide plunging in an artificial reservoir like the Vaiont landslide of October 9, 1963 (←Box 7.1). Dam breaks may occur as a consequence of inadequate design or exceptional precipitation. This happened, e.g., at Sella Zerbino (Italy, 1935), where an unexpected high rainfall caused the dam overtopping and its subsequent collapse. This accident claimed about 100 lives.

The discharge of the sudden water wave traveling at high speed can potentially wipe off everything along its path. If the water release is small (e.g., Stava in northern Italy), water may also mix with soil and transform into a debris flow or a

Fig. 10.2 The remains of the 60-m high Malpasset dam in France that collapsed on 2nd December 1959. Huge blocks have been transported by the water to long distances. Original photo courtesy of Ole Mejlhede Jensen

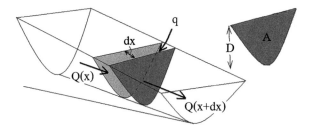

Fig. 10.3 Geometry of the system for the calculation of the de St. Venant equation

hyperconcentrated flow. The opposite process may also occur, where a debris flow traveling along a river bed dilutes to a flood with high solid load content.

The peak discharge following the dam collapse can be huge. Pilotti et al. (2006, 2010) provide a reliable methodology, along with a program, to compute the discharge hydrograph at the dam cross-section even in case of partial dam break. The mathematical description of the sudden water release along an open channel can be formulated in terms of the one-dimensional de Saint-Venant equations. Figure 10.3 shows the geometry of the system. Let us consider a volume element of length dx and cross section $A(x)$ and $A(x+dx)$ along a channel, defining the control volume. Water enters from the upstream end (located on the left in Fig. 10.3) and exits from the right.

Let $Q(x)$ be the volume of water entering the control volume per unit time from upstream, and $Q(x+dx)$ the volume of water per unit time exiting downstream. Some water may also enter or exit the control volume from the sides of the valley. We call q this net lateral inflow. The dimension of Q and q are m^3 s^{-1} and m^3 s^{-2}, respectively.

The equation for water balance is given by a simple balance as

$$\rho Q(x + dx) = \rho Q(x) + \rho q dx - \frac{\partial(\rho A(x) dx)}{\partial t} \qquad (10.18)$$

which simplifies to

$$\frac{\partial Q}{\partial x} + \frac{\partial A}{\partial t} = q \qquad (10.19)$$

where for simplicity we write Q and A instead of $Q(x)$ and $A(x)$. We also define a mean velocity as

$$U = \frac{Q}{A} \qquad (10.20)$$

so that the previous equation becomes

$$\frac{\partial(UA)}{\partial x} + \frac{\partial A}{\partial t} = q \qquad (10.21)$$

10.3 Catastrophic Flood Waves

We consider now the momentum equation. The elementary gravity force acting on the control volume is

$$dF_g = \rho g A dx \sin \beta \tag{10.22}$$

where β is as usual the slope angle while the lateral (longitudinal) variation of water pressure is

$$dF_p = -\rho g A \frac{\partial D}{\partial x} dx \tag{10.23}$$

and the resistance of the walls can be accounted for in a Chezy-like model

$$dF_F = \rho g A C^2 U^2 dx / R \tag{10.24}$$

where R is the hydraulic radius and C is Chezy's constant ($\leftarrow$Sect. 4.7.4) (in some formulations the resistance is expressed in terms of the slope of the energy line).

Newton's equation becomes so

$$\frac{1}{A}\frac{dQ}{dt} = \frac{dU}{dt} = -\frac{1}{\rho}\frac{dF_g + dF_p + dF_f}{dx} \tag{10.25}$$

Using the definition of the Eulerian derivative

$$\frac{d}{dt} = \frac{\partial}{\partial t} + U \frac{\partial}{\partial x} \tag{10.26}$$

it is found

$$\frac{\partial U}{\partial t} + U \frac{\partial U}{\partial x} = g \sin \beta - g \frac{U^2}{C^2 R} - g \frac{\partial D}{\partial x} + \frac{q}{A} U. \tag{10.27}$$

where the last term derives from the momentum carried by the lateral inflow. Equations 10.21 and 10.27 in combination are known as the de Saint-Venant equations. The section of the channel must be known along the transects of the channel. They can be used to calculate numerically the velocity and the water depth in time, called the hydrograph, at a certain position. The valley profile for a grid of longitudinal positions along the valley must be provided as an input together with the initial conditions at time $t = 0$ where the water is contained in the reservoir. The solution of these hyperbolic, non-linear partial differential equations is not immediate and requires appropriate algorithms, but can be worked out reliably with good approximation on modern computers.

Figure 10.4 shows the hydrograph calculated for the Gleno dam break (see opening figure) at a location about 7–8 km valley ward (Pilotti et al. 2011): The maximum level reached by water (approximately 15 m) is consistent with both

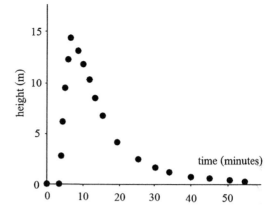

Fig. 10.4 The hydrograph calculated about 7 km from the point of the dam break. The data reported are an average of two calculations performed with different integration schemes (Simplified from Pilotti et al. 2011)

witness reports and water marks on the houses walls. In the same location, the critical water level lasted about 20–25 min. The calculations by Pilotti et al. (2011) also show that the hydrograph decreases in correspondence of widenings of the valley, and that the flood covered about 15 km in 45 min, traveling at an average velocity of 5–6 m/s. However, peak velocities immediately after the dam break were much higher, reaching 20 m/s.

10.4 Snow Avalanches

Snow avalanches deserve a short description, also considering their commonness in the mountain environment. The most distinctive characteristic of snow avalanches is the fact that they are composed of ice and water, which affects their behavior in many ways. Water is always present in the atmosphere and is easily transported by the winds and precipitation. This makes the accumulation rate fast and the return period for avalanching comparatively shorter than it is for landslides. As a consequence, some slopes may produce avalanches every year. Because temperatures in the mountain environment are often close to the melting point, snow may change behavior fast, which entails a modification of the characteristics of the resulting avalanche. In fact, an avalanche may be initiated simply when the sun warms locally a patch of snow, causing it to partially melt.

Based on the properties of snow, one recognizes two types of snow avalanches: non-cohesive and cohesive. The non-cohesive type is formed when snow lacking cohesion becomes unstable starting from a small area, often because of local melting. As the material starts to flow, it affects the snow at the sides. The result is that this kind of avalanche often extends in width downhill resembling a triangular shape similar to the superficial debris flows seen in (←Fig. 4.8). The lack of cohesion in snow may be due to two opposite situations. The first possibility is when snow is very dry and cold, and wind is absent. In this condition, snow

10.4 Snow Avalanches

crystals are incapable of forming bounds; snow remains powdery and crystals do not adhere. The resulting avalanches are usually not particularly large and dangerous, as they tend to be small and slow. The second situation resulting in non-cohesive avalanches is when the snow is completely wet due to intense heating or rain.

In contrast, cohesive snow produces avalanches that are particularly dangerous and responsible for most casualties. This is because the collapse involves instantaneously a large slab of hard snow; the resulting avalanche has large volume and high speed. The detachment is often provoked by an external perturbation like a skier passing nearby or overloading caused by intense precipitation. Failure often occurs in correspondence of a layer of low shear resistance, such as an ice sheet or bedrock.

The role of water in conferring non-cohesive and cohesive behavior to the snow is reminiscent of the building of sand castles. It is impossible to shape sand when it is either dry or completely wet, because in these conditions sand lacks the necessary cohesion, which is ensured only with intermediate amount of water in the pores.

Let us now briefly consider the dynamics of cohesive "slab" avalanches, those more similar to a rock avalanche. Immediately after detachment, the avalanche acquires speed and fragments into numerous ice particles. Mixed with air, ice develops a strong vertical density gradient. The avalanche can be so roughly divided into a dense part (density 0.2 and velocity up to 50 m/s) in contact with the terrain, and a less dense part (density less than 0.05 and velocity of 60 m/s). If the snow is dry, also a cloud of snow and air may travel suspended by the intense turbulence. This stratification may have an equivalent in both subaerial and subaqueous landslides; subaerial landslides often develop clouds of broken rock lifted up by turbulence; the subaqueous ones may also create a suspension in the form of a turbidity current ($\rightarrow$Sect. 10.6). The denser part of an avalanche is usually thinner than 5 m, while the suspension may be much thicker. Radar measurements are employed to study the internal structure and velocity distribution of snow avalanches.

Impact forces exerted by snow avalanches on walls or edifices can be surprisingly high. As seen for debris flows, the dynamical part of impact pressure is proportional to the density and to the velocity squared ($\leftarrow$Sect. 4.7.2)

$$P \approx \frac{1}{2} K \rho U^2. \qquad (10.28)$$

In a snow avalanche, the force is exerted by three component: air, powdery snow suspended in air, and larger agglomerates. The density ρ in Eq. 10.28 results from these three components. The constant K is usually estimated to be of the order two. This constant, however, depends on many variables like the Froude number, the type of avalanche, and the form of the obstacle; values as high as 6 have been suggested (e.g., Sovilla et al. 2008).

Maximum measured pressures attain values of the order 1 MPa, sufficient to displace structures in concrete. Smaller avalanches reach maximum pressures of 100 kPa, still sufficient to eradicate large trees. Measurements of pressures as a function of time show strong variation of pressure in time, especially for dry snow. Maximum pressures are measured at the front of the avalanche.

10.5 Slow Landslides and Soil Creep

10.5.1 Sackungs and Lateral Spreads

Some landslides may start as a slow creep before terminating catastrophically (←Box 6.2). Another kind of slow landslide, called sackung, is characterized by a large portion of rock flowing at a very slow rate. Volumes reached are often enormous (in the range of hundreds of million cubic meters) and involve areas of square kilometers. Several huge blocks may participate, resulting in a mass movement that resembles a slump in the form, and demonstrating the depth of the deformation (Fig. 10.5). Attempts have been made to describe these phenomena in terms of deep-seated micro-fracturing or viscous, non-Newtonian flow, but the rheological behavior remains largely speculative.

Lateral spreads are another class of landslides that normally move slowly. A typical example is the spreading of the La Verna in Tuscany, a monastery founded by St. Francis of Assisi (Fig. 10.6; see also Fig. 1.9). In this case, the intact rock splits in two series of perpendicular joints, resulting in a sub-division in prismatic blocks. Because the main faces of the blocks are perpendicular to the

Fig. 10.5 Example of sackung. Val di Fassa (Italy) (Photograph courtesy of Giovanni B. Crosta)

Fig. 10.6 The La Verna landslide in central Italy consists of sub-vertical rocky blocks resting on clayey materials

terrain, there is no geometric possibility of sliding or toppling and the sole instability results from the weakness of the clayey layers underneath the rocky blocks. Blocks thus tend to widen the gaps with time.

10.5.2 Soil Creep and Other Superficial Mass Movements

Soil creep is the slow movement of soil under the effect of gravity. Creep may be an important mechanism of mass movement and denudation; it may also evolve into a larger (and faster) superficial landslide (←Sect. 4.6.2).

The phenomenon of creep can be observed with periodical measurements separated by years or decades; creep is also noticed inadvertently because of the change in the position and deformation of shafts, pipes or fences. Typical velocities range between 0.002 and 0.3 m/a. Higher rates are more typical for arctic environments, while slow rates are normally measured in temperate zones. The rate of creep is not uniform, and becomes negligible in contact with the base.

Soil creep occurs also on slopes much gentler than the angle of repose of the composing grains. To explain the occurrence of creep, one has to consider the contraction and expansion of soil induced by cycles of heating–cooling and wetting–drying. When the soil swells as a consequence of warming and wetting conditions, it does so under the effect of the gravity pull. The gravity component along slope causes the soil to move slightly. Also during contraction will the soil move a little down slope. Thus, after many cycles of expansion and contraction, the particles of soil experience a zigzag path with a net component down the incline.

The simplest model to quantitatively describe soil creep is the viscous Newtonian model. The solution for the flow of an infinite Newtonian fluid is a parabolic velocity (←Sect. 3.7.2) $U(z) = U_0 - \frac{\rho g \sin \beta}{2\langle\mu\rangle} z^2$ where U_0 is the velocity at the surface, ρ is the average density of the soil and z is the downward height measured from this surface. This equation identifies a depth at which the velocity goes to zero $D = \sqrt{\frac{2\langle\mu\rangle U_0}{\rho g \sin \beta}}$, and so we can write $U(z) = U_0 \left[1 - \left(\frac{z}{D}\right)^2\right]$. Using typical values for $U_0 = 0.01$ m/a, $\beta = 10°$ and $\rho = 1,600$ kg/m³ and $D = 0.5$ m viscosities turn out to be of the order $\approx 10^{11}$ Pa s. Note that here the viscosity is a phenomenological parameter which allows us a first approximate calculation, but has no direct interpretation.

10.6 Suspension Flows: Turbidites and Turbidity Currents, and Relationship with Submarine Landslides

Suspension flows are made up of solid grains suspended in water or air. The gravity pull is maintained by the small density difference between the medium and the suspension. Pyroclastic flows, consisting of hot volcaniclastic particles in turbulent movement, are an example of suspension flows in air. Suspension flows in the ocean

10.6 Suspension Flows

are called turbidity currents (←Chap. 9). In the field, the products of turbidity currents appear as a succession of turbidite deposits like in Fig. 10.7. Although turbidites appear as horizontal deposits, the occurrence of gravitational instability and the presence of a slope during the emplacement is frequently confirmed by the presence of small slumps in turbiditic basins (Fig. 10.7b).

Fig. 10.7 (a) Turbidites in the Miocenic basin of Carboneras (southern Spain). (b) A small slump in the same locality appears in the lower part of this unit as the disturbed pack of sediments, demonstrating the unstable situation in turbiditic basins. The slump has been covered by new turbidites that appear horizontal

10.6.1 Turbiditic Basins

10.6.1.1 Where and When Turbidity Currents Occur

It has been discussed how the Grand Banks landslide mass was initially dense, and that it was probably transformed into a debris flow following water entrainment and remolding of the material (←Box 9.5). It is likely that when the landslide came to a halt, the suspended material continued for several hundred kilometers as a turbidity current, much like the snow suspension of a powder snow avalanche, which persists traveling after the bulk has come to rest. A turbidity current is thus likely coupled to a subaqueous debris flow (Hampton 1972). This is confirmed by laboratory experiments of subaqueous debris flows that show a thin suspension of fine material on top of the artificial debris flow.

A second possible mechanism for the formation of turbidity currents has already been mentioned. In correspondence of the continental rise, where slopes are steeper, some canyons may canalize high-speed turbidity currents. The resulting deposits are more abundant at the foot of the continental rise, where they form an underwater fan. The Ganges fan, about 2,600 km long and more than 1,000 km wide, is an example of huge fan formed by the deposition of turbidites. The slope gradient of the Ganges canyon is modest: only $0.27°$ (1:125). The gradient diminishes even more in correspondence of the fan to only $4'$ of degree (1:890), which explains the massive deposition in correspondence of the slope break.

Finally, turbidity currents may also form when rivers damp their solid content in the sea or in a lake. The first observed turbidity current, described by August Forel in 1885, was of this kind. Noticing the turbid, clay-rich waters of the Rhone stream disappearing into the Geneva lake, he deduced that the density contrast with the clear waters of the lake was causing the river to sink to the lake bottom.

10.6.2 Ancient Turbidites

Whereas the analysis of recent turbidites requires complex offshore studies, there are numerous examples of ancient turbidites directly accessible on land where they can be analyzed in detail.

A typical turbidite sequence is shown in Fig. 10.8a and b. It is usually made up of a sequence of layers of thickness variable from some mm to some meters. Each layer is often composed of sand at the bottom, gradually changing into more silty and muddy component at the top. The Dutch sedimentologist Arnould Bouma has suggested that the change from the arenaceous part in the bottom to muddy part at the top may follow an ideal sequence. Figure 10.9a and b shows a typical Bouma sequence and the standard interpretation. From the bottom to the top the sequence comprises:

(A) A normally graded bedding composed by sand or coarse silt devoid of laminations
(B) Sand and silt finer than in unit 1; the layer appears laminated horizontally
(C) Cross laminated sand or silt (finer than in unit B)

10.6 Suspension Flows

Fig. 10.8 (**a**) Turbidites in the Apennines, Italy. Miocene. (**b**) Eocenic turbidites in Broto, Pyrenees, northern Spain. Hedda Breien standing for scale

(D) Fine silt, horizontally laminated (finer than in unit C)
(E) Massive non-laminated muds, sometimes bioturbated (i.e., they present structures due to animal feeding or locomotion)

On top of the level E there may appear the level A of the next sequence.

In short, the most evident character of a turbidite is the gradation. When direct gradation occurs, particles have fallen in the water column freely which indicates a relatively low particle density. The standard interpretation of this sequence is as follows

(A′) The layer of graded coarser sand is deposited during the passage of the denser basal part of a turbidity current; the reason for the normal grading is that the large grains settle faster in a fluid.

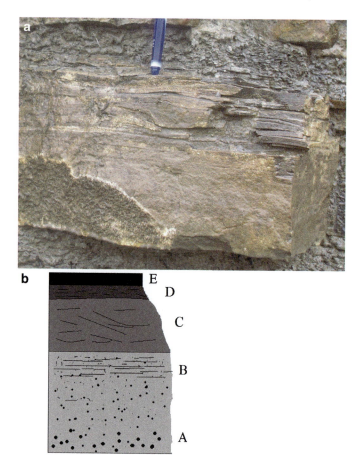

Fig. 10.9 (**a**) A Bouma sequence from the same locality of (Fig. 10.8b). (**b**) The classical Bouma sequence

(B′) The presence of a lamination is suggestive of a laminar current at the moment of deposition; granules are deposited quickly, but the current is sufficient to displace them further.
(C′) The presence of a cross lamination in the turbidites demonstrates a change of current direction during the waning phase of the turbidity current.
(D′) In this phase, the current becomes weak.
(E′) There is no current and the mud layer results from settling of clay particles; while the other layers are perhaps deposited in matter of hours, this layer may take much longer time to form, due to the low fall velocity of clay particles (→PhysApp).

A complete Bouma sequence is, however, uncommon in the sedimentological record. The Bouma concept thus represents an idealized situation, whose significance and generality has been the subject of controversy.

10.6 Suspension Flows

Fig. 10.10 (a) A long and wide tool mark is visible in the centre of the picture. It is due to the scouring of an object at the bottom of the sea, dragged by a strong current. Bioturbation is also visible. (b) Flute casts due to the effect of vortices on the sandy–silty sediment. The current was from the top right to bottom left. Such primary structures indicate strong bottom currents produced by turbidity currents, and are usually typical of the upper basins and of channels, where currents are stronger. Eocene of the Ainsa basin, northern Spain

Turbidite systems may reach enormous volumes and lengths of hundreds of kilometers. The coarse-grained component tends to be deposited close to the region of origin in the basin, whereas silt and clay may reach longer distances. In the proximal area of the basin, the high energy associated with the currents may form characteristic sedimentary structures such as flute casts or tool marks (Fig. 10.10), whereas such structures are less common in the deep basin where the energy of the current is lower.

10.6.3 Flow of a Turbidity Current

10.6.3.1 Anatomy and Speed of a Turbidity Current

Figure 10.11 shows schematically the experimental observation that the head of a turbidity current is thicker than the body. The experiments by Kuenen and Middleton have shown that the head moves with a velocity

Fig. 10.11 Highly schematic picture of a turbidity current

$$U_H \approx 0.7\sqrt{g\frac{\Delta\rho}{\rho}H_H} \qquad (10.29)$$

where H_H is the head thickness and $\frac{\Delta\rho}{\rho} = \frac{\rho - \rho_w}{\rho}$ is as usual the Archimedean buoyancy. Note that the slope angle is absent in this equation. The body, on the other hand, moves in response to the gradient with a velocity given approximately as

$$U \approx \sqrt{\frac{8}{f}g\frac{\Delta\rho}{\rho}\sin\beta H} = C\sqrt{\frac{\Delta\rho}{\rho}\sin\beta H} \qquad (10.30)$$

where β is as usual the slope angle and the constant is

$$C = \sqrt{\frac{8}{f}g} \qquad (10.31)$$

while f is a friction factor. Notice that the form is similar to the Chezy equation for a river or a debris flow (←Sect. 4.7.4). This is not surprising: a turbidity current is, loosely speaking, like a giant river moving within a submarine channel. A main difference is that the current is driven by a density contrast, hence the buoyancy term in (10.30). In a turbidity current, the resistance to flow occurs not only with the bottom and with the canyon walls, but also due to the steady water at the top. This implies that the friction factor f is made up of two contributions

$$f = f_0 + f_1 \qquad (10.32)$$

where f_0 (Darcy–Weisbach term) represents the friction with the canyon walls and f_1 the effect of steady water. Although reliable data are difficult to achieve, a reasonable range of values is $0.01 < f < 0.05$ and so $40\ \text{m}^{1/2}\ \text{s}^{-1} < C < 88\ \text{m}^{1/2}\ \text{s}^{-1}$, with the highest values referring to situation at high Froude and Reynolds numbers. It is also believed that of the two contributions the dominant one is the effect of the walls, whereas static water plays a minor role.

As an example, let us use values $\frac{\Delta\rho}{\rho} = 0.01$, $\beta = 1°$, $H = 10$ m, $H_{\text{HEAD}} = 50$ m so that $U \approx 39.6\sqrt{(\Delta\rho/\rho)\sin\beta H}$; velocities are 1.6 m/s for the body and 1.5 m/s for the head. Higher velocities may be reached for denser or thicker currents, or steeper slopes.

10.6.3.2 Physical Processes in a Turbidity Current

First, the settling time may be long enough compared to the flowing time that fine particles will remain suspended for the duration of the event ($\rightarrow$PhysApp). For example, a silt particle with fall velocity of 0.1 cm/s will be able to travel horizontally for 20 km before settling at the bottom of a 10-m high turbidity current moving at 2 m/s. This explanation may hold for the finest particle component like silt or clay. To explain why sand is also transported by a turbidity current, two additional mechanisms should be invoked: hindered settling and turbulence.

When a suspension settles, the particles displace water. If the suspension is sufficiently dense, this causes an upward water flow that opposes particle settling and reduces the downward velocity. The resulting falling velocity is (Richardson–Zaki law)

$$u_\downarrow = \frac{\eta_0}{\eta} u_\infty \varepsilon^2 f(\varepsilon) \qquad (10.33)$$

where η_0 is the viscosity of pure water, η is the viscosity of the water + clay suspension, ε is the void fraction $\varepsilon = 1 - C_S$ where C_S is the volume fraction of particles larger than clay. The function

$$f(\varepsilon) = \varepsilon^\alpha \qquad (10.34)$$

gives the increase of viscosity due to particles larger than clay and α is the Richardson–Zaki constant, which is 2.65 for low Reynolds numbers ($Re_p < 0.3$) about 0.4 for $Re_p > 500$, and acquires intermediate values between $0.3 < Re < 500$.

The second process is turbulence. Estimated values for the Reynolds number in a turbidity current may reach 10^6–10^7, which implies a full turbulent regime. Turbulence develops a water velocity component perpendicular to the terrain, which prevents particles from settling. In turn, particle-laden currents maintain a high velocity that generates turbulence, a feedback process called autosuspension. A criterion of autosuspension is Bagnold's

$$\frac{U \sin \beta e}{u_\infty} > 1 \qquad (10.35)$$

where e is an efficiency constant. Note the explicit dependence on the slope angle. Another autosuspension criterion is the one suggested by Stacey and Bowen

$$\frac{UC_D}{u_\infty \cos \beta} \approx \frac{UC_D}{u_\infty} > \frac{1}{S_0} \qquad (10.36)$$

where C_D, S_0 are, respectively, the drag coefficient for the grain and a dimensionless constant. Note that the autosuspension is favored by high velocity of the turbidity current.

General references Chapter 10: McClung and Schaerer (1993); Scheidegger (1975); Bridge and DeMicco (2008); Hsü (2002); Simpson (1997).

Appendix GeoApp (Geological and Geotechnical)

GeoApp1: Some Miscellaneous Properties of Rocks and Soils

Some properties of rocks

Rock	Friction coefficient	Friction angle	Shear strength (MPa)	Density
Sandstone	0.84–1	40–45	4	1.9–2.2
Limestone	0.70	35	5–30	2.3–2.6
Chalk	0.47	25	3	1.8
Granite	1.43	55	35	2.7
Basalt	1.19	50	40	2.9
Gneiss	0.58	30	30	2.7
Schist	0.47	25		2.7
Slate	0.47	25		2.7

Source: From Waltham 2004, reduced and modified

Unconfined Compressive Strength. It refers to the strength that a rock can support in unconfined setting and under uniaxial load. Typical values range from some hundreds of MPa for igneous rocks down to some tens of MPa for sedimentary rocks. The strength of rocks increases enormously in confined setting (triaxial strength).

Particle size classes. Particles building up clastic sediments can be classified according to their size as shown in the table.

	Clay	Silt	Fine sand	Coarse sand	Gravel	Pebbles	Boulders
Diameter	2 µm	2 µm	20 µm	200 µm	2 mm	2 cm	0.2 m

GeoApp2: List of Some Landslides on Mars and Earth

The following tables report a selection of landslides from the Earth and Mars. The data reported are: location of the landslide, volume V, area A, fall height H, runout R, runout ratio H/R. The apparent friction angle can be calculated from the runout ratio as $\alpha = \tan^{-1}(H/R)$. The comparison with Mars is interesting in revealing that: (1) both terrestrial and Martian landslides present a volume effect, and (2) the runout ratio is higher for Mars than for the Earth for a slide of the same volume. Data from different sources.

Name	Volume V ($\times 10^6 m^3$)	Area (km^2)	Height (m)	Runout R (m)	Runout ratio (H/R)
Mars (Olympus Mons)	14400000		8000	500,000	0.016
Mars (Olympus Mons)	2500000		8000	290,000	0.028
Mars	833000	1075	5400	56,250	0.096
Mars	688000	888	3600	45,000	0.08
Mars	668000	656	4400	30,985	0.142
Mars	655000	470	7600	53,900	0.141
Mars	321000	312	5400	36,000	0.15
Mars	157000	325	2800	32,941	0.085
Mars	32000	125	3600	29,032	0.124
Mars	29000	350	4000	20,000	0.2
Mars	14000	175	2000	18,018	0.111
Mars	11000	44	1200	8,000	0.15
Mars	5500	84	6400	20,983	0.305
Mars	5300	81	6200	20,000	0.31
Mars	4300	66	6200	19,018	0.326
Mars	3300	50	5000	15,974	0.313
Mars	1400	22	6200	16,986	0.365
Mars	900	13	2200	7,000	0.314
Mars	300	4	2200	5994	0.367
Mars	100	3	4200	7500	0.56
Felsberg	0.1	0.1	702	900	0.78
Haltenguet	0.15	0.075	252	300	0.84
Fidaz	0.4	0.2	520	1000	0.52
Airolo	0.5		192	300	0.64
Winkelmatten	3	0.25	224	700	0.32
Fionnay	4	0.4	210	300	0.7
Prayon	5	0.2	375	500	0.75
Elm	10	0.58	493	1700	0.29
Lago di Antrona	12	1	510	1500	0.34
Hintersee	13	0.95	1178	3100	0.38
Biasca	15	1	605	1100	0.55
Brione	16	0.7	238	700	0.34
Grand Clapier	16	0.41	686	1400	0.49
Mordbichl	20	0.45	200	500	0.4
Lago di Alleghe	20	0.5	408	800	0.51
St André	21	0.79	235	500	0.47
San Giovanni	25	1.2	272	1700	0.16
Haiming	29.5	1.7	510	1500	0.34
Haslensee	30	0.7	420	1000	0.42
Dobratsch	30	5			0.27

(continued)

Appendix

(continued)

Name	Volume V ($\times 10^6 m^3$)	Area (km^2)	Height (m)	Runout R (m)	Runout ratio (H/R)
Voralpsee	30	0.85	513	1900	0.27
Torbole	30	1.1	527	1700	0.31
Goldau	35	4	660	3000	0.22
Marquartstein	50	2.3	540	2000	0.27
Diablerets	50	2.2	1575	4500	0.35
Lac Lauvitel	68	1.6	1326	3900	0.34
Pletzachkogel	80	4	900	3000	0.3
Lofer	80	3.58	810	3000	0.27
Am Saum	100	1	280	700	0.4
Oberterzen	100	5			0.32
Mallnitz	100	2.4	875	3500	0.25
Cal de la Madeleine	115	1.84	900	2500	0.36
Oeschinensee	120	1.7	270	1000	0.27
Obersee	140	2.5	1131	3900	0.29
Abimes de Myans	150	15			0.22
Lago di Poschiavo	165	1	238	700	0.34
Masiere di vedane	170	7	966	4200	0.23
Dobratsch (1)	170	8	1053	3900	0.27
Tshirgant	210	13.2	720	4500	0.16
Lago di Tovel	250	5.2			0.25
Dobratsch (2)	360	16	1100	4400	0.25
Lago di Molveno	400	3.6			0.24
Eibsee	400	11	1311	6900	0.19
Monte Spinale	550	11.6			0.23
Totalp	600	4.3	748	4400	0.17
Kandertal	900	6.8	1314	7300	0.18
Siders(Sierre)	2000	28	1397	12700	0.11
Köfels	2100	12	954	5300	0.18
Engelberg	2500	9	1100	4400	0.25
Twin Slides (W)	7		887.3	4670	0.19
Twin Slides (E)	7		836	4400	0.19
Antronapiana	12		1634.1	4190	0.39
Huascarán (1962)	13		3569.6	15520	0.23
Sherman	13.3		1071	5950	0.18
Trilet Glac.	18		1863	6900	0.27
Steller 1	20		1206	6700	0.18
Allen 4	23		1309	7700	0.17
Fairweather	26		3300	10000	0.33
Schwan	27		1525	6100	0.25
Devastation Gl.	27		1220	6100	0.2
Diablerets	30		1210	5500	0.22
Rubble Cr.	33		1035	6900	0.15
Huascarán (1970)	75		3900	15600	0.25
Dusty Cr.	7		971.1	2490	0.39
Elm	10		598	2300	0.26

(continued)

(continued)

Name	Volume V ($\times 10^6 m^3$)	Area (km²)	Height (m)	Runout R (m)	Runout ratio (H/R)
Sasso Englar	13		369.6	1680	0.22
Drinov	18.5		405.6	1560	0.26
Mystery Cr.	35		1240	4000	0.31
Goldau	35		1098	6100	0.18
Frank	36.5		789.6	3290	0.24
Low. Gros Ventre	38		652.5	4350	0.15
Stalk Lakes	53		690	3000	0.23
Lavini di Marco	200		1186.5	5650	0.21
Mont Granier	210		1538	7690	0.2
Silver Reef	227		733.7	6670	0.11
Blackhawk	283		1084.6	9860	0.11
Martinez Mt.	380		1883.2	8560	0.22
Maligne Lake	667		984.6	5470	0.18
Mayunmarca	1600		1760	8000	0.22
Constantino	20		940.8	2240	0.42
Madison Canyon	28		436.8	1680	0.26
Monte Zandilla	40		1382.5	3950	0.35
Hope	47		1229.6	4240	0.29
Triple Slide	47		555.8	3970	0.14
Parpan	400		1310	6550	0.2
Lavini di Marco	100		950	4500	0.21
Flims	9000		1500	11000	0.14
Saidmarreh	20000		1000	12000	0.08
Aksu	≈1500	3.3	1900	4600	0.41
Kokomeren	≈1000	5.4	1800	4600	0.39
Beshkiol	≈10000	49	2500	10500	0.24
Karakudjur	≈10000	>32	1600	6000	0.27
Chukurchak	≈1000	5.7	1200	7500	0.16
Dead Lakes	2500	17	1800	7700	0.23
Djamantau	≈1000	10	1300	6000	0.22
Kugart	2250	20	700	7750	0.09
Sarychelek	2250	<30	1700	6250	0.25

Other Data

Name	Volume V ($\times 10^6 m^3$)	Runout ratio (H/R)
Val Lagone	0.65	0.44
Wengen (1)	2.5	0.45
Schächental	0.5	0.58
Tucketthütte	6	0.34
Val Brenta	8	0.6
Oberes	8.5	0.31
Voralpsee	40	0.32
Fadalto	95	0.25
Cima di Saoeseo	150	0.27
Bormio	180	0.29

(continued)

(continued)

Name	Volume V ($\times 10^6 m^3$)	Runout ratio (H/R)
Dejenstock	280	0.21
Almtal	300	0.11
Rockslide Pass	350	0.15
Fernpaß	1000	0.09
Parpan-Lenzerheide	400	0.31
Sierre	2000	0.11
Pandemonium Cr	5	0.22
Damocles	10	0.16
Twin E	10	0.2
Glärnisch-Guppen	730	0.36
Vallesinella	9	0.31
Fedaia	10	0.38
Melköde	10	0.47
Lago di Autrona	12	0.34
Monte Corno	17	0.31
Pamir	2000	0.24
Tamins	1300	0.095
Poschiavo	150	0.36
Gros Ventre	38	0.17
Corno di Dosté	20	0.32
Little Tahoma	11	0.29
Lecco	0.03	0.88
Airolo	0.5	0.65
Vaiont	230	0.23
Monbiel	0.75	0.42
Goldau	35	0.21

GeoApp3: List of Processes of Sediment Transport from the Shelf into the Deep Ocean

1. Sediments carried along canyons may be due either to tidal currents or to episodic surges, probably related to turbidity currents (←Sect. 10.6). Turbidity currents are capable of depositing thick sedimentary bodies in the deep ocean known as turbidites. Turbidity currents typically form fans in correspondence of the continental rise, often building stacked systems of coalesced units from different canyons.
2. Landslides and gliding of large bodies of gravitationally unstable sediments also contribute to sediment budget in the deep ocean. Sliding masses may reach enormous volumes and extreme distances.
3. Contour currents are generated by slight density gradients of surface water in the oceans, due to salinity and temperature differences. They give rise to enormous subaqueous rivers all around the globe, partly superficial (e.g., the Gulf current)

and partly completely submerged in deep waters. For example, the plunging of dense cold water of the North Atlantic gives rise to a southward deep sea current. Contour currents are influenced by Earth's rotation (Coriolis acceleration). They may reach velocities as high as 0.4 m/s especially around the continents. The sediments deposited, known as contourites, are typically muddy to gravel-rich units.

4. Explosive volcanism may be responsible for transport of ash, lapilli to the shelf and deep waters, particularly around island arches.
5. Pelagic rains (oozes) are due to the settling of planktic organisms and may be calcareous or siliceous depending on the skeletal composition of the dead organisms. Like contourites, they originate from mostly hemipelagic muds, i.e., muddy sediments partly composed of at least 5% of biogenic material and at least 40% of clastic or volcaniclastic component.
6. Marine sediments of glacial origin may be transported offshore by ice rafting.
7. Carbonate rocks of the shelf may be the source material of subaqueous rock avalanches.

Appendix PhysApp (Physics)

PhysApp1: Physical Quantities and Their Dimensions

Quantities	SI units	CGS units	Dimensional formulae
Length	m	cm	L
Particle diameter	m	cm	L
Water or channel depth	m	cm	L
Boundary layer thickness	m	cm	L
Area	m^2	cm^2	L^2
Volume	m^3	cm^3	L^3
Time	s	s	T
Velocity	m/s	cm/s	LT^{-1}
Acceleration	m/s^2	cm/s^2	LT^{-2}
Diffusivity (general)	m^2/s	cm^2/s	L^2T^{-1}
Kinematical viscosity (a diffusivity)	m^2/s	cm^2/s	L^2T^{-1}
Rate of fluid discharge	m^3/s	cm^3/s	L^3T^{-1}
Mass	kg	g	M
Point force	N (Newton)	dyn (dyne)	MLT^{-2}
Density	kg/m^3	g/cm^3	ML^{-3}
Specific weight	N/m^3	$g\ wt/cm^3$	$ML^{-2}T^{-2}$
Dynamic viscosity	Pa·s	P (poise)	$ML^{-1}T^{-1}$
Sediment transport rate	kg/m width s	g/cm width s	$ML^{-1}T^{-1}$
Pressure	Pa (Pascal)	dyn/cm^2	$ML^{-1}T^{-2}$
Shear stress	Pa	dyn/cm^2	$ML^{-1}T^{-2}$
Momentum	kg m/s	g cm/s	MLT^{-1}
Energy	J (Joule)	erg	ML^2T^{-2}
Work	J	erg	ML^2T^{-2}
Power	J/s	erg/s	ML^2T^{-3}

PhysApp2: Properties of Water and Air

Table 1 Properties of *water* at atmospheric pressure

Temperature (C)	Density (kg/m^3)	Viscosity (Pa s)	Kinematic viscosity (m^2/s)
0	999.87	0.001794	0.000001794
4	1000.00	0.001568	0.000001568
5	999.99	0.001519	0.000001519
10	999.73	0.00131	0.00000131
15	999.13	0.001145	0.000001146
20	998.00	0.001009	0.000001011
30	996.00	0.0008	0.000000803

Table 2 Properties of *air* at atmospheric pressure

Temperature (°C)	Density (kg/m^3)	Viscosity (Pa s)	Kinematic viscosity (m^2/s)
0	1.293	0.00001709	0.00001322
50	1.093	0.00001951	0.00001785

Appendix MathApp (Mathematical)

MathApp1: Trigonometry

Trigonometric Identities

$$\sin x \equiv \sin(180 - x) \equiv -\sin(180 + x) \equiv -\sin(360 - x)$$
$$\sin x \equiv \cos(90 - x) \equiv -\cos(90 + x)$$

$$\cos x \equiv -\cos(180 - x) \equiv -\cos(180 + x) \equiv \cos(360 - x)$$
$$\cos x \equiv \sin(90 - x) \equiv \sin(90 + x)$$

$$\tan x \equiv -\tan(180 - x) \equiv \tan(180 + x) \equiv -\tan(360 - x)$$

$$\sin^2 x + \cos^2 x \equiv 1$$
$$1 + \tan^2 x \equiv \sec^2 x$$
$$1 + \cot^2 x \equiv \csc^2 x$$

$$\sin(x \pm y) \equiv \sin x \, \cos y \pm \cos x \, \sin y$$
$$\cos(x \pm y) \equiv \cos x \, \cos y \mp \sin x \, \sin y$$
$$\tan(x \pm y) \equiv \frac{\tan x \pm \tan y}{1 \mp \tan x \, \tan y}$$

$$\sin 2x \equiv 2 \sin x \, \cos x$$
$$\cos 2x \equiv \cos^2 x - \sin^2 x$$
$$\tan 2x \equiv \frac{2 \tan x}{1 - \tan^2 x}$$

$$\sin\frac{x}{2} \equiv \pm\sqrt{\frac{1-\cos x}{2}}$$

$$\cos\frac{x}{2} \equiv \pm\sqrt{\frac{1+\cos x}{2}}$$

$$\tan\frac{x}{2} \equiv \pm\sqrt{\frac{1-\cos x}{1+\cos x}} = \frac{\sin x}{1+\cos x}$$

$$\sin x \pm \sin y \equiv 2\sin\tfrac{1}{2}(x\pm y)\cos\tfrac{1}{2}(x\pm y)$$

$$\cos x + \cos y \equiv 2\cos\tfrac{1}{2}(x+y)\cos\tfrac{1}{2}(x-y)$$

$$\cos x - \cos y \equiv -2\sin\tfrac{1}{2}(x+y)\sin\tfrac{1}{2}(x-y)$$

Natural Trigonometric Functions

Degrees	Radians	Sin	Cos	Tan	Degrees
0°00′	0.0000	0	1	0	90°00′
1°00′	0.0175	0.0175	0.9998	0.0175	89°00′
2°00′	0.0349	0.0349	0.9994	0.0349	88°00′
3°00′	0.0524	0.0523	0.9986	0.0524	87°00′
4°00′	0.0698	0.0698	0.9976	0.0699	86°00′
5°00′	0.0873	0.0872	0.9962	0.0875	85°00′
6°00′	0.1047	0.1045	0.9945	0.1051	84°00′
7°00′	0.1222	0.1219	0.9925	0.1228	83°00′
8°00′	0.1396	0.1392	0.9903	0.1405	82°00′
9°00′	0.1571	0.1564	0.9877	0.1584	81°00′
10°00′	0.1745	0.1736	0.9848	0.1763	80°00′
11°00′	0.1920	0.1908	0.9816	0.1944	79°00′
12°00′	0.2094	0.2079	0.9781	0.2126	78°00′
13°00′	0.2269	0.2250	0.9744	0.2309	77°00′
14°00′	0.2443	0.2419	0.9703	0.2493	76°00′
15°00′	0.2618	0.2588	0.9659	0.2679	75°00′
16°00′	0.2793	0.2756	0.9613	0.2867	74°00′
17°00′	0.2967	0.2924	0.9563	0.3057	73°00′
18°00′	0.3142	0.3090	0.9511	0.3249	72°00′
19°00′	0.3316	0.3256	0.9455	0.3443	71°00′
20°00′	0.3491	0.3420	0.9397	0.3640	70°00′
21°00′	0.3665	0.3584	0.9336	0.3839	69°00′
22°00′	0.3840	0.3746	0.9272	0.4040	68°00′
23°00′	0.4014	0.3907	0.9205	0.4245	67°00′
24°00′	0.4189	0.4067	0.9135	0.4452	66°00′

(continued)

(continued)

Degrees	Radians	Sin	Cos	Tan	Degrees
25°00′	0.4363	0.4226	0.9063	0.4663	65°00′
26°00′	0.4538	0.4384	0.8988	0.4877	64°00′
27°00′	0.4712	0.4540	0.8910	0.5095	63°00′
28°00′	0.4887	0.4695	0.8829	0.5317	62°00′
29°00′	0.5061	0.4848	0.8746	0.5543	61°00′
30°00′	0.5236	0.5000	0.8660	0.5774	60°00′
31°00′	0.5411	0.5150	0.8572	0.6009	59°00′
32°00′	0.5585	0.5299	0.8480	0.6249	58°00′
33°00′	0.5760	0.5446	0.8387	0.6494	57°00′
34°00′	0.5934	0.5592	0.8290	0.6745	56°00′
35°00′	0.6109	0.5736	0.8192	0.7002	55°00′
36°00′	0.6283	0.5878	0.8090	0.7265	54°00′
37°00′	0.6458	0.6018	0.7986	0.7536	53°00′
38°00′	0.6632	0.6157	0.7880	0.7813	52°00′
39°00′	0.6807	0.6293	0.7771	0.8098	51°00′
40°00′	0.6981	0.6428	0.7660	0.8391	50°00′
41°00′	0.7156	0.6561	0.7547	0.8693	49°00′
42°00′	0.7330	0.6691	0.7431	0.9004	48°00′
43°00′	0.75049	0.6820	0.7314	0.9325	47°00′
44°00′	0.76794	0.6947	0.7193	0.9657	46°00′
45°00′	0.7854	0.7071	0.7071	1.0000	45°00′

MathApp2: Exponential Function

Values of e^x and e^{-x}

x	e^x	e^{-x}	x	e^x	e^{-x}
0	1	1	1	2.718	0.368
0.1	1.105	0.905	2	7.39	0.135
0.2	1.221	0.819	3	20.09	0.0498
0.3	1.35	0.741	4	54.6	0.0183
0.4	1.492	0.67	5	148.4	0.00674
0.5	1.649	0.607	6	403.4	0.00248
0.6	1.822	0.549	7	1,097	0.000912
0.7	2.014	0.497	8	2,981	0.000335
0.8	2.226	0.449	9	8,103	0.000123
0.9	2.46	0.407	10	22,026	0.000045

MathApp3: Derivatives and Indefinite Integrals

General Derivation Rules

Let $u = u(x)$ and $v = v(x)$ be two functions

Function	Derivative
$u + v$	$u' + v'$
$u - v$	$u' - v'$
uv	$u'v - uv'$
$\dfrac{u}{v}$	$\dfrac{u'v - uv'}{v^2}$
$f(u)$	$f'(u) \cdot u'$

Special rules

a, b, c, and r are arbitrary constants.

$f(x)$	$f'(x)$	
c	0	
x^r	rx^{r-1}	(arbitrary r)
$\sqrt{x}$	$\dfrac{1}{2\sqrt{x}}$	(the case $r = \frac{1}{2}$)
$\dfrac{1}{x}$	$-\dfrac{1}{x^2}$	(the case $r = -1$)
$\sin x$	$\cos x$	
$\cos x$	$-\sin x$	
$\tan x$	$\dfrac{1}{\cos^2 x}$	
a^x	$a^x \cdot \ln a$	($a > 0$)
$\log_a x$	$\dfrac{1}{x \ln a}$	
e^x	e^x	
$\ln x$	$\dfrac{1}{x}$	
$\sin(ax + b)$	$a\cos(ax + b)$	
$a\cos(ax + b)$	$-a\sin(ax + b)$	
e^{cx}	ce^{cx}	
$|x|$	$\dfrac{|x|}{x}$	($x \neq 0$)
$\tan^{-1} x$	$\dfrac{1}{1 + x^2}$	
$\sin^{-1} x$	$\dfrac{1}{\sqrt{1 - x^2}}$	
$\cos^{-1} x$	$-\dfrac{1}{\sqrt{1 - x^2}}$	

Appendix

Table of Integrals

The constant of integration is to be added in each case.

---

1. $\displaystyle\int u^n\,du = \frac{u^{n+1}}{n+1}$

2. $\displaystyle\int \frac{du}{u} = \log|u|$

3. $\displaystyle\int a^u\,du = \frac{a^u}{\log a}$ $\hspace{4em}(a>0,\ \ a\neq 1)$

4. $\int e^u\,du = e^u$

5. $\int \sin u\,du = -\cos u$

6. $\int \cos u\,du = \sin u$

7. $\int \tan u\,du = \log|\sec u|$

8. $\int \cot u\,du = \log|\sin u|$

9. $\displaystyle\int \sec u\,du = \log|\sec u + \tan u| = \log\left|\tan\left(\frac{u}{2}+\frac{\pi}{4}\right)\right| = \cosh^{-1}(\sec u)$

10. $\displaystyle\int \csc u\,du = \log|\csc u - \cot u| = \log\left|\tan\frac{u}{2}\right| = -\sinh^{-1}(\cot u)$

11. $\displaystyle\int \frac{du}{u^2+a^2} = \frac{1}{a}\tan^{-1}\frac{u}{a}$ $\hspace{4em}(a\neq 0)$

12. $\displaystyle\int \frac{du}{a^2-u^2} = \frac{1}{2a}\log\left|\frac{u+a}{u-a}\right| = \begin{cases}\dfrac{1}{2a}\log\dfrac{a+u}{u-a} = \dfrac{1}{a}\coth^{-1}\dfrac{u}{a} & (u^2>a^2)\\[6pt] \dfrac{1}{2a}\log\dfrac{a+u}{a-u} = \dfrac{1}{a}\tanh^{-1}\dfrac{u}{a} & (u^2<a^2)\end{cases}$

13. $\displaystyle\int \frac{du}{\sqrt{u^2+a^2}} = \log\left|u+\sqrt{u^2+a^2}\right| = \sinh^{-1}\frac{u}{a}$

14. $\displaystyle\int \sin^n u\,du = -\frac{\sin^{n-1} u\,\cos u}{n} + \frac{n-1}{n}\int \sin^{n-2} u\,du$ $\hspace{2em}n$ integral and ≥ 2

15. $\displaystyle\int \cos^n u\,du = \frac{\cos^{n-1} u\,\sin u}{n} + \frac{n-1}{n}\int \cos^{n-2} u\,du$ $\hspace{2em}n$ integral and ≥ 2

MathApp4: Coordinate Systems

Coordinate Systems

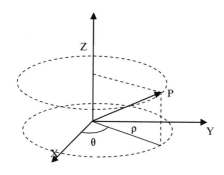
Cylindrical coordinates

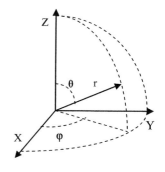
Spherical coordinates

Cylindrical coordinates

$$X = \rho \cos \theta$$
$$Y = \rho \sin \theta$$
$$Z \equiv Z$$

Spherical coordinates

$$X = r \sin \theta \cos \varphi$$
$$Y = r \sin \theta \sin \varphi$$
$$Z = r \cos \theta$$

MathApp5: Hyperbolic Functions

hyperbolic sinus: $\quad \sinh(x) = \dfrac{e^x - e^{-x}}{2}$

hyperbolic cosinus: $\quad \sinh(x) = \dfrac{e^x + e^{-x}}{2}$

hyperbolic tangent: $\tanh(x) = \dfrac{\sinh(x)}{\cosh(x)} = \dfrac{e^x - e^{-x}}{e^x + e^{-x}}$

MathApp6: Taylor Expansion of a Real Function

Consider a function $f(x)$ which is differentiable an infinite number of times in a certain open set. Then, if x_0 is a point belonging to this region, we can write

$$f(x) = \sum_{n=0}^{\infty} f^{(n)}(x_0) \frac{(x - x_0)^n}{n!}$$

where $f^{(n)}(x_0)$ is the nth derivative of the function calculated at the point x_0.

Taylor expansion allows us to write a transcendent function as a sum of polynomials progressively approximating the value of the function in the point. For example, the exponential function can be written as the following series

$$e^x = \sum_{n=0}^{\infty} \frac{x^n}{n!}$$

A second-order approximation is then

$$e^x \approx 1 + x + \frac{1}{2}x^2$$

which, depending on the degree of precision needed, can be a reasonably good approximation for $x < 0.1$.

MathApp7: The Symbol "Nabla"

It is common to find in the literature the "nabla" vector symbol $\vec{\nabla}$. In Cartesian coordinates, the nabla is an operator of this form

$$\vec{\nabla} = \hat{i}\frac{\partial}{\partial x} + \hat{j}\frac{\partial}{\partial y} + \hat{k}\frac{\partial}{\partial z}$$

It is easy to operate with this symbol, as it acts exactly like a vector (just remember that it is an operator, i.e., it acts on a space-dependent variable whose symbol should always stay to the right of the nabla symbol). Taking the scalar product of nabla with a vector in this order (the nabla first, and the vector after) one finds

$$\vec{\nabla} \cdot \vec{v} = \left[\hat{i}\frac{\partial}{\partial x} + \hat{j}\frac{\partial}{\partial y} + \hat{k}\frac{\partial}{\partial k} \right] \cdot \vec{v}$$

$$= \left[\hat{i}\frac{\partial}{\partial x} + \hat{j}\frac{\partial}{\partial y} + \hat{k}\frac{\partial}{\partial k} \right] \cdot [\hat{i}u_x + \hat{j}u_y + \hat{k}u_z] = \frac{\partial u_x}{\partial x} + \frac{\partial u_y}{\partial y} + \frac{\partial u_z}{\partial z}$$

This is called the *divergence* of the vector. Thus, the equation of continuity can be expressed simply as $\vec{\nabla} \cdot \vec{v} = 0$ where $\vec{v}$ is the velocity of the fluid.

Another operation consists in applying nabla to a scalar (for example, the height of a terrain h).

$$\vec{\nabla} h = \left[\hat{i}\frac{\partial}{\partial x} + \hat{j}\frac{\partial}{\partial y} + \hat{k}\frac{\partial}{\partial z} \right] h = \hat{i}\frac{\partial h}{\partial x} + \hat{j}\frac{\partial h}{\partial y} + \hat{k}\frac{\partial h}{\partial z}$$

This is called the *gradient* of h.

The divergence of the gradient of a scalar is also called the Laplacian and is indicated with the symbol $\nabla^2 h$. In Cartesian coordinates

$$\nabla^2 h \equiv \vec{\nabla} \cdot \vec{\nabla} h = \left[\hat{i}\frac{\partial}{\partial x} + \hat{j}\frac{\partial}{\partial y} + \hat{k}\frac{\partial}{\partial z} \right] \cdot \left[\hat{i}\frac{\partial h}{\partial x} + \hat{j}\frac{\partial h}{\partial y} + \hat{k}\frac{\partial h}{\partial z} \right] = \frac{\partial^2 h}{\partial x^2} + \frac{\partial^2 h}{\partial^2 y} + \frac{\partial^2 h}{\partial^2 z}$$

MathApp8: Leibnitz's Rule

Let us consider the following mathematical problem. A function $u(y,t)$ depending on two variables y and t is integrated with respect to the variable y from two points A and B: $\int_A^B u(y,t) dy$. What is the derivative of this integral with respect to the other variable, t? If the extremes of integration A and B do not depend on t, the result is obvious: the derivative can be carried out inside the integral

$$\frac{\partial}{\partial t} \int_A^B u(y,t) dy = \int_A^B \frac{\partial u(y,t)}{\partial t} dy.$$

However, if A and B depend on t, the final result is a little more complicated. Leibnitz's rule provides the result

$$\frac{\partial}{\partial t} \int_{A(t)}^{B(t)} u(y,t) dy = \int_{A(t)}^{B(t)} \frac{\partial u(y,t)}{\partial t} dy + u(B(t),t) \frac{\partial B(t)}{\partial t} - u(A(t),t) \frac{\partial A(t)}{\partial t}$$

Thus, in addition to the previous term, there appear two more addenda containing the derivative with respect to t of the two extremes of integration, multiplied by the integrand calculated at the same extremes.

References

Abrakhmatov K, Strom A (2006) Dissected rockslide and rock avalanche deposits. Tien Shan, Kyrgyzstan
Adushkin VV (2006) Mobility of rock avalanches triggered by underground nuclear explosions. In: Evans SG, Scarascia Mugnozza G, Strom A, Hermanns R (eds) Landslides from massive rock slope failure, NATO Science Series. Springer, Dordrecht
Ancey C (2006) Plasticity and geophysical flows: a review. Earth Sci Rev 45:230–287
Archard JF (1957) Elastic deformations and the laws of friction. Proc R Soc Lond Ser A 243:190–205
Azzoni A, de Freitas MH (1995) Experimentally gained parameters, decisive for rock fall analysis. Rock Mech Rock Eng 28:111–124
Bagnold RA (1954) Experiments on a gravity-free dispersion of large solid spheres in a Newtonian fluid under shear. Proc R Soc Lond Ser A 225(1160):49–63
Barla G, Dutto F, Mortara G (2000) Brenva glacier rock avalanche of 18 January 1997 on the Mount Blanc Range, Northwest Italy. Landslide News 13:2–5
Barnes HA, Hutton JF, Walters K (1989) An introduction to rheology. Elsevier, Amsterdam
Barton NR (1973) Review of a new shear strength criterion for rock joints. Eng Geol 7:287–332
Bianchi Fasani G (2007) La frana del 22 Agosto 2006: Cinematismo e possibili scenari future (in Italian). In: Pecci M, Scarascia Mugnozza G (eds) Il Gran Sasso in Movimento. Istituto Nazionale della Montagna, Bologna
Bondevik S, Løvholt F, Harbitz C, Mangerud J, Dawson A, Svendsen JI (2005) The Storegga Slide tsunami – comparing field observations with numerical simulations. Mar Petroleoum Geol 22:195–208
Bottino G, Chiarle M, Joly A, Mortara G (2002) Modelling rock avalanches and their relation to permafrost degradation in glacial environments. Permafrost Periglac Process 13:283–288
Bourgeois J (2009) Geological effects and records of tsunamis. In: Robinson A, Bernard EN (eds) The sea, vol 15, Tsunamis. Harvard University Press, Cambridge, pp 53–91
Bowden FP (1953) Proc R Soc Lond A 217:462
Bozzolo D, Pamini R (1986) Simulation of rock falls down a USA. Acta Mech 63:113–130
Breien H, De Blasio FV, Elverhøi A, Høeg K (2008) Erosion and morphology of a debris flow caused by a glacial lake outburst flood, Western Norway. Landslides 5:271–280. doi:10.1007/s10346-008-0118-3
Bridge JS, DeMicco RV (2008) Earth surface processes, landforms and sediment deposits. Cambridge University Press, Cambridge
Brilliantov NV, Poschel T (2000) Velocity distribution in granular gases of viscoelastic particles. Phys Rev E 61:2809
Brilliantov NV, Poschel T (2003) Kinetic theory of granular gases. Oxford University Press, Oxford
Broili L (1967) New knowledges on the geomorphology of the vaiont slide slip surfaces. Rock Mech Eng Geol 5:38–88

Brunetti MT, Guzzetti F, Rossi M (2009) Probability distributions of landslide volumes. Nonlinear Process Geophys 16:179–188

Byerlee JD (1978) Friction of rocks. Pure Appl Geophys 116:615–626

Campbell CS (1990) Rapid granular flows. Annu Rev Fluid Mech 22:57–92

Campbell CS, Cleary PW, Hopkins M (1995) Large scale landslide simulations: global deformations, velocities and basal friction. J Geophys Res 100:8267–8283

Cancelli A, Granata E, Griffini L (1991) Esempio di recupero all'uso insediativo di un'area interessata da fenomeni di scendimento massi. Acta Geol 68(2):105–121

Chiocci FL, Romagnoli C, Tommasi P, Bosman A (2008) The Stromboli 2002 tsunamigenic submarine slide: characteristics and possible failure mechanisms. J Geophys Res 113:B10102. doi:10.1029/2007JB005172

Colbeck SC (1995) Pressure melting and ice skating. Am J Phys 65:488–492

Collins GS, Melosh HJ (2003) Acoustic fluidization and the extraordinary mobility of sturzstroms. J Geophys Res 108:B10. doi:10.1029/2003JB002465

Corominas J (1996) The angle of reach as mobility index for small and large landslides. Can Geotech J 33:260–271

Costard F, Forget F, Jomelli V, Mangold N, Peulvast J-P (2007) Debris flows in greenland and on Mars. In: Chapman M (ed) The geology of Mars: evidence from earth-based analogs. Cambridge University Press, Cambridge

Coussot PC (1997) Mudflow rheology and dynamics: IAHR-AIRH monographs. AA Balkema Publishers, Rotterdam, p 272

Crandell DR (1971) Postglacial lahars from Mount Rainier Volcano, Washington. USGS professional paper 677. US Geological Survey, Reston

Crosta GB (2001) Failure and flow development of a complex slide: the 1993 Sesa landslide. Eng Geol 59:173–199

Crosta GB, Agliardi F (2004) Parametric evaluation of 3D dispersion of rockfall trajectories. Nat Hazards Earth Syst Sci 4:583–598

Crosta GB, Frattini P, Fusi N (2007) Fragmentation in the Val Pola rock avalanche, Italian Alps. J Geophys Res 112:23. doi:10.1029/2005JF000455

Cruden DM, Krahn J (1973) A reexamination of the geology of the Frank Slide. Can Geotech J 10:581–591

Cui P, Chen X, Wang Y, Hu K-H, Li Y (2005) Jiangjia Ravine debris flows in south-western China. In: Jakob M, Hungr O (eds) Debris-flow hazards and related phenomena. Springer-Praxis, Berlin

Curry R (1966) Observation of alpine mudflows in the Tenmile Range, central Colorado. Geol Soc Am Bull 77(7):771–776

Dade WB, Huppert HE (1998) Long-runout rockfalls. Geology 26:803–806

Danes ZF (1972) Dynamics of lava flows. J Geophys Res 77:1430–1432

Dantu P (1968) Etude statistique des forces intergranulaires dans un milieu pulverulent. Geotechnique 18:50–55 (in French)

Datei C (2005) Vaiont. La storia idraulica. Libreria Internazionale Cortina, Padua (in Italian)

Davies TRH (1982) Spreading of rock avalanche debris by mechanical fluidization. Rock Mech Rock Eng 15:9–24

Davies TRH, McSaveney MJ (1999) Runout of dry granular avalanches. Can Geotech J 36:313–320

Dawson AG, Stewart I (2007) Tsunami deposits in the geological record. Sed Geol 200:166–183

De Blasio FV (2007) Production of frictional heat and hot vapor in a model of self-lubricating landslides. Rock Mech Rock Eng 219–216. doi:10.1007/s00603-007-0153-8

De Blasio FV (2009) Rheology of a wet, fragmenting granular flow and the riddle of the anomalous friction of large rock avalancher. Granular Matter 11:179–184

De Blasio FV (2011) The aureole of Olympus Mons (Mars) as the compound deposit of submarine landslides. Submitted to: Earth and Planetary Science Letters

De Blasio FV, Elverhøi A (2008) A model for frictional melt production beneath large rock avalanches. J Geophys Res 113:F02014

De Blasio FV, Sæter M-B (2009a) Rolling friction on a granular medium. Phys Rev E 79:1

De Blasio FV, Sæter M-B (2009b) Small-scale experimental simulation of talus evolution. Earth Surf Process Land 34:1685–1692

De Blasio FV, Engvik L, Harbitz CB, Elverhøi A (2004) Hydroplaning and submarine debris flows. J Geophys Res 109:C01002. doi:10.1029/2002JC001714

De Blasio FV, Elverhøi A, Issler D, Harbitz CB, Bryn P, Lien R (2005) On the dynamics of subaqueous clay rich gravity mass flow – the giant Storegga slide, Norway. Mar Petrol Geol 22:179–186

De Blasio FV, Elverhøi A, Engvik LE, Issler D, Gauer P, Harbitz C (2006) Understanding the high mobility of subaqueous debris flows. Norw J Geol 86:275–284

Deganutti AM (2008) The hypermobility of rock avalanches. Tesi di Dottorato, Universitaa di Padova

Deparis J, Jongmans D, Cotton F, Baillet L, Thouvenot F, Hantz D (2008) Analysis of rock-fall and rock-fall avalanche seismograms in the French Alps. Bull Seismol Soc Am 98(4): 1171–1796

Dickens GR, O'Neil JR, Rea DK, Owen RM (1955) Dissociation of oceanic methane as a cause of the carbon isotope excursion at the end of the Paleocene. Paleoceanography 10:965–971

Dikau R, Brunsden D, Schrott L, Ibsen M-L (eds) (1996) Landslide recognition. Identification, movement and causes. Wiley, Chichester

Dorren LKA (2003) A review of rockfall mechanics and modelling approaches. Prog Phys Geogr 27:69–87

Dufresne A, Davies TR (2009) Longitudinal ridges in mass movement deposits. Geomorphology 105:171–181

Duncan JM (1996) Soil slope stability analysis. In: Turner AK, Schuser RL (eds) Landslides. Investigation and mitigation. Special Report 247. National Academy Press, Washington, DC

Duran J (2003) Powders and grains. Springer, Berlin

Edgers L, Karlsrud K (1982) Soil flows generated by submarine slides: case studies and consequences. Norwegian Geotech Inst Bull 143:1–11

Elverhøi A, De Blasio FV, Butt FA, Issler D, Harbitz C, Engvik L, Solheim A, Marr J (2002) Submarine mass-wasting on glacially influenced continental slopes-processes and dynamics. In: Dowdeswell JA, CÓ Cofaigh (eds) Glacier-influenced sedimentation on high-latitude continental margins. Geological Society, London, pp 73–87, Special publication 203

Elverhøi A, Breien H, De Blasio FV, Harbitz CB, Pagliardi M (2010) Submarine landslides and the importance of the initial sediment composition for run-out length and final deposit. Ocean Dyn 60:1027–1046. doi:10.1007//s10236-010-0317-z

Erismann TH (1979) Mechanisms of large landslides. Rock Mech Rock Eng 12:5–46

Erismann TH (1985) Flowing, rolling, bouncing, sliding: synopsis of basic mechanisms. Acta Mech 64:101–110

Erismann TH, Abele G (2001) Dynamics of rockslides and rockfalls. Springer, Berlin

Evans SG (2006) Single-event landslides resulting from massive rock slope failure: characterising their frequency and impact on society. In: Evans SG, Scarascia Mugnozza G, Strom A, Hermanns R (eds) Landslides from massive rock slope failure, NATO Science Series. Springer, Dordrecht

Evans SG, Clague JJ (1988) Catastrophic rock avalanches in glacial environments. In: Proceedings 5th international symposium on landslides. Balkema, Rotterdam, pp 1153–1158

Evans SG, Clague JJ, Woodsworth GJ, Hungr O (1989) The Pandemonium Creek rock avalanche, British Columbia. Can Geotech J 26:427–446

Evans SG, DeGraff JV (eds) (2002) Catastrophic landslides: effects, occurrence, and mechanisms. The Geological Society of America, Boulder. JV. Geol Soc Am Rev Eng Geol 15

Evans SG, Hungr O (1993) The assessment of rockfall hazard at the base of talus slopes. Can Geotech J 30:620–636

Fahnestock RK (1978) Little Tahoma Peak rockfall and avalanches, Mount Rainier, Washington, U.S.A. In: Voight B (ed) Rockslides and avalanches, 1. Elsevier, Amsterdam, pp 181–196

Federico A, Elia G, Fidelibus C (2009) Previsione del tempo di ricorrenza di una frana. Eng Hydro Environ 11:3–13 (in Italian)

Fine IV, Rabinovich AB, Bornhold BD, Thomson RE, Kulikov EA (2005) The Grand Banks landslide-generated tsunami of November 18, 1929: preliminary analysis and numerical modeling. Mar Geol 215:45–57

Francis PW, Wadge G (1983) The Olympus Mons aureole: formation by gravitational spreading. J Geophys Res 88:8333–8344

Galgaro A, Tecca PR, Genevois R, Deganutti AM (2005) Acoustic module of the Acquabona (Italy) debris flow monitoring system. Nat Hazards Earth Syst Sci 5:211–215

Gardner J (1969) Observations of surficial talus movement. Geomorholpol Z 13:317–323

Gauer P, Elverhøi A, Issler D, De Blasio FV (2006) On numerical simulations of subaqueous slides: back-calculations of laboratory experiments of clay-rich slides. Norw J Geol 86:295–300

Gauer P, Kvalstad TJ, Forsberg CF, Bryn P, Berg K (2005) The last phase of the Storegga slide: simulation of retrogressive slide dynamics and comparison with slide-scar morphology. Mar Petrol Geol 22:171–178

Gee MJR, Gawthorpe RL, Friedmann JS (2005) Giant striations at the base of a submarine landslide. Mar Geol 214:287–294

Genevois R, Ghirotti M (2005) The 1963 Vaiont landslide. G Geol Appl 1:41–52

Genevois R, Armento C, Tecca PR (2002) Failure mechanisms and runout behavior of three rock avalanches in the North-Eastern Italian Alps. In: Evans SG, Scarascia Mugnozza G, Strom A, Hermanns R (eds) Landslides from massive rock slope failure, NATO Science Series, Springer, Dordrecht, pp 431–444

Giani GP (1992) Rock slope stability analysis. Balkema, Rotterdam

Goguel J (1978) Scale-dependent rockslide mechanisms, with emphasis on the role of pore fluid vaporization. In: Voight B (ed) Rockslides and avalanches. Elsevier, Amsterdam, pp 693–706 (a cura di)

Griswold J, Bulmer MH, Beller D, McGovers PJ (2008) An examination of Olympus Mons aureoles: 39th lunar and planetary science conference, LPI contribution No. 1391, 2239

Guyon E, Hulin J-P, Petit L, Mitescu CD (2001) Physical hydrodynamics. Oxford University Press, Oxford

Guzzetti F, Crosta G, Detti R, Aglietti F (2002) STONE: a computer program for the three-dimensional simulation of rock-falls. Comput Geosci 28:1079–1093

Habib P (1975) Production of gaseous pore pressure during rock slides. Rock Mech Rock Eng 7:193–197

Haefeli R (1966) Note sur la classification, le mechanisme et le contrôle des avalanches de glacies et des crues glaciaires extraordinaires. IAHS-AISH Publ 69:316–325, in French

Haff PK (1983) Grain flow as a fluid-mechanical problem. J Fluid Mech 134:401–430

Haflidason H, Lien R, Sejrup HP, Forsberg CF, Bryn P (2005) The dating and morphometry of the Storegga slide. Mar Petrol Geol 22:123–136

Hampton AA (1972) The role of subaqueous debris flow in generating turbidity currents. J Sed Petr 42:775–793

Hampton MA, Lee HJ, Locat J (1996) Submarine slides. Rev Geophys 34:33–59

Harbitz CB (1992) Model simulation of tsunamis generated by the Storegga slides. Mar Geol 105:1–21

Harbitz CB, Pedersen G, Gjevik B (1993) Numerical simulations of large water waves due to landslides. J Hydraul Eng-ASCE 119:1325–1342

Harbitz CB, Løvholt F, Pedersen G, Masson DG (2006) Mechanisms of tsunami generation by submarine landslides: a short review. Norw J Geol 86:249–258

Harrison KP, Grimm RE (2003) Rheological constrains on martial landslides. Icarus 163:347–362

Heim A (1932) Bergsturz und Menschenleben. Fretz und Wasmuth, Zürich

Hendron AJ, Patton FD (1985) The Vaiont slide, a geotechnical analysis based on new geologic observations of the failure surface. I, II technical reports GL-85-5, U.S. Army Eng. Waterways Experiment Station, Vicksburg, Massachusetts

Hergarten S (2002) Self-organized criticality in earth systems. Springer, Berlin

Hewitt K (2009) Rock avalanches that travel onto glaciers and related developments, Karakoram Himalaya, Inner Asia. Geomorphology 103:66–79

Hodges CA, Moore HJ (1979) The subglacial birth of Olympus Mons and its aureoles. J Geophys Res 84:8061–8074

Hoek E (1987) Rockfall – a program in Basic for the analysis of rockfalls from slopes. Unpublished notes, University of Toronto, Ontario

Hsü KJ (1978) Albert Heim: observations on landslides and relevance to modern interpretation. In: Voight B (ed) Rockslides and avalanches, 1. Amsterdam, Elsevier, pp 71–93

Hsü KJ (1991) Catastrophic debris streams (Sturzstroms) generated by rockfalls. Geol Soc Am Bull 86:129–140

Hsü KJ (2002) Physics of sedimentology: textbook and reference. Springer, Berlin

Huang X, Garcia MH (1998) A Herschel-Bulkley model for mud flow down a slope. J Fluid Mech 374:305–333

Hübl J, Suda J, Proske D, Kaitna R, Scheidl C (2009) Debris flow impact estimation. In: International symposium on water management and hydraulic engineering, paper A56, Ohrid/Macedonia, 1–5 Sept 2005

Hungr O (2005) Classification and terminology. In: Jakob M, Hungr O (eds) Debris-flow hazards and related phenomena. Praxis, Springer, Berlin/Hiedelberg, pp 9–23

Hungr O, Evans SG, Bovis MV, Hutchinson JN (2001) A review of the classification of landslides of the flow type. Environ Eng Geosci 7(3):221–238

Hunt ML, Zenit R, Campbell CS, Brennen CE (2002) Revisiting the 1954 suspension experiment of R.A. Bagnold. J Fluid Mech 452:1–24

Hürlimann M, Rickenmann D, Graf C (2003) Field and monitoring of debris-flow events in the Swiss Alps. Can Geotech J 40:161–175

Hutchinson JN (1988) General report: morphological and geotechnical parameters of landslides in relation to geology and hydrogeology. In: Bonnard C (a cura di) Proceedings of the fifth international symposium on landslides, Balkema, Rotterdam, pp 3–35

Hutchinson JN (2002) Chalk flows from the coastal cliffs of northern Europe. In: Evans SG, DeGraaf JV (eds) Catastrophic landslides: effects, occurrence, and mechanisms. Geol Soc Am Rev Eng Geol vol 55

Hwang H, Hutter K (1995) A new kinetic model for rapid granular flow. Continuum Mech Thermodyn 7(3):357–384

Ilstad T, De Blasio FV, Elverhøi A, Harbitz CB, Engvik L, Longva O, Marr J (2004) On the frontal dynamics and morphology of submarine debris flows. Mar Geol 213:481–497

Imran J, Harff P, Parker G (2001) A numerical model of submarine debris flows with graphical user interface. Comput Geosci 27(6):721–733

Iverson RM (1997a) The physics of debris flows. Rev Geophys 35:245–296

Iverson R (1997b) Hydraulic modelling of unsteady debris-flow surges with solid-fluid interactions. In: Chen C-I (ed) Debris-flow hazards mitigation: mechanics, prediction, and assessment. New York, ASCE, pp 550–560

Iverson RM (2005) Debris-flow mechanics. In: Jakobs M, Hungr O (eds) Debris-flow hazards and related phenomena. Springer-Praxis, Berlin

Iverson RM, Denlinger RP (2001) Flow of variably fluidized granular masses across three-dimensional terrain 1. Coulomb mixture theory. J Geophys Res 106:537–552

Jakob M, Hungr O (eds) (2005) Debris-flow hazards and related phenomena. Praxis, Springer, Berlin, pp 9–23

Johnson AM (1970) Physical processes in geology. Freeman, Cooper & Company, San Francisco

Johnson AM (1984) Debris flow. In: Brunsden D, Prios DB (eds) Slope stability. Wiley, New York, pp 257–290

Johnson KL (1987) Contact mechanics. Cambridge University Press, Cambridge
Julien PY (1998) Erosion and sedimentation. Cambridge University Press, Cambridge
Keefer DK (2002) Investigating landslides caused by earthquakes – a historical review. Surv Geophys 23:473–510
Kent PE (1996) The transport mechanism in catastrophic rock falls. J Geol 74:79–83
Kozak J, Rybar J (2003) Pictorial series of the manifestations of the dynamics of the Earth. 3. Historical images of landslides and rock falls. Stud Geophys Geod 47:221–232
Kuijpers A, Nielsen T, Akhmetzhanov A, de Haas H, Kenyon NH, van Weering TCE (2001) Late quaternary slope instability on the Faeroe margin: mass flow features and timing of events. Geo-Mar Lett 20:149–159
Lan H, Martin CD, Lim CH (2009) Rockfall analyst: A GIS extension for three-dimensional and spatially distributed rockfall hazard modelling. Comput Geosci 33:262–279, in stampa
Legros F (2002) The mobility of long-runout landslides. Eng Geol 63:301–331
Legros F, Cantagrel J-M, Devouard B (2000) Pseudotachylyte (Frictionite) at the base of the Arequipa Volcanic Landslide Deposit (Peru): implications for emplacement mechanisms. J Geol 108:601–611
Lejeunesse E, Mangeney-Castelnau C, Vilotte JP (2006) Spreading of a granular mass on a horizontal plane. Phys Fluids 16:2371–2381
Levin B, Nosov M (2009) Physics of tsunami. Springer, Berlin
Li J, Yanmo Y, Cheng B, Defu L (1983) The main features of the mudflow in Jiang-Jia ravine. Z Geomorph NF 27:325–341
Locat J, Demers D (1988) Viscosity, yield stress, remolded shear strength, and liquidity index relationship for sensitive clays. Can Geotech J 25:799–806
Locat J, Lee HJ (2002) Submarine landslides: advances and challenges. Can Geotech J 39:193–212
Locat J, Lee HJ (2005) Subaqueous debris flows. In: Jakobs M, Hungr O (eds) Debris-flow hazards and related phenomena. Springer-Praxis, Berlin
Longva O, Janbu N, Blikra LH, Bøe R (2003) The 1996 Finneidfjord slide; seafloor failure and slide dynamics. In: Locat J, Mienert J (eds) Submarine mass movements and their consequences. Kluwer, Dordrecht, pp 531–538
Lopes RMC, Guest JE, Wilson CJ (1980) Origin of the Olympus Mons aureole and perimeter scarp. Moon Planets 22:221–234
Lorenzini G, Mazza N (2004) Debris flow – phenomenology and rheological modelling. WIT Press, Southampton, p 216
Lucas A, Mangeney A (2007) Mobility and topographic effects for large Valles Marineris landslides on Mars. Geophys Res Lett 34:L1021
Lucchitta BK (1979) Landslides in Vallis Marineris, Mars. J Geophys Res 84:8097–8113
Lucchitta BK, McEwen AS, Clow GD, Geissler PE, Singer RB, Schultz RA, Squyres SW (1992) The canyon system on Mars. In: Kieffer HH, Jakosky BM, Snyder CW, Matthews MS (eds) Mars. The University of Arizona Press, Tucson
Major JJ, Pierson TC, Scott KM (2005) Debris flows at Mount St. Helens Washington, USA. In: Jakob M, Hungr O (eds) Debris flow hazards and related phenomena. Springer Praxis, Chichester, pp 685–731
Makse HA, Johnson DL, Schwartz LM (2000) Packing of compressible granular materials. Phys Rev Lett 84:4160
Masch L, Wenk HR, Preuss E (1985) Electron microscopy study of hyalomilonites: evidence for frictional melting in landslides. Tectonophysics 115:131–160
Masson DG, Canals M, Alonso B, Urgeles R, Hühnerbach, V (1998) The Canary debris flow: source area morphology and failure mechanisms. Sedimentology 45:411–432
Mazzanti P (2008) Sapienza. PhD thesis, Universita di Roma
McCauley JM, Carr MH, Cutts JA, Hartmann WK, Masurski H, Milton DJ, Sharp RP, Wilhelms DE (1972) Preliminary mariner 9 report on the geology of Mars. Icarus 45:264–303
McClung D, Schaerer P (1993) The avalanche handbook. The Mountaineers Books, Seattle

References

McEwen AS (1989) Mobility of large rock avalanches: evidence from Valles Marineris, Mars. Geology 17:1111–1114

McGovern PJ, Smith JR, Morgan JK, Bulmer MH (2004) Olympus Mons aureole deposits: new evidence for a flank failure origin. J Geophys Res 109:E08008

McSaveney M (1978) Sherman glacier rock avalanche, Alaska, U.S.A. In: Voight B (ed) Rockslides and avalanches, 1. Elsevier, Amsterdam, pp 197–258

McSaveney M (2002) Recent rockfalls and rock avalanches in Mount Cook National Park, New Zealand. In: Evans SG, DeGraff JV (eds) Catastrophic landslides: effects, occurrence, and mechanisms. The Geological Society of America, Boulder, pp 35–70

McSaveney M, Davies TD (2006) Rapid rock mass flow with dynamic fragmentation: inferences from the morphology and internal structure of rockslides and rock avalanches. In: Evans SG, Scarascia Mugnozza G, Strom A, Hermanns R (eds) Landslides from massive rock slope failure. Springer, Dordrecht, pp 285–304

Melosh HJ (1979) Acoustic fluidization: a new geological process? J Geophys Res 84:7513–7520

Middleton GV, Wilcock PR (1994) Mechanics in the earth and environmental sciences. Cambridge University Press, Cambridge

Mohrig D, Whipple KX, Hondzo M, Ellis C, Parker G (1998) Hydroplaning of subaqueous debris flows. Geol Soc Am Bull 110:387–394

Morris EC (1981) Structure of Olympus Mons and its basal scarp. In: Paper presented to the 3rd international colloquium on Mars (abstract). Pasadena, pp 161–162

Morton DM, Campbell RH (1974) Spring mudflows at Wrightwood, Southern California. Q J Eng Geol 7:377–384

Mueth DM, Jaeger HM, Nagel SR (1998) Force distribution in a granular medium. Phys Rev E 57:3164

Müller L (1964) The rock slide in the Vaiont valley. Rock Mech Eng Geol 2(3/4):148–212

Murray B, Malin MC, Greeley R (1981) Earthlike planets. Freeman and Company, San Francisco

Nash D, Brundsen DK, Hughes RE, Jones DKC, Whalley BF (1985) A catastrophic debris flow near Gupis, northern areas, Pakistan. In: Proceedings of the 11th international conference on soil mechanics and foundation engineering, vol 3, San Francisco, 12–16 Aug 1985, pp 1163–1166

Nicoletti PG, Sorriso-Valvo M (1991) Geomorphic controls of the shape and mobility of rock avalanches. Geol Soc Am Bull 193(10):1365–1373

Nissen SE, Haskell NL, Steiner CT, Coterill KL (1999) Debris flow outrunner blocks, glide tracks, and pressure ridges identified on the Nigerian continental slope using 3-D seismic coherency. The Leading Edge Soc Explor Geophysicists 18(5):550–561

Pelletier JD, Malamud BD, Blodgett T, Turcotte DL (1997) Scale-invariance of soil moisture variability and its implications for the frequency-size distribution of landslides. Eng Geol 48:255–268

Pengcheng Z (1992) A discussion on the velocity of debris flows. Erosion, debris flows and environment in mountainuous regions. In: Proceedings of the international symposium, Chengdu, 5–9 July 1992

Penner RA (2001) The physics of sliding cylinders and curling rocks. Am J Phys 69:332–339

Pérez FL (1985) Surficial talus movement in an Andean Paramo of Venezuela. Geogr Ann 67A:221–237

Pérez FL (1998) Talus fabric, clast morphology, and botanical indicators of slope processes on the Chaos Crags (California Cascades), U.S.A. Géogr Phys Quatern 52:1–22

Persson BNJ (2000) Sliding friction. Physical principles and applications. Springer, Berlin

Petrovich JJ (2003) Mechanical properties of ice and snow. J Mater Sci 38:1–6

Pfeiffer T, Bowen T (1989) Computer simulation of rockfalls. Bull Assoc Eng Geol XXVI (1):185–196

Pierson TC (1980) Erosion and deposition by debris flows at Mt Thomas, North Canterbury, New Zealand. Earth Surf Proc 5:227–247

Pilotti M, Maranzoni A, Tomirotti M (2006) Modellazione matematica della propagazione dell'onda di piena conseguente al crollo della diga del Gleno. XXX Conv. Idr. Costr. Idr (in Italian)

Pilotti M, Tomirotti M, Valerio G, Bacchi B (2010) Simplified method for the characterization of the hydrograph following a sudden partial dam break. J Hydraul Eng 136(10):693–704

Pilotti M, Tomirotti M, Valerio G, Bacchi B (2011) The 1923 Gleno dam-break: case study and numerical modelling, Journal of Hydraulic Engineering, ASCE, pp 137, 480

Plafker G, Ericksen GE (1978) Nevados Huascaran avalanches, Peru. In: Voight B (ed) Rockslides and avalanches, 1. Elsevier, Amsterdam, pp 277–314

Pollet N, Schneider J-L (2004) Dynamic disintegration processes accompanying transport of the Holocene Flims sturzstrom (Swiss Alps). Earth Planet Sci Lett 221:433–448

Prior DB, Borhold BD, Coleman JM, Bryant WR (1982) Morphology of a submarine slide, Kitimat Arm, British Columbia. Geology 10:588–592

Quantin C, Allemand P, Mangold N, Delacourt C (2004) Ages of Valles Marineris (Mars) landslides and implications for canyon history. Icarus 172:555–572

Rabinowicz E (1995) Friction and wear of materials. Wiley-Interscience, New York

Reiche P (1937) The Toreva-block-a distinctive landslide type. J Geol 45:538–548

Rose ND, Hungr O (2007) Forecasting potential rock slope failure in open pits mines using the inverse-velocity method. Int J Rock Mech Min Sci 44:308–320

Sæter M-B (2008) Dynamics of talus formation. Master Thesis, Oslo Universitetet, A.A. 2007–2008

Saltzmann N, Kaab A, Huggel C, Allgower B, Haeberli W (2004) Assessment of the hazard potential of ice avalanches using remote sensing and GIS-modelling. Norsk Geogr Tidsskr (Norw J Geogr) 58:74–84

Sass O (2006) Determination of the internal structure of alpine talus deposits using different geophysical methods (Lechtaler Alps, Austria). Geomorphology 80:45–58

Sass O, Krautblatter M (2007) Debris flow-dominated and rockfall-dominated talus slopes: genetic models derived from GPR measurements. Geomorphology 86:176–192

Savage SB, Hutter K (1989) The motion of a finite mass of granular material down a rough incline. J Fluid Mech 199:177–215

Scheidegger AE (1973) On the prediction of the release and velocity of catastrophic rockfalls. Rock Mech 5(4):231–236

Scheidegger AE (1975) Natural catastrophes. Wiley, New York

Schenk PM, Bulmer MH (1998) Origin of mountains on Io by thrust faulting and large-scale mass movements. Science 279:1514–1517

Schlichting W (1960) Boundary layer theory. Wiley, New York

Scholz CH (2002) The mechanics of earthquakes and faulting, 2nd edn. Cambridge University Press, Cambridge

Scott KM, Vallance JM (1995) Debris flows, debris avalanches, and flood hazards at and downstream from Mt. Rainier, Washington (USGS hydrologic investigations atlas HA-729). US Geological Survey, Reston

Selby MJ (1993) Hillslope materials and processes. Oxford University Press, Oxford

Semenza E (2001) La storia del Vaiont raccontata dal geologo che ha scoperto la frana. K-Flash, Ferrara (in Italian)

Sharp RH, Nobles LH (1953) Mudflow of 1941 at Wrightwood, southern California. Geol Soc Am Bull 64:547–560

Shekesby RA, Matthews JA (2002) Sieve deposition by debris flows on a permeable substrate, Leirdalen, Norway. Earth Surf Process Land 27:1031–1041

Shreve RL (1966) Sherman landslide, Alaska. Science 154:1639–1643

Shreve RL (1968) The Blackhawk landslide. Geol Soc Am Spec Pap 108:47

Sidle CR, Ochiai H (2006) Landslides. Processes, prediction, and land use. AGU Books, Washington, DC

Simpson JE (1997) Gravity currents. Cambridge University Press, Cambridge

References

Sosio R, Crosta GB, Hungr O (2008) Complete dynamic modelling calibration of the Thurwieser rock avalanche (Italian Central Alps). Eng Geol 100:11–26

Sovilla B, Schaer M, Kern M, Bartelt P (2008) Impact pressures and flow regimes in dense snow avalanches observed at the Valee de la Sionne test site. J Geophys Res 113:F01010. doi:10.1029/2006JF000688

Spera FJ (2000) Physical properties of magmas. In: Sigurdsson H, Houghton B, Rymer H, Stix J (eds) Encyclopedia of volcanoes. Academic, New York

Statham I (1972) Scree slope development under conditions of surface particle movement. Trans Inst Br Geogr 59:41–53

Statham I (1976) A scree slope rockfall model. Earth Surf Processes 1:43–62

Straub S (2001) Bagnold revisited; implications for the rapid motion of high concentration sediment flows. In: McCaffrey WD, Kneller BC (eds) Particulate gravity currents. Blackwell Publishing Ltd, Oxford, pp 91–109, International association of sedimentologists special publication, 31

Tanaka KL (1985) Ice-lubricated gravity spreading of the Olympus Mons aureole deposits. Icarus 62:191–206

Tianchi L (1983) A mathematical model for predicting the extent of a major rockfall. Z Geomorphol Neue Folge 27:473–482

Tinti S, Pagnoni G, Zaniboni F, Bortolucci E (2003) Tsunami generation in Stromboli island and impact on the south-east Tyrrhenian coasts. Nat Hazards Earth Syst Sci 3:299–309

Tinti S, Pagnoni G, Zaniboni F (2005) The landslides and tsunamis of the 30th of December 2002 in Stromboli analysed through numerical simulation. Bull Volcanol 68:462–479

Tommasi P, Baldi P, Chiocci FL, Coltelli M, Marsella M, Pompilio M, Tomagnoli C (2006) The landslide sequence induced by the 2002 eruption at Stromboli volcano. Landslides 27:342–356

Turnbull JM, Davies TRH (2006) A mass movement origin for cirques. Earth Surf Process Land 31(9):1129–1148

Turner AK, Schuster RL (1996) Colluvium and talus. In Landslides – investigaton and mitigation, special report 247:525–549. Transportation Research Board, National Research Council. National Academy Press, Washington DC

Urgeles R, Canals M, Masson DG, Gee MJR (2003) El Hierro: shaping of oceanic island by mass wasting. In: Mienert J, Weaver P (eds) European margin sediment dynamics. Springer, Berlin

Vanneste M, Harbitz CB, De Blasio FV, Glimsdal S, Mienert J, Elverhøi A (2010) Hinlopen-Yermak landslide, Arctic Ocean-geomorphology, landslide dynamics and tsunami simulation. In: C Shipp (ed), SEPM 34 vol

Van Steijn H, Bertran P, Francou B, Hetu B, Texier JP (1995) Models for genetic and environmental interpretation of stratified slope deposits: review. Permafrost Periglac 6:125–146

Vardoulakis I (2002) Dynamic thermo-poro-mechanical analysis of catastrophic landslides. Geotechnique 52(3):157–171

Varnes DJ (1978) Slope movements: type and processes. In: Eckel AE (ed) Landslides Analysis and Control. Transp. Res. Board, Spec. Rep. 176, pp 11–33

Vischer DL, Hager WH (1998) Dam hydraulics. Wiley, Chichester

Voellmy A (1955) Über die Zerstörungskraft von Lawinen. Schweiz, Bauzeitung, 73, 159–165, 212–217, 246–249, 280–285

Voight BE, Faust C (1982) Frictional heat and strength loss in some rapid slides. Geotechnique 32(1):43–54

Von Porchinger A (2002) Large rockslides in the Alps: a commentary on the contribution of G. Abele (1937-1994) and a review of some recent developments. In: Evans SG, DeGraff JV (eds) Catastrophic landslides, vol XV. Geological Society of America Reviews in Engineering Geology, Boulder, pp 237–255

Von Porchinger A, Kippel T (2008) Alluvial deposits liquefied by the flims rock slide. Geomorphology 103:50–56. doi:10.1016/j.geomorph.2007.09.016

Wagner NJ, Brady JF (2009) Shear thickening in colloidal dispersions. Phys Today 62:27–32

Waltham T (2004) Foundations of engineering geology, 2nd edn. Spon Press, London

Watts P, Grilli ST (2003) Underwater landslide shape, motion, deformation and tsunami generation. Int Soc Offshore Polar Eng 27:364–371

Weast RC (ed) (1989) Handbook of chemistry and physics, 56th edn. CRC, Cleveland

Wieczorek GF, Snyder JB, Waitt RB, Morissey MM, Uhrhammer RA, Harp EL, Norris RD, Bursik MI, Finewood LG (2000) Unusual July 10, 1996, rock fall at Happy Isles, Yosemite National Park, California. GSA Bull 112:75–85

Zanuttigh B, Lamberti A (2007) Instability and surge development in debris flows. Review of geophysics, 45, RG3006, doi:10.1029/2005RG000175, AGU Publ., USA

Index

In **boldface** the names of localities

A
Acheron rock avalanche (New Zealand), 51
Acoustic fluidization, 211–212
Acquabona (Italy), 122, 126
Added mass, 314–318
Adhesion, 28, 29
Altels (Switzerland), 359
Amontons, 16, 26
Apparent friction angle, 189, 199, 208, 209, 377
Apparent friction coefficient, 179, 180, 188, 189, 194, 196, 197, 199, 239, 287, 359
Armero (Colombia), 5, 6
Aspect ratio, 156, 209, 312, 319, 320
Aureole (Mars), 258
Autosuspension, 375

B
Baga Bogd (Mongolia), 166
Bagnold, criterion for autosuspension, 375
Bagnold number, 155
Bandai (Japan), 33
Barton formula, 29
Bear Islands, 297, 299, 301
Bernoulli equation, 62–65, 85, 327, 331
Betze-Post, 32
Bingham (rheology, model), 97, 100–103, 122, 124, 130, 210, 215, 348, 349
BING model, 347–349
Blackhawk (USA), 163, 164, 172, 178, 212, 213, 380
Bouma, sequence, 370, 372
Brazil nuts effect, 153–158, 202, 288
Brenva (Italy), 247
Bulganuc (USA), 241
Buoyancy, 295, 303, 304, 347, 374

C
Canary Islands (Spain), 234, 300
Chezy (equation), 127–129, 374
Clay, 15, 18, 31, 77, 83, 89, 93–100, 105, 107, 108, 111–117, 227, 231, 239, 243, 297, 301, 302, 307–309, 319, 338, 340–343, 345, 349, 370, 372, 373, 375, 377
Coefficient of restitution, 17, 142–148, 150–152, 202, 206, 225, 271, 273, 277–282
Cohesion (definition), 31, 43
Colluvium, 107, 109, 302
Compliance function, 148
Compressive strength, 29, 377
Continuity equation, 73–75, 82, 328, 335, 347
Coulomb (friction law), 26, 187
Coulomb (scientist), 26
Creep, 18–20, 23, 24, 32, 85, 173, 175, 180–183, 226–228, 231, 283, 286, 288, 359, 366–368
CRSP model, 275–277

D
Dam, 3, 60, 164, 182, 223, 224, 226, 228, 231, 353, 361, 362
Dam break, 6, 53, 353, 361–364
Dante Alighieri, 49, 159, 161, 171
Debris flow, 89–130, 297, 341–342, 347–349
Densimetric Froude number, 350
De Saint-Venant equations, 362, 363
Diagonalization, 38, 39
Dilute suspensions, 91–93
Dispersive pressure, 153–157
Donnas (Switzerland), 161
Drag, 67, 69, 101, 122, 123, 127, 183, 185, 187, 191–192, 274, 295, 303, 304, 309–315, 319–321, 344, 350, 375

405

E

Earth pressure force, 49, 123–124, 211, 214, 239, 250
Earthquake, 3, 6, 23, 33, 47, 49–51, 159, 161, 166, 169, 173, 180, 212, 239, 245, 247, 251, 295, 301, 304, 305, 309, 324, 325, 334, 336, 338
Einstein notation, 81
Elastic collision, 142, 144–146, 149
El Hierro (Canary Islands), 298–299
Elm (Switzerland), 151, 161, 170, 172, 173, 175–176, 178, 193–194, 201, 378, 380, 381
Energy (height), 274
EPOCH, 12, 13, 15
Eulerian derivative, 77, 328, 363
Eulerian description, 76
Euler number, 85
Evapotranspiration, 33
Extensometer, 181
Extremely energetic rockfalls, 281–282

F

Factor of safety, 33, 34, 36, 40, 46–50
Fahrböschung, 113, 175, 196–201, 203, 209, 210, 215, 251, 266–267, 295, 345
Fellenius (method), 48, 49
Felsberg (Switzerland), 199, 378
Finneidfjord (Norway), 235, 343–346
Fjærland (Norway), 105, 157
Flank collapse, 2
Flims (Switzerland), 173, 177, 200, 202, 381
Flood waves, 164, 361–364
Fluidization, 211–212
Flux (in fluid mechanics), 61, 74
Force chains, 140–141, 180
Frank (Canada), 32, 164, 165, 168, 239, 380, 381
Friction, 9, 11, 16, 17, 23–52, 66, 78, 89, 122–124, 134, 135, 138, 139, 152, 159, 172, 179, 180, 183–191, 193, 194, 196–197, 199, 203, 204, 208–211, 214, 217–221, 231, 237–239, 247, 249, 250, 252, 253, 257, 267–269, 271, 278–281, 286, 287, 304, 314, 319, 321–323, 338, 340, 341, 349, 359, 360, 374, 377
Friction (rolling), 17, 269, 271, 274, 280, 286
Frictionite, 215–222
Froude number, 85, 118, 120, 121, 128, 231, 245, 336–337, 350, 365, 374

G

Gas hydrates, 305, 307
Glacial cirque, 50–52
Glacial deposits. *See* Moraines
Glacial lake outburst flood (GLOF), 6, 106, 361
Glacier, 1, 5, 7, 51, 105, 107, 151, 164, 180, 183, 190, 200, 212, 223, 227, 245–253, 256, 257, 281, 297, 302, 305, 309, 358–361
Gleno dam (Italy), 353, 363
GLOF. *See* Glacial lake outburst flood (GLOF)
Grand Banks (Canada), 321, 338–341, 370
Gran Sasso (Italy), 282
Granular temperature, 148–152, 178
Green Lake (New Zealand), 166

H

Haff equation, 150–151
Hall–Petch law, 360
Hawaii (USA), 32, 234, 259, 261, 298, 299
Heim, Albert, 161
Herschel-Bulkley (rheology), 95–97, 124
Hertz theory, 28, 147
Hindered settling, 375
Hinlopen-Yermak landslide (Arctic), 296, 299–301, 337, 344
Honolulu, 32
Huascaran, 1, 2, 6, 21, 49, 178, 179, 245, 247, 281, 282, 358, 380
Hydroplaning, 239, 242, 341, 348–351
Hyperconcentrated flows, 16, 105, 362

I

Ice avalanche, 49, 353, 358–360
Incompressibility, 61–62
Inelastic collision, 142, 144–146, 149

J

Jigsaw-puzzle effect, 178, 202, 212
Jotulhogget (Norway), 285
Junctions, 28

K

Kinematic viscosity, 240, 384
Kinetic sieving, 157
Kinetic theory, 69, 153
Kofels (Austria), 168, 173, 174, 202, 215–221, 380
Krieger–Dougherty formula, 93

Index

L
Lagrangian derivative, 77, 81, 328
Lagrangian description, 77, 121
Lahar, 5, 6, 16, 31, 89, 90, 108–111, 113–116, 129, 130, 297, 304
Langtang (Nepal), 166, 170
Lateral spreading, 13–15, 111, 261, 366–368
Lava flow, 53, 151, 175, 261, 355–358
La Verna (Italy), 20, 181, 366, 367
Lavini di Marco (Italy), 7, 8, 49, 159, 161, 171, 380, 381
Leonardo da Vinci, 26
Lift, 62–65, 136, 221, 310, 324, 344, 348, 350, 365
Liquid limit, 99
Lituya Bay, 233, 235–238
Loess, 6, 16, 107, 108, 302
Lubrication, 53, 99, 164, 180, 209, 212–222, 239, 241, 242, 245, 246, 249–251, 253, 257, 261, 289, 348, 350
Lumped mass model, 183–192, 273–275

M
Malpasset dam (France), 361
Manning equation, 127
Mars, 198, 212, 223, 253–261, 377–381
Mechanical fluidization, 209
Modified Bishop (method), 49
Mohr circle, 41–43, 45, 46
Moment of inertia, 144, 270, 293
Møns Klint (Denmark), 242
Monte Paci (Italy), 239
Moraines, 5, 6, 16, 51, 105, 106, 201, 302, 346, 347, 361
Mount Cook (New Zealand), 247
Mount Granier (France), 160–162
Mount Rainier (USA), 89, 113, 115, 116, 247
Mudflow, 15, 16, 53, 55, 89–130, 151, 254, 301, 319, 349

N
Navier–Coulomb criterion, 43
Navier–Stokes equation, 67, 76–87, 152, 311, 327, 335, 336, 355
Nevado del Ruiz (Columbia), 5, 6, 297
Newtonian fluid (definition), 54, 55, 66, 82
Newton's laws, 10, 27, 78, 81, 191, 356
Normal stress (definition), 25, 31, 38
Norway, 3, 32, 105, 106, 109–112, 115, 157, 181, 182, 214, 234, 236, 245, 248, 282, 284, 285, 291, 305, 307, 308, 337, 338, 343, 345, 351

No-slip condition, 65–66, 87, 276, 315, 327
Novaya-Zemlia (Russia), 179, 194–196

O
Osceola (USA), 113, 115–117, 303, 304
Outrunner blocks, 242, 244, 301, 303, 343–347

P
Paleocene–Eocene thermal maximum (PETM), 306, 307
Pandemonium Creek (USA), 179, 247
Papua New Guinea, 295, 334
Paretone (Gran Sasso, Italy), 282
Pascal (unit for pressure), 25, 31, 56, 66, 240
Penetration hardness, 28
Permeability, 35, 221, 222, 233, 305, 341
PETM. *See* Paleocene–Eocene thermal maximum (PETM)
Photoelastic response, 140
Piezometric height, 35
Piuro (Italy), 161, 162
Planets, 223, 253–255, 259, 305
Plasticity index, 99
Plastic limit, 55, 99, 135
Plug layer, 102–104
Poiseuille flow, 61
Poisson coefficient, 39
Pore water pressure, 32, 49, 58, 124, 214, 221, 303, 304, 351
Power law, 50, 93, 98, 135, 168, 169, 199
Prandtl–Schlichting, 314
Pressure force (fluid mechanics), 78–81
Protalus rampart, 289
Pseudotachylyte. *See* Frictionite
Pyroclastic (flows, materials), 3, 16, 19, 23, 53, 108, 297, 355, 368

Q
Quasi-periodicity, 126–127
Quick clay, 15, 108, 111–117, 345

R
Ramnefjell (Norway), 234, 236
Randa (Switzerland), 181
Rapid granular flows, 153, 202, 209
Refraction, 333–334
Residual soil, 6, 108
Restitution, coefficients of, 17, 142–148, 150–152, 202, 204, 206, 225, 271, 273, 277–281
Reynolds number, 62, 67–69, 85, 87, 129, 310–312, 314, 374, 375

Rheological flows, 15–16, 21, 53, 69, 89, 91, 104–117, 347
Rissa (Norway), 111
Rock avalanche, 6–8, 14, 15, 17–19, 21, 151, 152, 159–223, 226, 228, 231, 233, 234, 245–248, 251, 253, 257, 261, 263, 267, 269, 281, 293, 295, 298–299, 301, 303–305, 319, 350, 359, 365
Rock fall, 5, 13, 17–19, 49, 164, 168, 175, 200, 234, 236, 254, 263–293, 301
Rotational landslide. *See* Slump
Roughness coefficient, 29
Runout ratio, 180, 188, 197, 199, 203, 304, 351, 377, 378

S
Sackung, 13, 366–368
Saidmarreh (Iran), 166–168, 174, 199, 381
San Francisco (USA), 33
Sarno (Italy), 33
Sciara del Fuoco (Italy), 238
Scilla (Italy), 235, 239–241
Seamounts, 300, 301, 303
Seed–Idriss formula, 49
Self-organized criticality (SOC), 169
Shear layer, 101, 102, 159, 202–208
Shear rate (definition), 66
Shear strength (definition), 31, 55, 96
Shear stress (definition), 25, 38
Shear thickening, 95–97
Shear thinning, 95–97, 99
Sherman (Alaska), 7, 168, 172, 178, 212, 245, 247, 249–252, 256, 257, 380
Sidescan sonar, 297
Skin friction, 122–123, 191, 313–314, 319, 320, 349
Slump (rotational landslide), 13, 14, 40–49, 51, 170, 253, 257, 366, 369
Snow avalanches, 21, 109, 192, 353, 364–365, 370
St. Helens (USA), 33, 119, 253
Stokes (formula, regime), 67, 310, 315
Storegga (Norway), 299, 301, 303, 307–309, 337, 338, 348, 351
Streamlines, 60–61, 320
Streamtubes, 60–63, 78
Stress tensor, 36–39, 42, 44, 80–82, 91, 103
Stress trajectories, 38, 40, 44
Stromboli (Italy), 233, 235, 238, 239
Sturzstrom, 159, 161, 197, 201, 204, 207, 212, 216, 217, 219, 358

Superficial avalanching, 148, 149
Suspension flow, 19, 368–375

T
Tafjord (Norway), 235
Talus, 107, 253, 263–293, 302
Tensile stress (definition), 34, 60, 180
Tensor, 36, 37, 39, 40, 82, 318
Thixotropy, 98, 99
Thurwieser (Italy), 179, 245–247
Tidal flat, 233, 239–243
Tjelle (Norway), 234
Topple, 5, 13, 14, 17–18, 291–293
Transverse ridges, 12, 163, 170, 173, 175
Trenches, 300–301
Triolet (Italy), 247
Tsunami, 3, 53, 113, 181, 224, 225, 233–235, 237, 239, 243–245, 295, 299, 308, 324–342
Tungurahua (Ecuador), 114
Turbidites, 7, 308, 368–375, 382
Turbidity currents, 3, 19, 53, 156, 299, 300, 338–341, 353, 365, 368–375, 382

U
Urdbø (Norway), 248

V
Vaiont (Italy), 3, 168, 181, 223, 224, 226, 227, 231–232, 235, 361, 381
Valcamonica (Italy), 113
Val di Fassa (Italy), 366
Val di Scalve (Italy), 19
Valles Marineris (Mars), 255–258
Valpola/Val Pola (Italy), 164, 165, 168, 170, 171, 224
Val Venina (Italy), 265
Versor, 37, 74, 315
Virtual mass, 315–321, 323, 324
Viscosity (definition), 56, 66, 69, 71–73, 97, 219
Viscous force (fluid mechanics), 78, 80–82, 85
Voellmy model, 192

W
Walton model, 143, 144

Y
Yosemite (USA), 168
Young modulus, 133, 135